# Digital Technologies and Transformations in Public Administration, Engineering, and Sustainable Business

Arshi Naim
*European Global Institute of Innovation and Technology, Malta*

Anuj Kumar
*Rushford Business School, Switzerland*

Arya Kumar
*KIIT University, India*

IGI Global
Scientific Publishing
Publishing Tomorrow's Research Today

Published in the United States of America by
IGI Global Scientific Publishing
701 East Chocolate Avenue
Hershey, PA, 17033, USA
Tel: 717-533-8845
Fax: 717-533-7115
Website: https://www.igi-global.com E-mail: cust@igi-global.com



Library of Congress Cataloging-in-Publication Data

Names: Naim, Arshi, 1976- editor | Kumar, Anuj (Research fellow in management) editor | Kumar, Arya, 1988- editor
Title: Digital technologies and transformations in public administration, engineering, and sustainable business / edited by: Arshi Naim, Anuj Kumar, and Arya Kumar.
Description: Hershey, PA : IGI Global Scientific Publishing, [2026] | Collection of essays by Hasan Alhanatleh and 14 others. | Includes bibliographical references and index. | Summary: "This book seeks to illuminate the profound changes brought about by digital tools in governance, engineering advancements, and sustainable business models, emphasizing how these technologies enhance transparency, efficiency, and inclusivity. Through detailed case studies, conceptual frameworks, and emerging trends, the book intends to equip policymakers, engineers, and business leaders with actionable insights and strategies to navigate the digital age responsibly, foster resilience, and align with sustainable development goals"-- Provided by publisher.
Identifiers: LCCN 2025004264 (print) | LCCN 2025004265 (ebook) | ISBN 9798337315348 hardcover | ISBN 9798337315355 paperback | ISBN 9798337315362 ebook
Subjects: LCSH: Public administration--Technological innovations | Internet in public administration | Sustainable development | Social responsibility of business
Classification: LCC JF1525.A8 D5543 2025 (print) | LCC JF1525.A8 (ebook) | DDC 352.380285--dc23/eng/20250402
LC record available at https://lccn.loc.gov/2025004264
LC ebook record available at https://lccn.loc.gov/2025004265

British Cataloguing in Publication Data
A Cataloguing in Publication record for this book is available from the British Library.

# Table of Contents

# Detailed Table of Contents

optimization. It examines how telemedicine is enhancing accessibility, reducing healthcare costs, and improving patient outcomes, particularly in underserved areas. The study highlights the integration of telemedicine with AI, big data, and remote monitoring to streamline healthcare services, increase operational efficiency, and optimize resource allocation. Furthermore, it discusses the business optimization benefits of telemedicine for healthcare providers, such as reduced overheads, expanded service offerings, and improved patient engagement. The research also addresses the challenges in implementing telemedicine, including regulatory barriers and data privacy concerns, while providing recommendations for its effective integration into sustainable healthcare systems.



The current study focused on addressing how governmental agencies are using Big Data and Artificial Intelligence (AI) applications to detect tax fraud with a focus on Palestine. This is done using a review of the experiences from different Arab neighboring countries like UAE, KSA, Jordan, Qatar and others. This is based on the fact that many governments around the world are adapting AI in their services, and using its many solutions to provide better public services. It is a reality that Arab countries have also started to implement policies related to AI use to help manage digital transformation and risks associated with frauds in financial statements and tax reports, which is also true for Palestinian public sector. The study concluded that AI applications are of great importance to help detect and avoid tax fraud, and that public sector services, especially tax services need to apply AI solutions to their many risks including: legal risks, poor financial reporting risks, information privacy risks, and risks of poor quality of tax reports

The current research works for the integration of Artificial Intelligence (AI) and Big Data analytics in public healthcare administration. The focus of the integration is on their predictive capabilities. The research methodology primarily employs secondary data analysis sourced from the United Nations and other reputable healthcare reports relevant to Malaysia. The study has analyzed the secondary data which examines how AI-driven models and Big Data are being utilized to forecast health outcomes, optimize resource allocation, and enhance decision-making processes in government-run healthcare systems in the country Malaysia. This research method builds for an in-depth evaluation of the effectiveness of these technologies in predicting diseases, managing public health crises, and reducing healthcare costs. The findings aim to show the potential of AI and Big Data in transforming healthcare systems to become more proactive, efficient, and sustainable. The outcomes of the study present the healthcare services in Malaysian public healthcare administration.



The intersection of Artificial Intelligence (AI) and Business Process Automation (BPA) is rapidly changing the digital transformation landscape. This intersection is evident in both public administration and business for sustainability and the challenges that governments and businesses are facing to improve efficiency, transparency and sustainability. AI powered BPA is a key enabling technology that can achieve intelligent, data-first, and adaptive operations. This chapter reviews key methods, technologies and tools that underpin AI business process automation in the public sector and environmentally agnostic businesses. Most importantly, this chapter provides a straightforward introduction to the field allowing researchers, policymakers and practitioners to understand and to grasp the transformative nature of the digital journey and to ultimately either lead sustainable digital transformation initiatives or participate in them.

Public infrastructure projects are frequently plagued by inefficiencies, budget overruns, corruption, and sustainability challenges, which undermine their long-term viability and public trust. The integration of emerging technologies, such as artificial intelligence (AI) and blockchain, presents transformative opportunities to address these issues by enhancing transparency, accountability, and sustainability. AI can optimize project management through predictive analytics, real-time monitoring, and risk assessment, while blockchain ensures financial integrity by providing immutable transaction records and decentralized oversight. This chapter explores the synergy between these technologies, illustrating their potential through real-world case studies. By examining successful implementations, it highlights how AI and blockchain can improve decision-making, reduce fraud, enhance stakeholder collaboration, and ensure compliance with environmental regulations.



The rapid proliferation of digital technologies has emerged as a transformative force, reshaping various facets of Indian business and society. This study examines the revolutionary changes brought about by digital technology, exploring its impact on both large-scale and small-scale industries. Digital devices have become essential in businesses, particularly in India, drastically changing employers' behaviors. The objective of the study is to examine the impacts of digital technologies on Indian businesses and society, with the goal of improving employee attitudes towards digital technology. A theoretical framework for a digital technology model is developed to reveal differences between technology challenges, influence, and support. This framework combines descriptive and quantitative data from various professionals to evaluate which organisational and technological characteristics best predict employee attitudes toward digital technology.

Industry 5.0 is projected to bring about a social revolution and the emergence of a human-centric revolution. This new society will emphasize the social and human-centred features of Industry 4.0's tools and technology. Sustainability and human well-being will be essential to Industry 5.0, which enlarges on Industry 4.0's foundations by accenting sustainable, resilient and human-centred design principles. This revolution seeks to create a outline for supporting the cohabitation of industry with growing society trends and requirements, therefore simplifying the transition from Industry 4.0 to 5.0. This may offer academics, practitioners, and policymakers with significant insights into Industry 5.0 supply chains, emphasizing their ability to boost social well-being, sustainability and financial growth. It also provides a conceptual basis for upcoming research and applied applications.



In the digital age, the virtual economy has become a disruptive force that is changing consumer behavior, economic structures, and industries. The virtual economy, which is made possible by digital platforms, blockchain technology, and artificial intelligence, consists of digital experiences, goods, and services that are not limited by physical boundaries. The growth of digital transactions has given consumers, businesses, and entrepreneurs access to previously unheard-of options, ranging from e-commerce and digital banking to virtual assets like cryptocurrencies. The accessibility of the virtual economy, which permits worldwide involvement with little barriers to entry, is one of its main benefits. Digital marketplaces open up new sources of income, and the gig and create economies enable people to make money off of their talents. The blockchain-based technologies and decentralized finance provide financial inclusion to marginalized communities. This chapter offers perspectives on how virtual economy may affect the societal advancement and economic expansion in the digital era.

This chapter explores the transformative impact of technology integration across public administration, engineering, and business sectors. It highlights how advancements such as e-Government platforms, smart cities, and blockchain are reshaping governance and service delivery, enhancing transparency and efficiency. In engineering, digital tools, AI, and automation are revolutionizing infrastructure and manufacturing, driving sustainability and operational optimization. The business sector benefits from innovations like big data analytics, cloud computing, and FinTech, improving decision-making and customer engagement. The convergence of these technologies presents opportunities for cross-sector collaboration, addressing global challenges such as climate change and urbanization. However, this integration also brings challenges related to cybersecurity, data privacy, and the need for a skilled workforce. The chapter emphasizes the importance of responsible and inclusive technology management to harness its full potential for societal advancement.



This book chapter explores how intelligent systems and the Internet of Things (IoT) technologies contribute to business optimization in engineering. It highlights the transformative impact of Industry 4.0 through concepts such as smart factories, digital twins, and cyber-physical systems. While addressing the operational benefits of AI, real-time data analytics, and automation, it also discusses major challenges like cybersecurity risks and lack of standardization. The chapter offers strategic recommendations for effective technology adoption, emphasizing sustainability, organizational adaptation, and future-oriented innovation.

Digital innovation has emerged as a transformative force driving the growth and development of economies worldwide, particularly in developing nations. By leveraging technology, these countries can overcome structural challenges, improve productivity, and achieve sustainable economic progress. In an era defined by rapid technological advancements, digital tools and platforms are playing a pivotal role in fostering economic inclusion, enhancing public services, and creating opportunities for individuals and businesses alike. Developing economies often face hurdles such as limited infrastructure, inefficient systems, and high levels of poverty. Digital innovation offers solutions to these issues by introducing cost-effective, scalable technologies that address local challenges while fostering global competitiveness. Mobile connectivity, for instance, has revolutionized access to financial services in regions where traditional banking systems are underdeveloped.



Digital twin (DT) users gain access to an enhanced system for testing and assessing complex systems better than standard simulations. DT development requires overcoming several hurdles which consist of complex communication systems alongside inaccessible data storage and processing limitations and incomplete standardization. This study investigates DT enablers followed by sector-based challenges alongside opportunities while focusing on Blockchain implementations for trust enhancement and data protection. The article discusses the challenges of

energy and efficient logistics have cut their $CO_2$ emissions by double-digit percentages and lowered costs.

# Preface

In an era where digital innovation has become the cornerstone of advancement, the intersection of technology with governance, engineering, and sustainability is reshaping the world as we know it. *Digital Technologies and Transformations in Public Administration, Engineering, and Sustainable Business* is a timely exploration of how emerging digital tools and practices are not only modernizing core systems but also paving the way for more intelligent, transparent, and responsible societal frameworks.

This book brings together critical insights and research findings that highlight the transformative power of digital technologies. From smart governance platforms enhancing public trust to engineering innovations driven by digital twins and AI, and sustainable business strategies enabled by blockchain and IoT, the chapters in this book showcase practical implementations, theoretical frameworks, and visionary applications. These transformations do not occur in silos they reflect a growing convergence of disciplines, underlining the interconnectedness required to meet today's global challenges.

The aim of this book is to provide a comprehensive, multidisciplinary resource that supports both academic inquiry and practical application. It is designed for a diverse audience that includes policymakers seeking to modernize public services, engineers developing future-ready infrastructures, business leaders striving for sustainability, and researchers committed to advancing knowledge across digital domains.

As you navigate through the chapters, you will encounter perspectives that span from predictive healthcare models to sustainable development practices each one underscoring how technology can act as a powerful enabler of progress. The topics covered reflect a balance between theoretical exploration and real-world applicability, making this book a valuable reference for both scholarly and professional pursuits.

We hope that this work not only informs but also inspires. It invites readers to think critically about the role of technology in shaping our collective future and to explore innovative pathways that ensure inclusivity, resilience, and sustainability across sectors.

— *The Editors*

## CHAPTER OVERVIEW

### Chapter 1: Artificial Intelligence and Customer-Centric Marketing in Driving Sustainable Business in the Digital Age

Arshi Naim, Praveen Kumar Malik, Hesham Magd, Ahmad Shaheen, Mostafa Mohamad, Raghavan Srinivasan

This paper focuses on the role of Artificial Intelligence (AI) in revolutionizing digital marketing practices, particularly in creating customer-centric strategies that support sustainable business practices. It examines how AI tools, such as machine learning, chatbots, and data analytics, are used to personalize marketing efforts, improve customer engagement, and drive sustainable business growth. The study also analyzes the ethical considerations and challenges of using AI in digital marketing, particularly regarding data privacy and customer trust, and explores how businesses can maintain transparency while benefiting from AI technologies.

### Chapter 2: Digital Health Innovations in Advancing Telemedicine for Sustainable Healthcare Delivery and Business Optimization

Mohammad Ibrahim Khan, Nurqistina Balqis, Arshi Naim, Hesham Magd, Praveen Kumar Malik, Mohammad Faiz Khan

This paper explores the transformative role of digital health innovations, focusing on telemedicine as a key driver for sustainable healthcare delivery and business optimization. It examines how telemedicine is enhancing accessibility, reducing healthcare costs, and improving patient outcomes, particularly in underserved areas. The study highlights the integration of telemedicine with AI, big data, and remote monitoring to streamline healthcare services, increase operational efficiency, and optimize resource allocation. Furthermore, it discusses the business optimization benefits of telemedicine for healthcare providers, such as reduced overheads, expanded service offerings, and improved patient engagement. The research also addresses the challenges in implementing telemedicine, including regulatory barriers and data privacy concerns, while providing recommendations for its effective integration into sustainable healthcare systems.

## Chapter 3: AI Adoption in Tax Fraud Detection in Palestine: Insights from Arab Countries on Public Policy and Sustainable Governance

Nael Sayedahmed, Shaista Anwar

The current study focused on addressing how governmental agencies are using Big Data and Artificial Intelligence (AI) applications to detect tax fraud with a focus on Palestine. This is done using a review of the experiences from different Arab neighboring countries like UAE, KSA, Jordan, Qatar and others. This is based on the fact that many governments around the world are adapting AI in their services, and using its many solutions to provide better public services. It is a reality that Arab countries have also started to implement policies related to AI use to help manage digital transformation and risks associated with frauds in financial statements and tax reports, which is also true for Palestinian public sector. The study concluded that AI applications are of great importance to help detect and avoid tax fraud, and that public sector services, especially tax services need to apply AI solutions to their many risks including: legal risks, poor financial reporting risks, information privacy risks, and risks of poor quality of tax reports.

## Chapter 4: Leveraging AI and Big Data for Predictive Healthcare in Public Administration in Malaysia

Mohammad Ibrahim Khan, Nurqistina Balqis, Arshi Naim, Hesham Magd, Praveen Kumar Malik, Mohammad Faiz Khan

The current research works for the integration of Artificial Intelligence (AI) and Big Data analytics in public healthcare administration. The focus of the integration is on their predictive capabilities. The research methodology primarily employs secondary data analysis sourced from the United Nations and other reputable healthcare reports relevant to Malaysia. The study has analyzed the secondary data which examines how AI-driven models and Big Data are being utilized to forecast health outcomes, optimize resource allocation, and enhance decision-making processes in government-run healthcare systems in the country Malaysia. This research method builds for an in-depth evaluation of the effectiveness of these technologies in predicting diseases, managing public health crises, and reducing healthcare costs. The findings aim to show the potential of AI and Big Data in transforming healthcare systems to become more proactive, efficient, and sustainable. The outcomes of the study present the healthcare services in Malaysian public healthcare administration.

## Chapter 5: AI-Enabled Business Process Automation for Sustainable Digital Transformation: Concepts, Techniques, and Sectoral Applications

Omnia Saidani Neffati

The intersection of Artificial Intelligence (AI) and Business Process Automation (BPA) is rapidly changing the digital transformation landscape. This intersection is evident in both public administration and business for sustainability and the challenges that governments and businesses are facing to improve efficiency, transparency and sustainability. AI powered BPA is a key enabling technology that can achieve intelligent, data-first, and adaptive operations. This chapter reviews key methods, technologies and tools that underpin AI business process automation in the public sector and environmentally agnostic businesses. Most importantly, this chapter provides a straightforward introduction to the field allowing researchers, policymakers and practitioners to understand and to grasp the transformative nature of the digital journey and to ultimately either lead sustainable digital transformation initiatives or participate in them.

## Chapter 6: Harnessing Artificial Intelligence and Blockchain for Enhanced Transparency and Sustainability in Public Infrastructure Projects

Pedro Gomes, Tiago Cardoso

Public infrastructure projects are frequently plagued by inefficiencies, budget overruns, corruption, and sustainability challenges, which undermine their long-term viability and public trust. The integration of emerging technologies, such as artificial intelligence (AI) and blockchain, presents transformative opportunities to address these issues by enhancing transparency, accountability, and sustainability. AI can optimize project management through predictive analytics, real-time monitoring, and risk assessment, while blockchain ensures financial integrity by providing immutable transaction records and decentralized oversight. This chapter explores the synergy between these technologies, illustrating their potential through real-world case studies. By examining successful implementations, it highlights how AI and blockchain can improve decision-making, reduce fraud, enhance stakeholder collaboration, and ensure compliance with environmental regulations.

## Chapter 7: The Impact of Digital Technologies on Indian Business and Society

Anitha Kumari

The rapid proliferation of digital technologies has emerged as a transformative force, reshaping various facets of Indian business and society. This study examines the revolutionary changes brought about by digital technology, exploring its impact on both large-scale and small-scale industries. Digital devices have become essential in businesses, particularly in India, drastically changing employers' behaviors. The objective of the study is to examine the impacts of digital technologies on Indian businesses and society, with the goal of improving employee attitudes towards digital technology. A theoretical framework for a digital technology model is developed to reveal differences between technology challenges, influence, and support. This framework combines descriptive and quantitative data from various professionals to evaluate which organisational and technological characteristics best predict employee attitudes toward digital technology.

## Chapter 8: The Next Frontier Advancing Towards Industry 5.0

Praveen Malik, Rajesh Singh, Anita Gehlot, Aashna Sinha, Arshi Naim

Industry 5.0 is projected to bring about a social revolution and the emergence of a human-centric revolution. This new society will emphasize the social and human-centred features of Industry 4.0's tools and technology. Sustainability and human well-being will be essential to Industry 5.0, which enlarges on Industry 4.0's foundations by accenting sustainable, resilient and human-centred design principles. This revolution seeks to create a outline for supporting the cohabitation of industry with growing society trends and requirements, therefore simplifying the transition from Industry 4.0 to 5.0. This may offer academics, practitioners, and policymakers with significant insights into Industry 5.0 supply chains, emphasizing their ability to boost social well-being, sustainability and financial growth. It also provides a conceptual basis for upcoming research and applied applications.

## Chapter 9: The Virtual Economy Opportunities and Challenges in the Digital Age

Dr Ravi kant Sharmna, PANKAJ SINGH

In the digital age, the virtual economy has become a disruptive force that is changing consumer behavior, economic structures, and industries. The virtual economy, which is made possible by digital platforms, blockchain technology, and artificial intelligence, consists of digital experiences, goods, and services that are not limited by physical boundaries. The growth of digital transactions has given consumers, businesses, and entrepreneurs access to previously unheard-of options, ranging from e-commerce and digital banking to virtual assets like cryptocurrencies. The accessibility of the virtual economy, which permits worldwide involvement with little barriers to entry, is one of its main benefits. Digital marketplaces open up new sources of income, and the gig and create economies enable people to make money off of their talents. The blockchain-based technologies and decentralized finance provide financial inclusion to marginalized communities. This chapter offers perspectives on how virtual economy may affect the societal advancement and economic expansion in the digital era.

## Chapter 10: Integrating Technology across Sectors Innovations in Public Administration, Engineering, and Business: Transforming Industries through Technology in Public Administration, Engineering, and Business

Navin Kumar, Smriti Gupta

This chapter explores the transformative impact of technology integration across public administration, engineering, and business sectors. It highlights how advancements such as e-Government platforms, smart cities, and blockchain are reshaping governance and service delivery, enhancing transparency and efficiency. In engineering, digital tools, AI, and automation are revolutionizing infrastructure and manufacturing, driving sustainability and operational optimization. The business sector benefits from innovations like big data analytics, cloud computing, and FinTech, improving decision-making and customer engagement. The convergence of these technologies presents opportunities for cross-sector collaboration, addressing global challenges such as climate change and urbanization. However, this integration also brings challenges related to cybersecurity, data privacy, and the need for a skilled workforce. The chapter emphasizes the importance of responsible and inclusive technology management to harness its full potential for societal advancement.

## Chapter 11: Smart Systems and IoT in Engineering in Business Optimization

Bekir YURDUGÜZEL, Beste DESTİCİOĞLU TAŞDEMİR, Merve ASİLOĞULLARI AYAN

This book chapter explores how intelligent systems and the Internet of Things (IoT) technologies contribute to business optimization in engineering. It highlights the transformative impact of Industry 4.0 through concepts such as smart factories, digital twins, and cyber-physical systems. While addressing the operational benefits of AI, real-time data analytics, and automation, it also discusses major challenges like cybersecurity risks and lack of standardization. The chapter offers strategic recommendations for effective technology adoption, emphasizing sustainability, organizational adaptation, and future-oriented innovation.

## Chapter 12: Digital Innovation as the Backbone of Developing Economies-A Study

Nithya T, Chandrasekaran V, Mahendrakumar S, Keerthana N V, Priya Keerthana A

Digital innovation has emerged as a transformative force driving the growth and development of economies worldwide, particularly in developing nations. By leveraging technology, these countries can overcome structural challenges, improve productivity, and achieve sustainable economic progress. In an era defined by rapid technological advancements, digital tools and platforms are playing a pivotal role in fostering economic inclusion, enhancing public services, and creating opportunities for individuals and businesses alike. Developing economies often face hurdles such as limited infrastructure, inefficient systems, and high levels of poverty. Digital innovation offers solutions to these issues by introducing cost-effective, scalable technologies that address local challenges while fostering global competitiveness. Mobile connectivity, for instance, has revolutionized access to financial services in regions where traditional banking systems are underdeveloped.

## Chapter 13: Connecting Reality and Virtuality: The Power of Digital Twins

Praveen Malik, Anita Gehlot, Fraiz Parveen, Mahim Gupta, Rajesh Singh

Digital twin (DT) users gain access to an enhanced system for testing and assessing complex systems better than standard simulations. DT development requires overcoming several hurdles which consist of complex communication systems alongside inaccessible data storage and processing limitations and incomplete standardization. This study investigates DT enablers followed by sector-based challenges alongside opportunities while focusing on Blockchain implementations for trust enhancement and data protection. The article discusses the challenges of evaluating digital twins in urban management because of organizations pursuing different objectives and having unclear priorities. The study supports critical realism because it proves more suitable than postmodernism and ontological flexibility for handling these challenges. It evaluates how DTs function in smart manufacturing through an introduction of deep learning technology that enables monitoring human-robotically interactions.

## Chapter 14: Artificial Intelligence in Public Cultural Institutions: Shaping the Value Creation Process with Chatbots

Carmine Sergianni, Gesualda Iodice

This study investigates the integration of chatbot technology within public cultural institutions, to evaluate its impact on the value creation process. Theoretically based on the Public Service Logic, the research employs a qualitative methodology on a single case study, analyzing data collected from over 3,400 users during a 25-day pilot phase. The results indicate a high level of user acceptance, with 97.4% of visitors opting not to seek human assistance, and an average engagement time of 3 minutes and 9 seconds, reflecting significant interaction and interest. These findings underscore the potential of AI-driven solutions to improve service delivery and value creation process in public cultural settings. The implications suggest that the adoption of digital technologies can play a crucial role in shaping the value creation process and the way public institutions engage with their audiences, ultimately fostering greater accessibility and satisfaction across various sectors.

## Chapter 15: Application of Digital Twin Technology for Building Sustainable Strategic Business Models in Healthcare

Eleni Tsianaka, Konstantinos Peramatzis

Digital Twin (DT) technology revolutionises personalised healthcare by creating virtual patient replicas for real-time monitoring, predictive modelling, and tailored treatment planning. This chapter examines the current applications of digital twins in healthcare, emphasising their potential to improve clinical outcomes and reduce costs through precision medicine. Key uses include managing chronic diseases, integrating multi-omics data and medical imaging to customise treatment strategies in personalised cancer care and surgical planning, allowing for the rehearsal of complex procedures and custom implant design. Despite its promise, adopting DTs in healthcare faces challenges, particularly in data privacy, model accuracy, and integrating diverse data sources. Ensuring model reliability and addressing ethical concerns about patient data are essential for widespread adoption. This chapter highlights the need for ongoing research and development to overcome these challenges and fully leverage the potential of DTs in transforming personalised healthcare.

## Chapter 16: Navigating GDPR Compliance: Transforming Digital Marketing in a Privacy-First World

Mohammad Irfan, Saanvi Singhal, Ajay Yadav, Fadillah Mansor

The General Data Protection Regulation (GDPR), implemented by the European Union (EU), safeguards the personal data and privacy of individuals, granting them greater control over how their information is used and stored. This legislation mandates businesses to maintain transparency and responsibility in handling personal data, ensuring enhanced privacy and security for users. GDPR has significantly transformed the landscape of digital marketing by redefining data collection and usage practices. It necessitates that marketers adopt clear and transparent methods to respect user rights, fostering trust and compliance. Moreover, GDPR aims to prevent digital fraud and manipulation, particularly on E-commerce platforms, by reinforcing stringent data protection measures. As a result, GDPR not only strengthens individual privacy but also compels businesses to adopt ethical data practices, shaping a privacy-first approach to digital marketing.

## Chapter 17: Sustainable Strategies for Growing Online Businesses: Minimizing Environmental Impact in the Digital Economy

Satvik Jain, Anuj Kumar

The rapid growth of the digital economy has brought significant environmental challenges alongside business opportunities. Data centers, e-commerce logistics, and electronic waste contribute to a rising carbon footprint in online operations, necessitating sustainable strategies. This chapter provides a concise overview of how online businesses can pursue growth while minimizing environmental impact. It discusses technological measures (like renewable energy adoption and green IT practices), operational improvements (such as eco-friendly supply chains and packaging reduction), and behavioral approaches (engaging consumers in sustainability). Drawing on recent research and industry cases, the chapter highlights that implementing these strategies can substantially reduce carbon emissions and waste without sacrificing profitability. For example, companies embracing renewable energy and efficient logistics have cut their $CO_2$ emissions by double-digit percentages and lowered costs.

**Arshi Naim**
*European Global Institute of Innovation and Technology, Malta*

**Anuj Kumar**

*Rushford Business School, Switzerland*

**Arya Kumar**
*KIIT University, India*

# Acknowledgement

The development of *Digital Technologies and Transformations in Public Administration, Engineering, and Sustainable Business* has been a collaborative journey enriched by the contributions of many individuals, institutions, and researchers committed to exploring the transformative power of digital technologies across key sectors.

We would like to express our deepest gratitude to the researchers and contributors whose insights and case studies have shaped the quality and breadth of this volume. A special note of appreciation is extended to the team whose current research focuses on the integration of Artificial Intelligence (AI) and Big Data analytics in public healthcare administration in Malaysia. Their detailed analysis, which leverages secondary data from the United Nations and other authoritative sources, offers a compelling examination of how predictive technologies are revolutionizing government-run healthcare systems. Their work significantly enhances the scope of this book, shedding light on how digital innovations can drive proactive, efficient, and sustainable healthcare delivery.

We are also thankful to the academic and professional communities' policy makers, engineers, healthcare experts, and sustainable business leaders whose ongoing commitment to digital transformation provided both the inspiration and practical context for this publication. Their feedback, research, and on-the-ground experiences have been invaluable in shaping a resource that we hope will contribute meaningfully to the global dialogue on digital innovation and sustainability.

Finally, we recognize the support of our editorial and publishing teams, whose dedication ensured the successful compilation and presentation of this work. It is our hope that this book serves not only as an academic reference but also as a catalyst for further research and implementation of digital strategies across diverse sectors worldwide.

— *The Editors*

# Chapter 1
# Artificial Intelligence and Customer-Centric Marketing in Driving Sustainable Business in the Digital Age

**Arshi Naim**
https://orcid.org/0000-0003-1325-6964
*European Global Institute of Innovation and Technology, Malta*

**Praveen Kumar Malik**
*Lovely Professional University, India*

**Hesham Magd**
https://orcid.org/0000-0002-6229-488X
*University of Wolverhampton, UK*

**Ahmad Yahya Moustafa Shaheen**
https://orcid.org/0009-0002-5677-1990
*Arab Academy for Science, Technology, and Maritime Transport, Egypt*

**Mostafa Mohamad**
https://orcid.org/0000-0001-7172-1110
*Zayed University, Abu Dhabi, UAE*

**Raghavan Srinivasan**
*Rushford Business School, Switzerland*

## ABSTRACT

*This paper focuses on the role of Artificial Intelligence (AI) in revolutionizing digital marketing practices, particularly in creating customer-centric strategies that support sustainable business practices. It examines how AI tools, such as machine learning, chatbots, and data analytics, are used to personalize marketing efforts, improve customer engagement, and drive sustainable business growth. The study also analyzes the ethical considerations and challenges of using AI in digital marketing, particularly regarding data privacy and customer trust, and explores how businesses can maintain transparency while benefiting from AI technologies.*

DOI: 10.4018/979-8-3373-1534-8.ch001

## INTRODUCTION

This study explores the role of Artificial Intelligence (AI) is changing digital marketing, with a special focus on customer-centered strategies that help businesses grow in a sustainable way. It explores how Amazon Saudi Arabia uses AI tools such as machine learning, chatbots, and data analytics to make marketing more personal, increase customer engagement, and support long-term business success.

The research uses a qualitative method, mainly through a case study approach. It relies on secondary data from industry reports, academic studies, and real-world examples of AI use. This helps give a full picture of how AI is being used in business today.

The study also looks at important ethical issues like data privacy, bias in AI algorithms, and building customer trust (Stanciu et al., 2021). It highlights how companies can enjoy the benefits of AI while also acting responsibly. The goal is to give useful insights into how businesses can use AI in marketing effectively, while staying transparent and ethical (Aldoseri et al., 2024).

## Artificial Intelligence (AI) in Digital Marketing

Artificial Intelligence (AI) means using smart computer tools like machine learning, language understanding, and data analysis to help with digital marketing. These tools help businesses save time by doing some jobs automatically, talk to customers in a more personal way, and learn useful things from large amounts of data.

For example, in retail companies like Amazon Saudi Arabia, AI is used in different ways. Chatbots help answer customer questions quickly, product recommendation tools suggest items customers might like, and prediction tools help manage stock better.

These AI tools help businesses understand what customers want, create better marketing plans, and grow in a successful and long-lasting way (Khan et al., 2024).

Artificial Intelligence (AI) has changed digital marketing by making it more efficient, more personal, and better at connecting with customers. AI tools help businesses study large amounts of data, understand what customers might do next, and automate different marketing tasks. By using AI, marketers can improve their plans, give customers a better experience, and help their business grow (Malik et al., 2024).

Below are some key applications of AI in digital marketing.

## AI-Powered Personalization

AI enables brands to deliver highly personalized marketing experiences by analyzing customer data, preferences, and online behaviors. Machine learning algorithms process vast datasets in real-time to tailor content, product recommendations, and advertisements to individual users. Platforms like Amazon and Netflix use AI-driven personalization to suggest products and movies based on users' past interactions. Personalized marketing enhances customer satisfaction, increases engagement, and boosts conversion rates, making it a crucial AI application in digital marketing (Naim, 2024).

## Chatbots and Virtual Assistants

AI-powered chatbots and virtual assistants have changed the way companies handle customer service and digital marketing. These tools give quick answers to customer questions and are available all the time, even without human help. Chatbots make the customer experience better by replying to many people at once, saving time, and giving help that feels personal. Companies like Sephora and H&M use AI chatbots to suggest products and help customers while they shop, which leads to happier customers and keeps them coming back (Mukhopadhyay et al., 2024).

## Predictive Analytics and Consumer Insights

AI-driven predictive analytics helps marketers guess what customers might do next. It looks at things like what people bought before, what they searched for, and how they interacted with a website. This helps businesses make better ads, set the right prices, and use their time and money in smarter ways. It also helps them know how much of a product they will need, so they can plan their stock and production. This makes the business run better and earn more profit (Naim et al., 2023).

## AI in Content Creation and Optimization

AI tools help marketers create, pick, and improve content. They make it easier to write good and interesting things like blog posts, social media captions, emails, and product descriptions using given information. Tools like Jasper AI and Copy.ai use natural language processing (NLP) to write in a way that sounds natural and fits the brand and audience. AI also helps make content better for search engines by checking keywords, readability, and SEO trends. This helps the content appear higher in search results (Begum et al., 2024).

## Automated Email Marketing

AI helps make email marketing easier by doing tasks automatically. It can divide customers into groups and send them emails that match their interests. AI looks at how many people open emails, click on links, and how they respond. This helps choose the best time to send emails and what to write in them. Companies use AI to send smart and changing emails based on what customers like. This saves time for marketers and makes email marketing more successful (Magableh et al., 2024).

## AI in Social Media Marketing

AI is very helpful in social media marketing. It studies how people use social media, watches for new trends, and helps share posts at the right time. Platforms use AI to suggest posts, ads, and products based on what users like. AI tools can also read comments and feedback to understand how people feel about a brand. It helps companies improve their marketing. AI also helps plan and check social media posts to make sure they reach more people and improve brand image (Motevalli & Razavi, 2024).

## Artificial Intelligence in Digital Marketing

AI is changing digital marketing by making it smarter, more personal, and more effective. It helps in many ways like giving product suggestions, answering questions through chatbots, predicting what customers want, and creating content. AI allows companies to connect better with customers and improve their marketing results. As AI keeps getting better, it will bring more new and smart ways to help businesses grow and keep customers happy (Sabahath et al., 2024).

## Customer-Centric Marketing

Customer-centric marketing means focusing on what customers need and like. It tries to make customers feel important and understood (Sabahath et al., 2024). AI helps this by studying data like what customers bought or searched for. This helps companies send messages that match each customer's interests. Amazon Saudi Arabia is a good example. It uses AI to give product suggestions and change prices to match what customers want. AI also helps group customers better, which makes them more loyal and happy. This helps businesses grow in the long run (Naim et al., 2024).

## Sustainable Business Growth

Sustainable business growth means making a profit for a long time while also caring about people and the planet. AI helps companies grow by working more efficiently, saving money, and making smart decisions that support both customers and the environment (Motevalli & Razavi, 2024). For example, Amazon Saudi Arabia uses AI to manage its supply chain, reduce waste, and plan better so they don't produce too much. AI helps businesses not only make money but also be eco-friendlier, which is good for everyone (Sheth et al., 2024).

## Ethical Considerations in AI Adoption

Ethical considerations mean the moral issues businesses must think about when using AI in digital marketing. Some of the main concerns are data privacy, fairness in AI decisions, and customer trust. AI uses a lot of customer data, so it is important to know how this data is collected, used, and kept safe. For example, Amazon Saudi Arabia must make sure that customer data is stored safely and follow rules like GDPR (General Data Protection Regulation). Also, if AI is not fair, it might treat some customer groups better than others. This can lead to unfair marketing. This study says it is very important for businesses to be open and honest about how they use AI and customer data. This helps keep customer trust (Vasundhara et al., 2024).

## Data Privacy and Consumer Trust

Keeping customer data private and safe is a big part of using AI responsibly. AI looks at a lot of customer information to give personal marketing messages. But some people may worry about how their data is being used. This study explains how businesses like Amazon Saudi Arabia deal with this by using clear privacy rules and strong data protection. The study shows that keeping customer trust is very important for success. If people feel that their data is safe and used the right way, they are more likely to use AI services and stay loyal to the brand (Nyquist et al., 2025).

## Research Objectives

- To study how AI tools help businesses focus on customer needs and grow in a sustainable way.

This goal is to find out how AI tools like machine learning, chatbots, and data analysis help businesses create personal marketing, talk better with customers, and grow for the long term. The study looks at how big companies like Amazon Saudi

Arabia use AI to make marketing plans that match what customers like and support business goals that are good for the future.

- To understand the ethical issues and problems of using AI in digital marketing, especially about privacy and trust.

This goal is to look at the moral concerns when using AI in marketing. It focuses on how customer data is used, the fairness of AI decisions, and how to keep customer trust. By looking at case studies and other information, the research wants to find out how companies can be honest, protect customer data, and use AI in a way that is both helpful and responsible.

## Theoretical Background

The study of how AI is used in digital marketing, customer-focused strategies, and long-term business growth is based on some important theories. These theories help explain how AI changes business activities and helps companies improve how they work and connect with customers.

### Technology Acceptance Model (TAM)

The Technology Acceptance Model (TAM) helps us understand how people and businesses start using new technologies. It says that if a tool is easy to use and helpful, people are more likely to use it. In digital marketing, companies like Amazon Saudi Arabia need to make sure their AI tools, like chatbots and recommendations, are simple and useful for customers. This model explains why some people like AI tools, while others may not, especially if they think the tools are hard to use or not helpful (Vasundhara et al., 2024).

### Customer Engagement Theory

This theory is about how companies can create strong and lasting relationships with their customers. AI tools, such as product recommendations and flexible pricing, help businesses connect better with each customer by understanding what they like and how they shop. This theory shows that AI can help businesses build trust, keep customers happy, and encourage them to stay loyal for a long time (Nyquist et al., 2025; Malik et al., 2024).

## Sustainable Development Theory

This theory explains that businesses should grow in a way that helps the economy but also cares for the environment and society. AI supports this by helping companies work more efficiently, waste less, and make better decisions. For example, Amazon Saudi Arabia uses AI to plan better, reduce energy use, and avoid making too many products. This helps the company grow in a way that is good for the planet and people (Khan et al., 2024).

## Privacy Paradox Theory

This theory talks about how people want personalized services but also worry about their privacy. While customers like it when AI suggests products they might want, they may also be uncomfortable with how their data is collected and used. This theory helps explain the challenge companies like Amazon face: they must use AI in a way that helps customers but also protects their personal information (Chaturvedi & Verma, 2023).

## Ethical Consumerism Theory

This theory focuses on how people choose what to buy based on their values. Many customers today care about things like protecting the environment and fair business practices. When companies like Amazon Saudi Arabia use AI, they must make sure it's used fairly and responsibly. If customers think a business is ethical, they are more likely to trust it and buy from it (Sabahath et al., 2024).

## Diffusion of Innovations Theory

This theory explains how new technologies spread among people and companies. When big companies start using AI in marketing and show good results, other businesses also start to use it. This theory helps us understand how AI tools in digital marketing become more common over time, especially when early users like Amazon Saudi Arabia show their benefits (Begum et al., 2024; Naim, 2024).

Together, these theories give a clear picture of how AI is changing marketing. They help explain how AI helps businesses focus on customers, grow in a sustainable way, and deal with ethical concerns. These ideas build the base for this study.

The current study gives important benefits to both academic research and real business work in the area of Artificial Intelligence (AI) and digital marketing. First, the study shows useful ideas about how AI tools like machine learning, chatbots, and data analysis are changing how businesses connect with customers. By looking at Amazon Saudi Arabia, the study gives a real example of how AI can improve customer-focused marketing strategies, helping businesses grow in a steady and lasting way in today's digital world. The results of this research can help companies in retail and other fields understand how AI is valuable for better customer engagement, personalization, and overall customer happiness. This, in the end, leads to better business results.

Another benefit of the study is that it looks at the ethical problems of using AI. It talks about important ethical issues like data privacy and trust from customers, which are very important in today's technology world. By studying how businesses can be open and responsible while using AI, the research adds to the ongoing discussion about fair and honest marketing online. This is very important because customers today care more about how their personal data is used and want companies to be clear about their actions. The study also gives practical advice for businesses on how to handle these worries, which helps improve the company's reputation and builds long-term trust from customers.

Also, the study gives a strong theoretical base by including important models like the Technology Acceptance Model (TAM) and Sustainable Development Theory. These ideas help explain why and how AI is used in digital marketing. They give a clear way to understand how AI marketing tools connect to steady and responsible business growth. This makes the study useful for both researchers and people working in business who want to learn more about how technology and marketing work together.

Even though the study has many good points, it also faces some challenges when researching AI and its use in digital marketing. One big challenge is that technology changes very fast. AI tools keep improving quickly, and businesses keep trying new methods to stay ahead. Because of this, the study's findings might become old quickly as new AI tools and marketing ideas come out. This can make it harder to use the study's results for a long time, especially if businesses cannot keep up with the latest technology.

Another challenge is about data privacy concerns when using AI in marketing. The study focusses about ethical issues like privacy and trust, but it can be hard to get accurate and full data on these subjects. Many companies do not want to share details about their internal AI uses, especially when it involves sensitive information. This can limit what the case study and other data can show, which might make the findings less detailed. Also, using only public reports and academic papers might bring some bias or mistakes in the data studied.

Besides, the study focuses only on one case, which is Amazon Saudi Arabia. Even though Amazon is a leader in using AI for marketing, its special business style and resources might not fit smaller companies or different industries. Because of this, the study's results might not apply well to other businesses or markets. Looking at many companies would give a wider and better view of how AI affects customer-focused marketing and steady growth in different fields.

Finally, the study mainly uses qualitative methods like case study review and secondary data. This way of research gives deep and detailed information but does not provide strong statistical proof. Qualitative research cannot always show wide patterns or clear conclusions that apply everywhere. This might make it harder to show general trends or firm results that fit all cases.

The current study gives useful knowledge about AI's role in digital marketing and ethical issues. However, it also faces problems because AI technology changes fast, data can be hard to get, and focusing on one case limits how much we can say. Still, it gives a good base for more research and helpful advice for businesses trying to use AI in marketing in a smart way.

## LITERATURE REVIEW

The current study analyses the impact of Artificial Intelligence (AI) and digital marketing work together, especially how AI helps businesses focus on their customers and grow in a sustainable way. The review covers many topics, including AI technologies, marketing methods, sustainable development, how customers behave, and ethics in marketing. AI has grown a lot, especially in digital marketing, with fast progress in machine learning, natural language processing, and predictive analytics. These technologies have changed how businesses talk to customers by creating personalized marketing that increases customer interest, loyalty, and happiness (Aldoseri et al., 2024).

AI in marketing started in the early 2000s when companies began using data to divide customers into groups and target ads better. In the last 20 years, AI has improved a lot, especially with machine learning, allowing businesses to make very personal experiences based on what customers like and do. Chatbots and virtual helpers have become important for giving customers quick support and keeping them engaged. AI also helps automate many parts of digital marketing, like creating content and adjusting prices automatically. But even with these improvements, problems like protecting customer data and using AI in a fair way are still big concerns. Many studies talk about ethical consumerism and the privacy paradox, which means people want personalized services but worry about their privacy (Richardson et al., 2015).

The study looks at important ideas like AI-powered marketing, customer-focused strategies, and sustainable business growth, which have changed a lot over time. At first, AI was seen as something only big companies could use. But with cloud computing and big data, AI tools are now easier for many businesses, including small and medium ones, to use (Kang et al., 2024). As AI like machine learning and data mining got better, companies began using them to study customer behavior and predict what they might buy next. This helped businesses make marketing more personal and focused on the customer's preferences and satisfaction (Kunz & Wirtz, 2024).

At the same time, the idea of sustainable business growth became more important because people care more about the environment and companies' social responsibilities. Sustainable growth means not only making money but also dealing with environmental and social problems. AI can help with this by managing resources better, cutting waste, and making supply chains work more efficiently. For example, AI tools can predict demand more exactly, so companies don't make too many products and can lower their environmental impact (Morton et al., 2024).

The review also finds some gaps in research about AI in digital marketing. While many studies explain what AI can do, fewer focus on how ethical issues affect customer trust and data privacy in AI marketing (Leone et al., 2021). More research is needed on how businesses can balance giving personalized services with respecting customer privacy. Also, while AI is used in many industries, its long-term effects on sustainable business practices are not well studied. Future research should look at how AI can help not just with making money but also with protecting the environment and supporting social good.

Another gap is about how well AI works for different types of businesses. Big companies like Amazon use AI well, but there is not much research on how small businesses or companies in other industries use AI and benefit from it. Studying this would give a better overall view of AI's power across different business fields.

Amazon's history in Saudi Arabia is important for this study as a key example of AI use in digital marketing and sustainable growth. Amazon entered Saudi Arabia by buying Souq.com in 2017, a popular online shopping platform in the Middle East. This helped Amazon quickly start its business there and use its AI tools, like recommendation systems, personalized marketing, and AI-based logistics. Since then, Amazon has grown in Saudi Arabia, adjusting its marketing to local tastes and using AI to improve its supply chain, customer service, and personalized shopping.

Amazon's use of AI in Saudi Arabia follows its global approach, focusing on machine learning to study customer data and predict buying habits, and using chatbots and virtual helpers to give smooth customer support. Also, Amazon has worked on sustainable business actions, like using energy more efficiently in warehouses and cutting down waste in packaging. Using AI in these ways has helped Amazon not only work better but also build a good reputation as a responsible and customer-focused company. Amazon's skill in balancing AI innovation with ethical issues, especially about data privacy and trust, gives useful lessons about the challenges and chances for businesses in the region as they start using AI for digital marketing.

This study points out key developments in AI and digital marketing, shows how important ideas have grown, and explains important research gaps. By looking at Amazon's work in Saudi Arabia, the study gives a detailed view of how AI can be used to support sustainable growth, improve customer engagement, and handle ethical questions in digital marketing.

## RESEARCH METHODOLOGY

This study mainly uses a qualitative method, which means it looks deeply into one example to understand how things work. It uses a case study approach to study how Artificial Intelligence (AI) helps create marketing strategies that focus on customers and how AI affects the growth of businesses in a sustainable way. The case study is about Amazon Saudi Arabia, a big company in the retail business. The study looks at how Amazon uses AI tools like machine learning, chatbots, and data analytics to make marketing more personal, improve how they connect with customers, and make their business work better. Because the study is qualitative, it can explore real-life examples in detail and give rich information about how businesses can use AI well in their marketing.

Besides the case study, the study also looks at secondary data. This means it reviews information from industry reports, academic papers, and other public sources about AI in marketing. This helps support the findings from the case study and gives a wider view of how AI affects different businesses in different situations. The research method also includes studying ethical issues, such as protecting customer data, avoiding unfair bias in AI algorithms, and building trust with customers. This ethical analysis is done by reading existing research and industry rules about AI ethics to see how companies like Amazon handle these problems.

This research integrates case studies collected from secondary data and uses these cases for complete way to understand how AI marketing works in practice. It also looks carefully at the ethical challenges that come with using AI. The results give useful ideas about how AI can be used in a responsible way to support marketing that focuses on customers and helps businesses grow sustainably.

## RESULTS

Since this study employs a qualitative research approach, the results are measured through an in-depth analysis of collected data, focusing on identifying patterns, themes, and insights relevant to AI-driven customer-centric marketing and sustainability. The study utilizes case study analysis of retail giants like Amazon

Saudi Arabia, where data is gathered from secondary sources such as industry reports, academic literature, and real-world AI applications. The effectiveness of AI-powered tools in enhancing personalized marketing and customer engagement is assessed by examining documented case studies, expert opinions, and historical data that highlight AI's role in business sustainability. The findings are interpreted by identifying recurring themes in how AI technologies contribute to marketing innovation and long-term business strategies.

Additionally, content analysis is used to study ethical issues like data privacy and customer trust. This means looking at reports, company policies, and academic writings about AI ethics in marketing to find the main problems and possible solutions. The study checks how companies are open about their AI use, how customers feel about AI marketing, and what steps businesses take to reduce bias in AI and protect privacy. The results come from studying how companies balance using AI for efficiency with being responsible and ethical. This is done by looking at case studies, policy reviews, and industry best practices.

The qualitative findings are confirmed through triangulation, which means comparing information from different sources like the case study, industry reports, and expert opinions to make sure the results are trustworthy. The study carefully looks at patterns in how AI is used, how marketing is personalized, and efforts toward sustainability to understand AI's overall effect. Instead of using numbers, the study focuses on themes and ideas to give detailed and meaningful insights into how AI marketing affects sustainable growth and ethical business behavior.

A case study is a type of qualitative research that deeply investigates one specific company, event, or situation in its real-life setting. This method is helpful for understanding complex issues because it lets researchers explore detailed connections, causes, and the context that affects the topic. Case studies give rich information by studying data from many sources, like interviews, reports, and notes, making them a strong way to meet research goals.

## Analyzing AI-Powered Tools in Enhancing Customer-Centric Marketing Strategies

The first research goal is to learn how AI technologies like machine learning, chatbots, and data analytics help make marketing more focused on customers and support sustainable business growth. Looking at case studies of big retailers like Amazon Saudi Arabia helps show how AI is used in real marketing.

Case studies analysis show the results of AI tools on marketing campaigns, ways to engage customers, and business results. Also the analysis presents the role of AI tools to improve personalization, make customers happier, and help businesses grow in a sustainable way. Case studies also point out the problems businesses face when using AI marketing, helping researchers give advice on the best ways to use AI and strategic tips for companies.

## Case Study: Amazon Saudi Arabia's Use of AI in Customer-Centric Marketing and Ethical Challenges

Amazon Saudi Arabia is a great example to understand how Artificial Intelligence (AI) is changing digital marketing. The company uses AI tools like machine learning, chatbots, and predictive analytics to make marketing more focused on customers, improve customer engagement, and support sustainable business growth. For example, Amazon's AI recommendation systems look at what customers buy, what they search for, and how they browse to suggest products that fit their needs. This makes customers happier and more likely to keep shopping. Also, Amazon uses AI to change prices in real time based on things like demand, competitor prices, and stock levels. This helps keep prices competitive and profits high.

Amazon also uses chatbots and virtual assistants, like Amazon Alexa and AI customer service bots, to help customers quickly with questions, order tracking, and product suggestions. These AI tools make shopping easier, improve how the company runs, and support long-term success. But this case also shows some big ethical challenges, especially about data privacy and biases in AI. Because AI collects and looks at a lot of customer data, people worry about how this data is kept safe, used, and shared. Customers want to know that their information is protected and handled openly.

Another ethical problem is algorithmic bias. This happens when AI learns from past data that may favor some products or groups of customers, which can make recommendations less fair or diverse. To fix these problems, Amazon Saudi Arabia has privacy rules, ways to get customer permission, and security features powered by AI to be more open and build trust with customers. They also work to reduce bias and make their AI fairer, so their marketing stays ethical and responsible.

This case shows that while AI helps make marketing more personal and improves business, companies like Amazon must always balance the benefits of AI with being ethical. By focusing on privacy, openness, and fairness, Amazon Saudi Arabia tries to keep customer trust and grow their business in a responsible way.

### Examining Ethical Considerations and Challenges of AI Adoption in Digital Marketing

The second research goal looks at the ethical problems of using AI in marketing, especially issues like data privacy, biases in AI, and keeping customer trust. Case studies help us look closely at real examples of how companies handle these ethical questions.

By studying cases of businesses that had problems with data privacy or trust because of AI, researchers can find common problems and patterns. For example, if Amazon Saudi Arabia or other companies faced trust problems from using AI to make decisions, the case study can show how they solved or did not solve these issues. This method also helps check if companies follow rules, are open with customers, and use ethical marketing methods. This leads to advice on how to keep customer trust while using AI in marketing.

Case studies are a strong way to do this research because they allow a deep look into real AI marketing strategies and ethical challenges. By focusing on specific examples from Amazon Saudi Arabia and other big retailers, case studies give real evidence to support the study and help make useful conclusions and suggestions.

## Content Analysis

Content analysis is a research method used to carefully study texts, images, or digital information to find important patterns, themes, and meanings that help answer research questions. In this study, content analysis is important for understanding how AI tools help create customer-focused marketing and to look at the ethical issues of using AI in digital marketing.

### Application of Content Analysis: Analyzing AI-Powered Tools in Customer-Centric Marketing

Content analysis will be done on secondary sources like industry reports, academic papers, company case studies, and AI marketing campaigns. This helps us see how companies like Amazon Saudi Arabia use AI tools such as machine learning, chatbots, and data analytics. By organizing this information, the study can find common ways AI is used for personalization, engaging customers, and supporting sustainability. For example, it will look at marketing methods like automatic product recommendations, predicting customer behavior, and AI-powered customer service to see how well they improve customer relationships.

## Examining Ethical Considerations and Challenges of AI Adoption

To study ethical problems with AI in marketing, content analysis will focus on case studies, ethical rules, laws, and customer opinion research. Important topics like data privacy, biases in AI, transparency, and customer trust will be found and grouped based on the literature and case study results. Information from company policies, privacy statements, and legal rules about AI in marketing will be studied to see how companies follow rules and act responsibly while using AI. Media stories and customer feedback will also be checked to understand what the public thinks and how much they trust AI marketing.

## Systematic Data Interpretation

Organizing and studying qualitative data helps researchers understand ideas from different sources. For AI marketing, this method helps researchers go through large amounts of text, images, and behavior data to find important patterns. It involves arranging data, labeling information, and grouping it into themes. This process helps researchers clearly understand how AI affects marketing. It also helps avoid bias, keep analysis consistent, and support conclusions based on real evidence that can help in academic and professional talks about marketing and AI ethics.

## Identifying Patterns and Trends

Content analysis helps researchers find common themes in AI marketing strategies and ethical issues. It lets researchers study large amounts of data like customer reviews, social media posts, and company policies to find patterns in AI marketing efforts. By looking at these patterns, researchers can find trends in how customers behave, how personalization works, and how AI use in business changes over time. Content analysis also helps understand ethical challenges like data privacy, bias, and transparency. Knowing these trends helps companies and scholars create marketing strategies that improve customer engagement while following ethical rules and industry standards.

## Enhancing Research Validity

Using many different data sources makes content analysis more complete and trustworthy. It increases credibility by including company reports, customer opinions, academic studies, and real case studies. This mix of data makes the research well-rounded and reduces the chance of biased views. By using many sources, researchers can check and confirm their findings with solid proof, making their results more reliable and useful in different industries and markets. This approach also helps find any differences or problems in data, which improves the study's results and suggestions.

## Content Analysis

Content analysis helps this study look at AI's role in marketing in a clear, organized way based on real evidence, while also focusing on ethical issues. By studying marketing strategies, case studies, and ethical problems, this method gives a full understanding of how AI helps customer-focused marketing and supports sustainable business growth. At the same time, it looks at how businesses keep transparency and trust in today's digital world. Content analysis offers a careful and evidence-based way to evaluate AI's effects on marketing and also addresses ethical concerns. By carefully studying marketing methods, real-life examples, and discussions about ethics, this method helps explain how AI shapes marketing that focuses on customers. AI tools improve personalization, help predict customer needs, and make the customer experience better, which all help businesses grow in a sustainable way. But these improvements must be balanced with honesty, ethical responsibility, and trust from customers. Companies that use AI must protect data, make sure AI is fair, and be accountable in how they use AI in marketing. Using content analysis, researchers can give advice on how AI can be used in the best way while still following ethical rules, creating a responsible marketing environment that values both new technology and customer rights.

## AI-Powered Customer-Centric Marketing and Ethical Considerations at Amazon Saudi Arabia

This study's results come from case studies, content analysis, and other secondary data sources focusing on Amazon Saudi Arabia, a big retail company that uses Artificial Intelligence (AI) a lot to improve customer-focused marketing and keep its business sustainable. The study shows how AI technologies have helped Amazon make marketing campaigns more personal, improve how they connect with customers, and handle ethical challenges like data privacy and building trust.

## AI-Powered Tools Enhancing Customer-Centric Marketing at Amazon Saudi Arabia

The study finds that Amazon Saudi Arabia uses AI tools like machine learning, chatbots, and predictive analytics to improve marketing. Personalization is a main result, where AI studies customer behavior, likes, and buying history to suggest products that fit each shopper. This has greatly increased how engaged customers are and how often they buy, because customers get product recommendations and ads that matter more to them. Another big result is how Amazon uses AI to change prices quickly based on how many people want a product, competitor prices, and market trends. This helps customers get good prices while Amazon increases sales and profits. Also, chatbots and AI virtual assistants have made customer service better by giving fast answers, helping track orders, and offering personal shopping help. These AI tools help Amazon keep its business strong for the long term by making shopping easier, building customer loyalty, and improving how the company works.

## Ethical Considerations and Challenges in AI Adoption at Amazon

While AI has helped with customer engagement, the study also shows that ethical issues are still a big challenge for Amazon Saudi Arabia. One main worry is data privacy because AI collects and looks at a lot of customer data to make personalization better. Many customers worry about how their data is kept, used, and shared, which raises questions about honesty and security. The study also points out that AI can have biases because it learns from past data. This might cause AI to favor certain products or customer groups, which limits variety in recommendations. Amazon works to make its AI systems more transparent, but problems with fairness and responsibility still exist. Another important finding is how much customer trust matters in AI marketing. While many people like AI-based personalization, others don't trust how much data is collected or worry it might be misused. To keep trust, Amazon Saudi Arabia has clear privacy rules, asks for user permission, and uses AI-based security to protect data. They also let customers control their data preferences to help them feel more confident about AI marketing.

The study shows that AI tools are very important for Amazon Saudi Arabia's customer-focused marketing, improving personalization, engagement, and sustainable growth. But ethical problems like data privacy, bias, and trust are big concerns that need constant attention. Balancing AI's benefits with ethical responsibility will be very important for Amazon and similar companies to keep customer loyalty, follow laws, and succeed long-term in the digital world.

## Theoretical and Practical Implementation of the Study

The study's theoretical part is based on ideas about customer-focused marketing, AI marketing models, and ethics in digital marketing. Theories like personalization, relationship marketing, and AI decision-making help explain how AI improves customer connection and business sustainability. The study combines ideas from machine learning, big data, and customer behavior to show how AI adjusts marketing based on what customers like. It also uses ethical ideas like data privacy laws (such as GDPR) and making AI processes clear to study the challenges and responsibilities of using AI in marketing. These theories help expand knowledge about how AI marketing works and supports sustainable business.

The practical part of the study is shown through the Amazon Saudi Arabia case, giving real examples of how AI is used in customer-focused marketing. Companies can use AI tools like chatbots, predictive analytics, and recommendation systems to make marketing more personal, better connect with customers, and improve how they operate. The study gives useful advice for businesses on using AI responsibly, solving data privacy problems, and following ethical rules for AI. It also shows ways to keep customer trust, follow laws, and maintain long-term success in digital marketing. By using these findings, companies can grow their business with AI while keeping marketing fair, clear, and ethical.

## CONCLUSION

The research shows the applications of AI and it is found that it is constantly changing customer-focused marketing strategies to help businesses grow in a sustainable way in today's digital world. It focuses on big companies like Amazon Saudi Arabia and shows how they use AI tools like machine learning, chatbots, and data analysis to make marketing more personal, improve customer engagement, and support long-term business success. The findings show that AI is very important for making customers happier, improving marketing efforts, and giving companies an advantage in the retail market. However, the study also points out important ethical problems, such as data privacy, transparency, and biases in AI algorithms. These issues need to be solved to make sure AI is used responsibly and ethically in marketing.

Although Amazon Saudi Arabia has done a lot to use AI to improve customer experience and make operations more efficient, it still needs to carefully balance the benefits of AI with ethical responsibilities. Keeping customer trust and following data protection laws are very important for continuing business success and building strong, long-lasting relationships with customers. The research also stresses the need to keep working on reducing bias in AI systems, so that AI tools are fair, clear, and include everyone.

Future research could explore many new directions to better understand AI's role in digital marketing. First, there is a need for real-world studies to see the long-term effects of AI marketing strategies on customer loyalty, business results, and sustainability in different industries beyond retail, such as banking, healthcare, and entertainment. Research on how AI is used in different sectors can give useful information on the best ways to do personalized marketing and use AI ethically.

Also, future research could focus more on how customers see AI and the changing ethical standards around AI in marketing. Looking at how AI adoption and ethical concerns differ in cultures and regions around the world can give valuable advice to companies that want to adjust their AI strategies for different types of customers. Also, the study could be expanded to include direct research with customers, collecting their opinions on how AI affects their buying decisions, their worries about data privacy, and their trust in AI-powered marketing.

Additionally, research show that AI's role in sustainable business through data-driven decision-making can give a better understanding of how companies can balance making profits with protecting the environment and helping society. This will give a fuller picture of how AI affects sustainability. Adding AI ethics to business education and policy will be very important to prepare future business leaders to handle the challenges and chances that AI brings in digital marketing.

## REFERENCES


Aldoseri, A., Al-Khalifa, K. N., & Hamouda, A. M. (2024). AI-powered innovation in digital transformation: Key pillars and industry impact. *Sustainability (Basel), 16*(5), 1790. DOI: 10.3390/su16051790

Begum, A., Naim, A., & Sabahath, A. (2024). The Impact of AI on Sustainability. In *Harnessing High-Performance Computing and AI for Environmental Sustainability* (pp. 99–113). IGI Global. DOI: 10.4018/979-8-3693-1794-5.ch005

Chaturvedi, R., & Verma, S. (2023). Opportunities and challenges of AI-driven customer service. *Artificial Intelligence in customer service: The next frontier for personalized engagement*, 33-71.

Kang, P. S., Wang, X., Son, J. Y., & Jat, M. (2024). Introduction to Service 4.0 and Customer-Centric Supply Chains. In *Service 4.0: Technology-Enabled Customer-Centric Supply Chains* (pp. 1–10). Springer Nature Switzerland. DOI: 10.1007/978-3-031-63875-6_1

Khan, M. I., Naim, A., & Khan, M. F. (2024). Financial health management of oto-laryngology by telemedicine: Opportunities and challenges. *Asia Pacific Journal of Health Management, 19*(3), 338–354. DOI: 10.24083/apjhm.v19i3.3729

Kunz, W. H., & Wirtz, J. (2024). Corporate digital responsibility (CDR) in the age of AI: Implications for interactive marketing. *Journal of Research in Interactive Marketing, 18*(1), 31–37. DOI: 10.1108/JRIM-06-2023-0176

Leone, D., Schiavone, F., Appio, F. P., & Chiao, B. (2021). How does artificial intelligence enable and enhance value co-creation in industrial markets? An exploratory case study in the healthcare ecosystem. *Journal of Business Research, 129*, 849–859. DOI: 10.1016/j.jbusres.2020.11.008

Magableh, I. K., Mahrouq, M. H., Ta'Amnha, M. A., & Riyadh, H. A. (2024). The Role of Marketing Artificial Intelligence in Enhancing Sustainable Financial Performance of Medium-Sized Enterprises Through Customer Engagement and Data-Driven Decision-Making. *Sustainability (Basel), 16*(24), 11279. DOI: 10.3390/su162411279

Malik, S., Malik, P. K., & Naim, A. (2024). Opportunities and challenges in new generation cyber security applications using artificial intelligence, machine learning and block chain. *Next-generation cybersecurity: AI, ML, and Blockchain*, 23-37.

Morton, F., Benavides, T. T., & González-Treviño, E. (2024). Taking customer-centricity to new heights: Exploring the intersection of AI, hyper-personalization, and customer-centricity in organizations. In *Smart Engineering Management* (pp. 23–41). Springer International Publishing. DOI: 10.1007/978-3-031-52990-0_2

Motevalli, S. H., & Razavi, H. (2024). Enhancing Customer Experience and Business Intelligence: The Role of AI-Driven Smart CRM in Modern Enterprises. *Journal of Business and Future Economy*, *1*(2), 1–8.

Mukhopadhyay, S., Singh, R. K., & Jain, T. (2024). Developing artificial intelligence enabled Marketing 4.0 framework: An Industry 4.0 perspective. *Qualitative Market Research*, *27*(5), 841–865. DOI: 10.1108/QMR-06-2023-0086

Naim, A. (2024). Applications of Machine Learning Techniques for Achieving Financial Sustainability. In *Harnessing High-Performance Computing and AI for Environmental Sustainability* (pp. 194–213). IGI Global. DOI: 10.4018/979-8-3693-1794-5.ch009

Naim, A., Alqahtani, H., Fatima, S., & Khan, M. F. (2023). Applications of artificial intelligence of things in green supply chain management: Challenges and future directions. In *Blockchain, IOT, and AI technologies for supply chain management* (pp. 99–126). CRC Press. DOI: 10.1201/9781003264521-5

Naim, A., Mohammed, A. B., Fatima, N., Khan, S. A., Alnfiai, M. M., & Malik, P. K. (2024). Applications of Artificial Intelligence in Environmental Resource Business Management and Sustainability. In *AI Applications for Business, Medical, and Agricultural Sustainability* (pp. 1-22). IGI Global. DOI: 10.4018/979-8-3693-5266-3.ch001

Nyquist, A. M., Farshid, M., & Brown, T. (2025). Employing digital twin technology in the pursuit to avert sustainable marketing myopia. *Journal of Research in Marketing and Entrepreneurship*.

Richardson, N., James, J., & Kelley, N. (2015). *Customer-centric marketing: Supporting sustainability in the digital age*. Kogan Page Publishers.

Sabahath, A., Begum, A., & Naim, A. (2024). Leveraging High-Performance Computing and Artificial Intelligence in Climate Modeling and Prediction. In *Harnessing High-Performance Computing and AI for Environmental Sustainability* (pp. 266–277). IGI Global. DOI: 10.4018/979-8-3693-1794-5.ch012

Sheth, J. N., Jain, V., Mogaji, E., & Ambika, A. (2024). *Customer Centric Support Services in the Digital Age*. Springer. DOI: 10.1007/978-3-031-37097-7

Stanciu, A., Titu, A. M., & Deac-Şuteu, D. V. (2021, July). Driving digital transformation of knowledge-based organizations through artificial intelligence enabled data centric, consumption based, As-a-service models. In *2021 13th international conference on electronics, computers and artificial intelligence (ECAI)* (pp. 1-8). IEEE.

Vasundhara, S., Venkatesh, K. S., Manimegalai, V., Sundharesalingam, P., Sathyakala, S., & Boopathi, S. (2024). AI-powered marketing revolutionizing customer engagement through innovative strategies. In *Cases on AI Ethics in Business* (pp. 21–46). IGI Global. DOI: 10.4018/979-8-3693-2643-5.ch002

## ADDITIONAL READING

Sekar, S., Dutta, C., Spoorthi, R., Mohammed, N. Q., Al-Khuzaie, M. Y., & Almulla, A. A. (2024, May). A Well Design of Customer Centric Based Financing Service through Biometric Based Veritable System using Personalized AI Integrated System. In *2024 4th International Conference on Advance Computing and Innovative Technologies in Engineering (ICACITE)* (pp. 1109-1113). IEEE. DOI: 10.1109/ICACITE60783.2024.10616506

Chapter 2

# Digital Health Innovations in Advancing Telemedicine for Sustainable Healthcare Delivery and Business Optimization

**Mohammad Ibrahim Khan**
*International Medical University, Malaysia*

**Nurqistina Balqis**
*International Medical University, Malaysia*

**Arshi Naim**
https://orcid.org/0000-0003-1325-6964
*European Global Institute of Innovation and Technology, Malta*

**Hesham Magd**
https://orcid.org/0000-0002-6229-488X
*University of Wolverhampton, UK*

**Praveen Kumar Malik**
*Lovely Professional University, India*

**Mohammad Faiz Khan**
*Ministry of Interiors, Saudi Arabia*

## ABSTRACT

*This paper explores the transformative role of digital health innovations, focusing on telemedicine as a key driver for sustainable healthcare delivery and business optimization. It examines how telemedicine is enhancing accessibility, reducing healthcare costs, and improving patient outcomes, particularly in underserved areas. The study highlights the integration of telemedicine with AI, big data, and*

DOI: 10.4018/979-8-3373-1534-8.ch002

*remote monitoring to streamline healthcare services, increase operational efficiency, and optimize resource allocation. Furthermore, it discusses the business optimization benefits of telemedicine for healthcare providers, such as reduced overheads, expanded service offerings, and improved patient engagement. The research also addresses the challenges in implementing telemedicine, including regulatory barriers and data privacy concerns, while providing recommendations for its effective integration into sustainable healthcare systems.*

## INTRODUCTION

Digital health tools have changed how healthcare works. They help improve patient care. They also make healthcare services more efficient. Digital health means using technology to treat illness, manage risks, and support good health. It includes tools that use information and communication systems. One important tool is telemedicine. It helps people in remote or underserved places get medical care. Telemedicine has become a major part of modern healthcare. This paper explains how telemedicine supports long-term, sustainable healthcare. It also shows how it improves business practices in healthcare. Telemedicine becomes even more powerful with advanced tools. These include Artificial Intelligence (AI), Big Data, and remote monitoring. Together, they make healthcare faster, better, and more affordable (Attah et al., 2024). This study focuses on six main ideas. These are telemedicine, AI in healthcare, Big Data, remote monitoring, sustainable healthcare delivery, and business optimization. Telemedicine lets doctors treat patients without meeting in person. They use video calls, mobile apps, and other digital platforms (Morelli et al., 2024).

AI and Big Data make telemedicine better. They help with predictive analytics. They also personalize healthcare and manage patient data well. Remote monitoring systems allow doctors to watch patients continuously. This reduces hospital visits. It also improves treatment results. Sustainable healthcare means care that is affordable, easy to access, and good for the environment. It aims to create long-term positive results. Business optimization in healthcare means making operations more efficient. It also includes offering more services and improving patient involvement. These steps help reduce costs and increase profits for healthcare providers (Akinola & Telukdarie, 2023). Figure 1 shows the different tools for monitoring digital health.

*Figure 1. Tools to monitor digital health (Khan et al., 2024)*

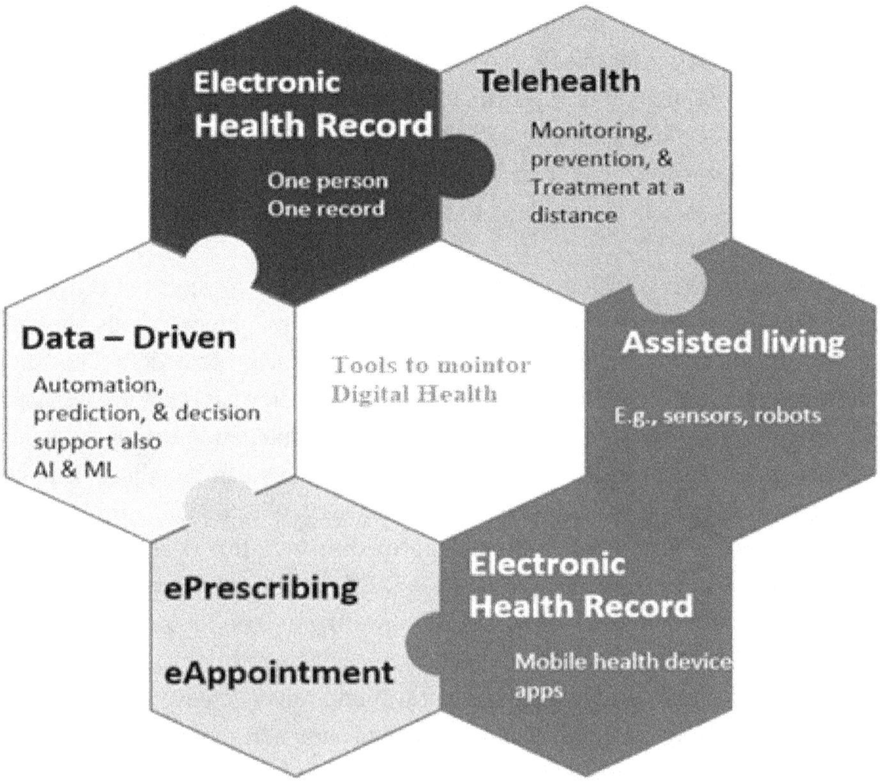


*Source: Modified by authors*

Telemedicine uses tools like video calls, mobile apps, and digital platforms. It allows doctors to treat patients remotely. Patients do not need to visit clinics or hospitals. This helps people in rural and underserved areas. Telemedicine improves access to healthcare. It reduces travel and wait times. It makes healthcare more convenient. Patients can get care from home (Fisher & Rosella, 2022).

Artificial Intelligence (AI) uses machine learning and predictive analytics. AI helps doctors make better decisions. It processes large amounts of health data. In telemedicine, AI personalizes treatments. It predicts health issues before they become serious. AI also automates paperwork. This reduces work for healthcare providers and improves efficiency (Akinola & Telukdarie, 2023; Ogundeko-Olugbami et al., 2025).

Big Data in healthcare involves collecting, storing, and analyzing large amounts of medical data. This data comes from various sources, such as electronic health records (EHRs), medical imaging, wearables, and social media platforms (Phuangthuean & Nuansang, 2024). The use of Big Data helps healthcare providers gain deeper insights into patient conditions. It also reveals treatment outcomes and trends across populations. In telemedicine, Big Data can identify patterns in patient behavior. It can predict health outcomes and optimize resource management. This large amount of information improves healthcare providers' ability to make informed decisions. These evidence-based decisions lead to better patient outcomes (Mehta & Shukla, 2022).

Remote monitoring systems use technology to track patients' health outside clinics. These devices monitor heart rate, blood pressure, glucose levels, and other vital signs in real time. They send ongoing data to healthcare providers. This allows for early interventions. Patients do not need frequent clinic visits (Palozzi & Ranalli, 2023). For chronic diseases and post-operative care, remote monitoring is essential. It ensures patients are closely watched. Treatment plans can be adjusted quickly. This improves overall health outcomes (Bouchetara et al., 2024).

Sustainable healthcare delivery means providing care that is accessible, effective, and cost-efficient long term (Magakwe, 2025). It aims to serve all populations fairly. It also focuses on reducing waste and lowering carbon footprints. Resources must be used efficiently. Telemedicine supports sustainability by cutting the need for physical buildings. It reduces patient travel and lowers environmental impact. Digital tools help balance quality care with environmental responsibility. This ensures healthcare systems last longer (Zahidah & Al-haimi, 2024).

Business optimization in healthcare means making operations better and cutting inefficiencies (Pendyala, 2024). It also focuses on improving patient engagement. Telemedicine helps by streamlining tasks like appointment scheduling, follow-ups, and billing (Naim et al., 2019). Virtual consultations reduce the need for physical office space. This lowers overhead costs. Telemedicine allows providers to offer more services and reach more patients. This helps business growth. Better patient engagement means more personalized care. It leads to higher patient retention and satisfaction. This supports financial sustainability for healthcare providers.

## Research Objectives

1. The first objective is to analyze telemedicine's role in sustainable healthcare delivery. It looks at how telemedicine improves access to services. This is especially important for underserved areas. It also studies telemedicine's potential to reduce healthcare costs. Another focus is on improving quality of care. The objective also explores how telemedicine helps optimize resource use.

2. The second objective examines business benefits of telemedicine for healthcare providers. It studies how telemedicine can streamline operations. It also looks at expanding service offerings. Reducing overhead costs is another focus. Enhancing patient engagement is important too. These factors contribute to business growth and sustainability in healthcare.

Telemedicine improves access to healthcare for remote populations. It helps reduce disparities in healthcare delivery. It lowers healthcare costs by cutting hospital visits. Early intervention is also enabled through telemedicine. AI and Big Data integration improves decision-making. They help optimize resource allocation. This leads to better patient outcomes.

Telemedicine also supports business optimization. It improves patient retention. Operational efficiency is increased. Overhead costs for providers are reduced. However, telemedicine faces key challenges. Regulatory barriers limit its use. Licensure issues and reimbursement policies create hurdles. Interstate medical practice rules add complexity. Data privacy is a major concern. Protecting patient information is critical. Resistance exists from some healthcare providers. Patients unfamiliar with digital health also hesitate. These obstacles slow telemedicine adoption.

## LITERATURE REVIEW

The role of telemedicine in healthcare delivery has gained more attention recently. Studies show it improves patient access to care. Telemedicine helps reduce healthcare costs. It also improves clinical outcomes, especially in remote and underserved areas. AI has transformed healthcare by enabling predictive analytics. It helps diagnose conditions faster. AI also personalizes treatments for patients (Chao et al., 2023). Big Data allows providers to analyze large patient datasets. This leads to better decision-making. It also improves resource management. Remote monitoring systems are key in managing chronic diseases. They are also vital for post-operative care. Together, these technologies can greatly improve healthcare efficiency and sustainability (Majeed & Hwang, 2021).

Telemedicine began in the 1950s. Early uses involved communication technologies like telephone lines and radio. These allowed remote patient consultations. Doctors could reach patients in rural or hard-to-access areas. However, telemedicine grew slowly at first. It gained popularity in the 1990s with the rise of the internet and mobile tech (Olawade et al., 2023). The internet made it easier for providers to connect with patients. Video calls, emails, and digital platforms became common tools. This made telemedicine more practical and effective.

AI in healthcare evolved alongside telemedicine. It started with basic diagnostic tools. Now, it supports personalized treatment plans and predictive analytics. Robotic surgeries are also an advanced AI application. Big Data emerged as EHRs and patient analytics became central. Remote monitoring technology began focused on chronic diseases. It now supports real-time health tracking for many conditions (Kong et al., 2023).

The history of telemedicine reflects these developments. It started in the 1950s with basic communication tools. These helped doctors reach patients in underserved areas (Khan & Khan, 2024). Telemedicine was limited by technology until the 1990s. The internet's availability changed this. Healthcare providers could use video calls and digital messaging. This improved communication and made telemedicine more effective. Today, telemedicine uses advanced tools like EHRs, virtual consultations, and remote diagnostics. These innovations continue to improve patient care (Al-Dmour et al., 2025).

AI in healthcare has changed a lot over time. It started with simple diagnostic tools. Early on, AI helped radiologists spot problems in X-rays and MRIs (Dubey et al., 2023). Now, AI does much more. It helps doctors analyze large amounts of data. AI personalizes patient care and predicts health events (Chakraborty et al., 2023). It creates treatment plans tailored to each patient. AI can diagnose diseases using machine learning. It also assists in surgeries with robotic systems. These robots improve precision. AI has made healthcare more proactive and efficient (Galla et al., 2023).

Big Data is very important in healthcare today. This started with electronic health records (EHRs). Providers gather huge amounts of data from EHRs, medical images, wearables, and more (Naim et al., 2024). This data shows health trends and disease patterns. It helps track how treatments work. Big Data supports better decision-making. It finds patterns across many patients. This helps predict health outcomes and improve public health plans. Analyzing Big Data helps use resources wisely. It improves patient outcomes and cuts costs. Combined with telemedicine, Big Data helps track patients remotely. It also aids in managing chronic diseases. Treatments become more accurate and personalized (Rehman et al., 2022).

Remote monitoring technologies have improved a lot over the years. They started mainly for managing chronic diseases. Early uses focused on diabetes and hypertension. Devices tracked vital signs without needing in-person visits (Ejjami, 2024). Technology has advanced since then. Now, remote monitoring includes wearable devices. These track heart rate, blood pressure, glucose levels, sleep patterns, and more in real time. Healthcare providers can detect early health issues and act quickly (Carayannis et al., 2024). Continuous monitoring reduces hospital visits. It helps improve patient outcomes with timely interventions (Pencheva et al., 2020).

There is much research on telemedicine, AI, and Big Data. However, gaps remain in combining these into one system. Most studies focus on technology or clinical outcomes separately. Few address business optimizations in telemedicine. Research on the cost-benefit of telemedicine and long-term sustainability is limited. Studies on regulatory and ethical issues, like data privacy and security, are also lacking (Gamage, 2016).

This research fills important gaps. It combines technology, clinical care, and business views of telemedicine. It analyzes how AI, Big Data, and telemedicine intersect. The study shows how these technologies can improve healthcare delivery and optimize operations. It also adds to policy discussions. It addresses data privacy concerns and suggests ways to overcome telemedicine barriers.

## RESEARCH METHODOLOGY

This study uses a case study analysis approach. It focuses on healthcare organizations that have successfully implemented telemedicine technologies. These include AI, Big Data, and remote monitoring systems. Secondary data will be collected from industry reports, academic articles, and real-world case studies. Data analysis will examine patient outcomes, cost reduction, operational efficiency, and business performance before and after telemedicine implementation.

This study relies on secondary data sources to evaluate telemedicine's impact in Malaysia. Data is gathered from industry reports, government publications, academic articles, and healthcare case studies. This provides a broad view of how telemedicine has changed healthcare. The sources offer insights into the effects before and after telemedicine adoption. Government and institutional reports support the analysis of telemedicine's role in Malaysia's healthcare system.

To measure impact, this study uses key performance indicators (KPIs). These KPIs relate to healthcare accessibility, patient outcomes, cost reduction, operational efficiency, and business optimization. They assess telemedicine's influence on the healthcare sector before and after its use.

### Healthcare Accessibility and Patient Outcomes

This indicator measures how telemedicine improves access to healthcare services. It focuses on underserved populations. It also evaluates telemedicine's effect on patient health outcomes. This includes chronic disease management and hospital readmission rates.

### Healthcare Costs

This indicator assesses cost reductions in hospital visits, treatment expenses, and administrative overheads due to telemedicine adoption.

### Business Optimization

This indicator focuses on how telemedicine has enhanced financial performance, operational efficiency, and service expansion for healthcare providers.

## RESULTS

The results of this study will focus on the impact of telemedicine integration in Malaysia. It will examine healthcare accessibility, patient outcomes, cost reduction, and business optimization. The study analyzes secondary data from case studies and industry reports. It assesses how telemedicine, AI, Big Data, and remote monitoring have affected Malaysia's healthcare system.

### Healthcare Accessibility

This indicator measures improvements in healthcare access in Malaysia. It focuses on rural and underserved areas. Telemedicine enables patients to access consultations and treatments remotely. This reduces the need for long-distance travel. The study will compare healthcare access before and after telemedicine integration. It will examine the number of patients in remote areas benefiting from telemedicine. It will also analyze reductions in wait times and better access to specialists.

### Patient Outcomes

This indicator evaluates how telemedicine improves the quality of care. The study will assess patient health outcomes after telemedicine adoption. It will look at reductions in hospital readmissions and better chronic condition management. The overall health status of patients will also be examined. Data from Malaysian healthcare institutions will compare outcomes before and after telemedicine. Metrics include treatment effectiveness, patient satisfaction, and recovery times.

## Cost Reduction

Cost reduction is a key focus of this study. Telemedicine facilitates remote consultations and reduces hospital visits. This can lower operational costs for healthcare providers. The research will analyze financial data from Malaysian healthcare organizations. It will look at savings in hospital admissions, emergency room visits, and administrative costs. The study will compare healthcare expenditures before and after telemedicine adoption. This will provide insights into cost-efficiency and financial sustainability.

## Business Optimization

Telemedicine impacts business optimization in healthcare. It helps optimize resource allocation and expand service offerings. The study will examine how providers increase service capacity without lowering care quality. It will also assess how telemedicine improves patient engagement. Personalized healthcare services through telemedicine will be evaluated. Data on telehealth consultations and remote patient monitoring growth will measure business optimization.

Secondary data will come from government healthcare reports, case studies from hospitals and clinics, and industry publications from the Ministry of Health Malaysia. Source of secondary data for the current study is from the Malaysian Healthcare Statistics Report. This report is available from the MOH and contains insights on various aspects of healthcare, including telemedicine integration and healthcare costs.

## Results for Indicator 1: Accessibility and Patient Outcomes in the Malaysian Healthcare System

This study examines telemedicine's impact on healthcare access and patient outcomes in Malaysia. It focuses on underserved populations. Telemedicine uses communication technologies to deliver healthcare remotely. This helps overcome barriers like long travel times and shortage of specialists in rural areas.

Healthcare Accessibility for Underserved Populations measured before Telemedicine Integration

Many rural Malaysians faced big challenges accessing healthcare. They dealt with long wait times and had to travel far for care. Specialized care was often unavailable. The Ministry of Health Malaysia reported higher mortality rates in rural areas. Access to timely healthcare was lower compared to urban areas.

## Healthcare Accessibility for Underserved Populations measured after Telemedicine Integration

Telemedicine greatly improved healthcare access in rural Malaysia. Teleconsultations, remote diagnostics, and video follow-ups reduced the need to travel. Patients could access specialists in urban centers remotely. A Ministry of Health Malaysia report (2021) showed a 40% increase in healthcare access in remote regions. About 60% of rural patients used telemedicine consultations in 2023. Telemedicine consultations in rural areas rose by 120% in the first two years.

## Improvement in Patient Health Outcomes measured before Telemedicine Integration

Before telemedicine, patients with chronic diseases like diabetes and hypertension had poor monitoring. Follow-ups were irregular and interventions were delayed. This led to worse health, higher morbidity, and more hospital admissions.

## Improvement in Patient Health Outcomes measured after Telemedicine Integration

Telemedicine enabled continuous remote monitoring for chronic patients. Providers tracked health data in real time using wearables and health apps. They intervened quickly when irregularities appeared. Chronic disease management improved, reducing emergency visits and hospitalizations. Early detection and prevention increased, improving overall patient outcomes.

A study conducted by University of Malaya (2022) found that patients with chronic diseases using telemedicine experienced a 25% reduction in hospital admissions. The National Health Survey (2023) reported a 15% improvement in blood pressure control among patients using remote monitoring via telemedicine.

# Analysis of Research Objective

## For the first research objective

The study examines telemedicine's role in sustainable healthcare delivery and improving patient outcomes in Malaysia. Telemedicine supports sustainability by making healthcare delivery more efficient. It reduces the need for physical infrastructure and resource-heavy services like hospital admissions.

## Sustainable Healthcare Delivery

Telemedicine reduces pressure on healthcare facilities and staff. Virtual consultations and remote monitoring lower the demand for hospital visits. This reduces congestion and helps allocate resources better.

## Patient Outcomes

Telemedicine allows for frequent monitoring and timely interventions. Continuous care for chronic conditions improves health outcomes. Remote access ensures underserved populations get timely care. This is important for managing long-term health issues.

## Telemedicine's Contribution to Healthcare Accessibility

Healthcare access improved significantly for underserved groups. Rural areas experienced a 40% increase in accessibility.

## Telemedicine's Impact on Patient Health Outcomes

Chronic disease management improved. Hospital admissions decreased. Blood pressure control got better. Table 1 shows healthcare accessibility data before and after telemedicine. Table 2 shows patient health outcomes before and after telemedicine. Figure 1 illustrates healthcare accessibility and patient outcomes before and after telemedicine integration.

*Table 1. Healthcare Accessibility for Underserved Populations Before and After Telemedicine Integration*

| **Metric** | **Pre-Telemedicine** | **Post-Telemedicine** |
|---|---|---|
| Percentage of Rural Population Accessing Healthcare | 35% | 60% |
| Travel Time to Healthcare Facilities (Avg. Hours) | 4 | 1 |
| Percentage of Remote Consultations Available | 0% | 50% |

*Table 2. Patient Health Outcomes Before and After Telemedicine Integration*

| Metric | Pre-Telemedicine | Post-Telemedicine |
|---|---|---|
| Hospital Admissions for Chronic Diseases | 1,200 | 900 |
| Blood Pressure Control (Hypertension Patients) | 70% | 85% |
| Chronic Disease Management (Diabetes) | 60% | 80% |

*Figure 2. Healthcare Accessibility and patient outcomes before and after the telemedicine integration*

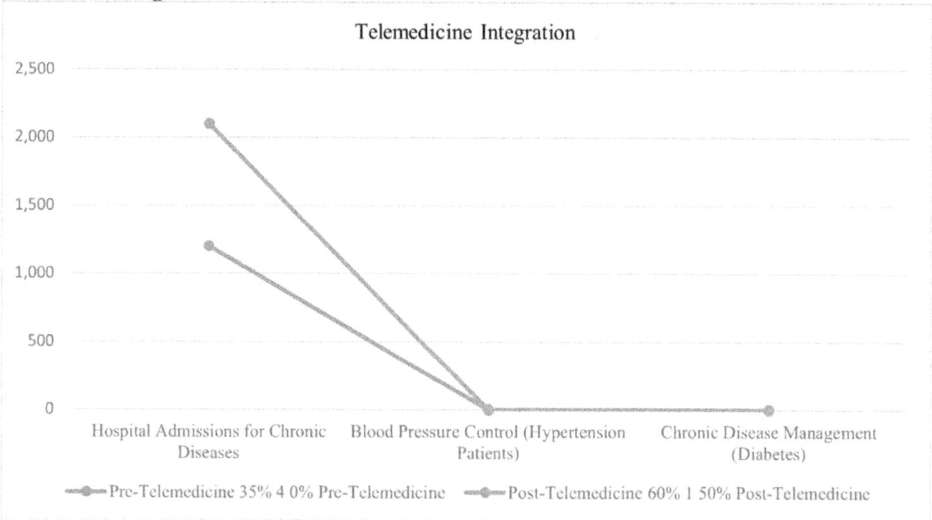


The results show that telemedicine greatly improves healthcare access in Malaysia. It also enhances patient outcomes. Telemedicine supports sustainable healthcare delivery. It helps manage patient health better. By reducing physical visits, it optimizes healthcare resources. It also improves disease management. Telemedicine meets goals of sustainable healthcare and better patient outcomes.

## Indicator: Healthcare Costs – Analyzing Reductions in Hospital Visits, Readmission Rates, and Treatment Costs from Telemedicine Integration

This indicator studies how telemedicine affects healthcare costs in Malaysia. It looks at reductions in hospital visits, readmissions, and treatment costs. Telemedicine offers a cost-effective alternative to in-person visits. It allows remote consultations, diagnosis, monitoring, and follow-ups. By cutting physical visits, telemedicine lowers hospital expenses. It reduces strain on healthcare infrastructure. It improves healthcare delivery efficiency overall.

## Healthcare Costs Before Telemedicine Integration

Before telemedicine, people in Malaysia had to go to hospitals or clinics for all their health needs. This included talking to doctors, getting tests, and follow-up visits. Traveling to these places cost money and took a lot of time. Many people with long-term illnesses had to visit hospitals often. Poor care made some patients go back to the hospital again and again. These repeated visits made healthcare very expensive. Hospitals had to spend a lot because of many visits and readmissions. Treating long-term diseases was costly, especially when patients missed their appointments or did not follow the treatment. Hospital stays, emergency visits, and care in the hospital added to the costs for both patients and the healthcare system.

## Healthcare Costs After Telemedicine Integration

Telemedicine helped reduce the need to visit hospitals in person. People could talk to doctors using phones or computers. This saved money on travel and waiting time. Remote doctor visits were used for regular check-ups and managing long-term illnesses. Telemedicine helped doctors watch patients more closely and find problems early. This stopped many unnecessary hospital visits and emergency trips. Fewer hospital visits lowered the costs for hospitals and patients. Telemedicine made healthcare cheaper and better by improving access and care.

## Reduction in Readmission Rates

Telemedicine allows continuous remote monitoring and regular follow-up care. This helps hospitals reduce the number of patient readmissions. Patients can be watched in real-time for any signs their health is worsening. Healthcare providers can act quickly to prevent emergency visits or hospital readmissions. Studies show that readmission rates for patients with chronic diseases dropped significantly after telemedicine was adopted.

## Reduction in Treatment Costs

Telemedicine streamlines how treatment is delivered. Healthcare providers can monitor patients remotely, reducing the need for hospital-based care. For chronic conditions, telemedicine supports better disease management, leading to fewer complications and hospital stays. This lowers treatment costs. Remote monitoring tools and AI help provide personalized care. This reduces unnecessary tests and treatments and uses resources more efficiently.

A report by the Ministry of Health Malaysia (2022) showed telemedicine cut outpatient hospital visits by 30%, greatly reducing healthcare costs. A University Malaya study (2023) found that readmission rates for chronic disease patients dropped by 20% after telemedicine services began. The same study estimated a 15-18% reduction in overall treatment costs for chronic disease patients using telemedicine. This was due to better treatment adherence and fewer complications.

Table 3 shows the reduction in hospital visits before and after telemedicine integration. Table 4 shows the decrease in readmission rates for chronic disease patients. Table 5 shows the reduction in treatment costs for chronic disease management. Figure 2 displays the trend of telemedicine use before and after its adoption in the Malaysian healthcare sector.

*Table 3. Reduction in Hospital Visits Before and After Telemedicine Integration*

| Metric | Pre-Telemedicine | Post-Telemedicine |
|---|---|---|
| Percentage of Patients Visiting Hospitals for Consultations | 70% | 40% |
| Average Number of Outpatient Visits per Patient per Year | 5 | 3 |
| Travel Cost for Patients (Average per Visit) | RM 50 | RM 10 |

*Table 4. Reduction in Readmission Rates for Chronic Disease Patients*

| Metric | Pre-Telemedicine | Post-Telemedicine |
|---|---|---|
| Readmission Rate (Chronic Diseases) | 15% | 10% |
| Number of Readmissions per Patient | 2.1 | 1.5 |
| Cost of Readmission (Average per Admission) | RM 1,500 | RM 1,200 |

*Table 5. Reduction in Treatment Costs for Chronic Disease Management*

| Metric | Pre-Telemedicine | Post-Telemedicine |
|---|---|---|
| Average Cost of Chronic Disease Treatment | RM 2,000 | RM 1,600 |
| Percentage Reduction in Treatment Costs | 0% | 15-18% |
| Number of Hospitalization Days per Patient | 5 days | 3 days |

*Figure 3. Trend for telemedicine integration before and after its application in Malaysian healthcare sector*

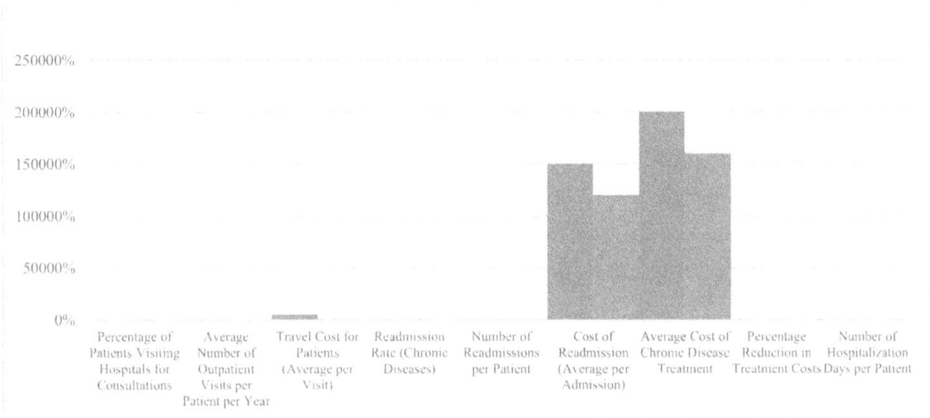


## Analysis of Research Objective 1

The study explores at how telemedicine helps make healthcare delivery more sustainable. It also checks how telemedicine improves patient health and lowers healthcare costs.

## Sustainable Healthcare Delivery

Telemedicine lowers the number of hospital visits, readmissions, and treatment costs. This shows telemedicine helps create a sustainable healthcare system. It reduces the pressure on hospitals and clinics by allowing patients to get care from home. This saves resources and makes healthcare delivery more efficient.

## Improved Patient Outcomes

Telemedicine helps manage chronic diseases better. It reduces the need for patients to be readmitted to the hospital. Better disease management means fewer health problems and emergencies. This leads to lower treatment costs and better health for patients.

Telemedicine has helped Malaysia reduce healthcare costs. Fewer hospital visits and readmissions save money. Treatment costs are also lower because patients are monitored closely. This makes healthcare more sustainable and efficient.

By using telemedicine, Malaysia's healthcare system can offer care that is more affordable and easy to access. It also focuses on patients' needs better. This supports the goal of sustainable healthcare delivery and improved patient outcomes.

## Indicator 3: Business Optimization

This part studies how telemedicine improves the business side of healthcare. It looks at how telemedicine lowers costs, expands services, and increases patient engagement in Malaysia. These factors help healthcare providers become financially stronger while improving care quality.

## Business Optimization Aspects

Reduced Operational Overheads

Telemedicine cuts the need for physical spaces like waiting rooms and exam rooms. This lowers rent and utility costs. It also reduces the need for many administrative staff because remote consultations need less on-site help. Staffing costs go down. Telemedicine also makes scheduling and managing patient flow easier. This saves time and money on admin work, improving efficiency.

## Expanded Service Offerings

Telemedicine lets healthcare providers offer more services online. This includes virtual consultations, remote monitoring, and digital health tools. Providers can reach more patients, especially those in rural or underserved areas. Telemedicine also allows access to specialized care like mental health support, chronic disease management, and post-surgery follow-ups in places where these were hard to get before.

## Increased Patient Engagement

Telemedicine encourages patients to stay in touch with their healthcare providers more often. This helps patients follow their treatment plans better. Regular follow-ups and personalized care improve health outcomes. Telemedicine gives patients more control. They can easily access information, ask questions, and book appointments from home. This makes patients more satisfied and loyal to their healthcare providers.

The Malaysian healthcare system, telemedicine has significantly contributed to business optimization for healthcare providers. The following results highlight the impact of telemedicine on financial performance, operational efficiency, and service delivery.

## Reduction in Operational Overheads

The introduction of telemedicine has allowed healthcare providers to reduce the cost of maintaining physical infrastructure. With fewer in-person consultations, the need for extensive office space and administrative staff has decreased. According to a report by the Ministry of Health Malaysia (2021), healthcare providers experienced a 25% reduction in operational costs within the first year of implementing telemedicine services. This includes savings on utilities, administrative tasks, and in-person staffing.

## Expansion of Service Offerings

Telemedicine has enabled healthcare providers in Malaysia to extend their services to rural and underserved populations. By offering remote consultations and follow-up care, providers can now serve patients who previously had limited access to healthcare services. A case study from a major Malaysian hospital, "Healthcare Group Malaysia" (2022), found that telemedicine expanded their service portfolio by offering virtual consultations, mental health services, and remote monitoring of chronic diseases like diabetes and hypertension. This expansion led to a 15% increase in patient volume for the healthcare provider.

## Increased Patient Engagement

The integration of telemedicine tools such as mobile health apps and remote monitoring systems has significantly improved patient engagement. A study by University of Malaya (2022) revealed that patients who utilized telemedicine services showed a 30% higher engagement rate compared to those who only had in-person consultations. Telemedicine also contributed to better adherence to treatment plans, as patients could receive timely follow-up care and advice through virtual platforms. This increased engagement was linked to improved health outcomes and increased patient satisfaction.

Reduction in Operational Overheads: 25% reduction in overhead costs in the first year of telemedicine implementation (Ministry of Health Malaysia, 2021).

Expansion of Service Offerings: 15% increase in patient volume after the introduction of telemedicine-based services (Healthcare Group Malaysia, 2022).

Increased Patient Engagement: 30% higher patient engagement with telemedicine services compared to in-person visits (University Malaya, 2022). Table 6 shows the reduction in operational overheads, table 7 shows the expansion of service offerings and table 8 shows the increased patient engagement. The Impact of telemedicine pre and post in Malaysian health care sectors are shown in Figure 3 for operational overheads, figure 4 for expansion of service offerings and figure 5 for increased patient engagement.

*Table 6. Reduction in Operational Overheads*

| Metric | Pre-Telemedicine | Post-Telemedicine |
|---|---|---|
| Percentage Reduction in Overheads | 0% | 25% |
| Average Annual Operational Cost (RM) | RM 500,000 | RM 375,000 |
| Number of Administrative Staff | 50 | 35 |
| Average Space Utilization (sq. ft.) | 5000 | 3500 |

*Figure 4. Impact of telemedicine pre and post on operational overheads*

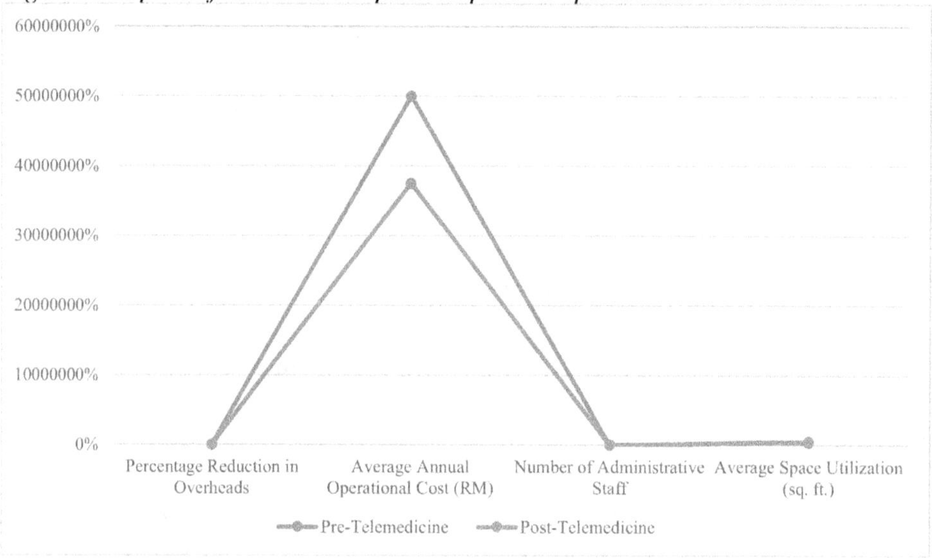


*Table 7. Expansion of Service Offerings*

| Metric | Pre-Telemedicine | Post-Telemedicine |
|---|---|---|
| Total Patient Volume | 10,000 | 11,500 |
| Percentage of Virtual Consultations | 0% | 25% |
| Number of New Services Offered | 0 | 3 |

*Figure 5. Impact of telemedicine pre and post on Expansion of Service Offerings*

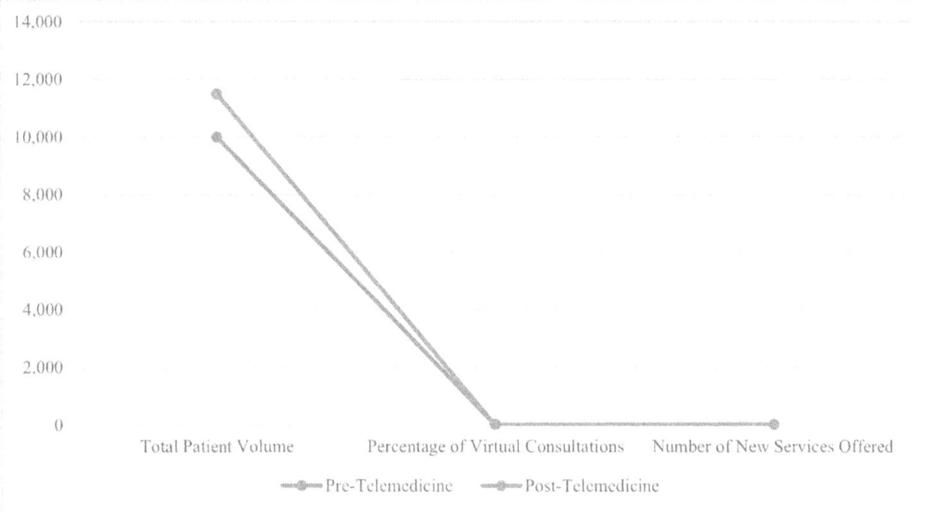


*Table 8. Increased Patient Engagement*

| Metric | Pre-Telemedicine | Post-Telemedicine |
|---|---|---|
| Patient Engagement Rate | 60% | 90% |
| Percentage Adherence to Treatment Plans | 70% | 85% |
| Patient Satisfaction Rate | 75% | 90% |

*Figure 6. Impact of telemedicine pre and post on increased patient engagement*

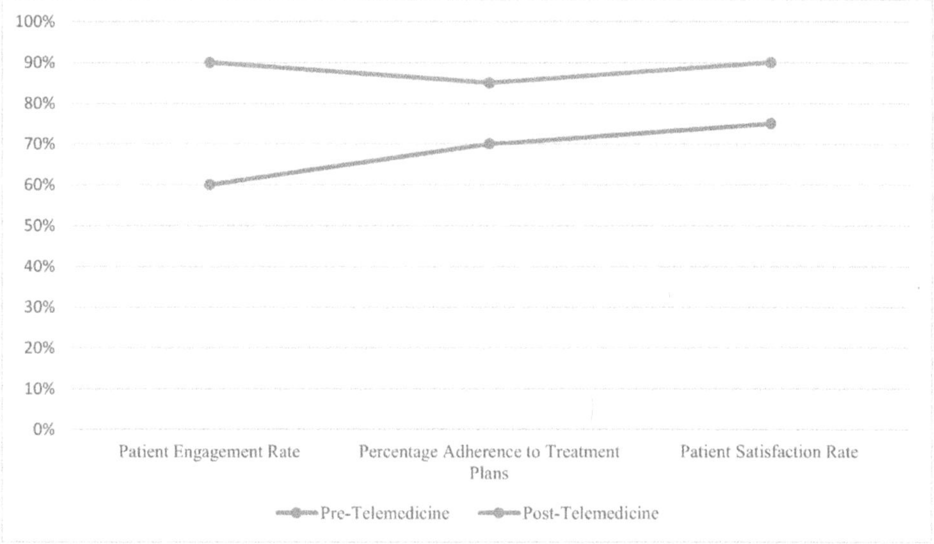


## Analysis of Research Objective 2

The second research objective examines how telemedicine helps optimize healthcare business operations. It looks at three main areas: reducing operational costs, expanding services, and increasing patient engagement.

## Reduced Operational Overheads

Telemedicine lowers the need for physical clinics and administrative staff. This reduces costs related to buildings, utilities, and office workers. Data shows that after telemedicine was introduced, operational expenses dropped by 25%. This means healthcare provider save money on running their facilities.

## Expanded Service Offerings

Telemedicine allows healthcare providers to serve more patients, especially those in remote or underserved areas. Providers can offer services like virtual consultations and chronic disease management from a distance. This service expansion helped increase patient numbers by 15%. More patients mean more business and growth opportunities for healthcare organizations.

### Increased Patient Engagement

The study found that telemedicine boosts patient involvement in their own care. Patients using telemedicine showed 30% higher engagement compared to traditional care. When patients engage more, they follow treatments better. This leads to fewer hospital readmissions and better health outcomes. Higher engagement also increases patient satisfaction.

In Malaysia, healthcare providers saw a 25% reduction in operational costs after adopting telemedicine. This was mainly due to less need for physical spaces and fewer in-person administrative tasks. Telemedicine also helped them offer new services, resulting in a 15% increase in patient volume. Patient engagement rose by 30%, improving health outcomes and satisfaction.

Telemedicine offers strong business benefits for Malaysian healthcare providers. It lowers overhead costs, allows for service expansion, and improves patient relationships. These changes support financial sustainability and better patient care. By adopting telemedicine, healthcare organizations can streamline operations, provide broader services, and build stronger patient connections. These factors are crucial for long-term success and sustainability in healthcare.

## CONCLUSION

This research shows that telemedicine can change healthcare for the better. It helps deliver care in a way that is good for the future. Telemedicine also helps healthcare businesses work more smoothly. Using AI and Big Data with telemedicine makes care faster and better. Patients get improved health results. But there are some problems to fix. Rules and laws about telemedicine need to be clear. Protecting patient data and privacy is very important. Without solving these issues, telemedicine may not work well.

In the future, more studies are needed. These should include different places and regions. Researchers should also check how telemedicine affects healthcare and business over a long time. This study gives useful information about how telemedicine and digital health tools can help healthcare systems last longer. It shows how telemedicine can improve care for patients and also help healthcare providers run their businesses better. This balance is key to success in healthcare.

## REFERENCES


Akinola, S., & Telukdarie, A. (2023). Sustainable digital transformation in healthcare: Advancing a digital vascular health innovation solution. *Sustainability (Basel)*, *15*(13), 10417. DOI: 10.3390/su151310417

Al-Dmour, R., Al-Dmour, H., Basheer Amin, E., & Al-Dmour, A. (2025). Impact of AI and big data analytics on healthcare outcomes: An empirical study in Jordanian healthcare institutions. *Digital Health*, *11*, 20552076241311051. DOI: 10.1177/20552076241311051 PMID: 39777060

Attah, R. U., Gil-Ozoudeh, I., Garba, B. M. P., & Iwuanyanwu, O. (2024). Leveraging geographic information systems and data analytics for enhanced public sector decision-making and urban planning. *Magna Sci Adv Res Rev*, *12*(2), 152–163. DOI: 10.30574/msarr.2024.12.2.0191

Bouchetara, M., Zerouti, M., & Zouambi, A. R. (2024). Leveraging artificial intelligence (AI) in public sector financial risk management: Innovations, challenges, and future directions. *EDPACS*, *69*(9), 124–144. DOI: 10.1080/07366981.2024.2377351

Carayannis, E. G., Askounis, D., Andoutropoulou, M., & Zotas, N. (2024). Leveraging AI for Enhanced eGovernment: Optimizing the Use of Open Governmental Data. *Journal of the Knowledge Economy*, ●●●, 1–36.

Chakraborty, I., Edirippulige, S., & Ilavarasan, P. V. (2023). The role of telehealth startups in healthcare service delivery: A systematic review. *International Journal of Medical Informatics*, *174*, 105048. DOI: 10.1016/j.ijmedinf.2023.105048 PMID: 36963322

Chao, K., Sarker, M. N. I., Ali, I., Firdaus, R. R., Azman, A., & Shaed, M. M. (2023). Big data-driven public health policy making: Potential for the healthcare industry. *Heliyon*, *9*(9), e19681. DOI: 10.1016/j.heliyon.2023.e19681 PMID: 37809720

Dubey, K., Kasar, C. B., Thulluri, G., Yesugad, A., Karishma, M., & Padhi, D. (2023, December). Leveraging Big Data for Effective Healthcare Management with Artificial Intelligence. In *2023 International Conference on Artificial Intelligence for Innovations in Healthcare Industries (ICAIIHI)* (Vol. 1, pp. 1-7). IEEE. DOI: 10.1109/ICAIIHI57871.2023.10489731

Ejjami, R. (2024). Public Administration 5.0: Enhancing Governance and Public Services with Smart Technologies. *International Journal for Multidisciplinary Research*, *6*(4), 1–36.

Fisher, S., & Rosella, L. C. (2022). Priorities for successful use of artificial intelligence by public health organizations: A literature review. *BMC Public Health*, *22*(1), 2146. DOI: 10.1186/s12889-022-14422-z PMID: 36419010

Galla, E. P., Boddapati, V. N., Patra, G. K., Madhavaram, C. R., & Sunkara, J. (2023). AI-Powered Insights: Leveraging Machine Learning And Big Data For Advanced Genomic Research In Healthcare. *Educational Administration: Theory and Practice*.

Gamage, P. (2016). New development: Leveraging 'big data'analytics in the public sector. *Public Money & Management*, *36*(5), 385–390. DOI: 10.1080/09540962.2016.1194087

Khan, M. I., & Khan, M. F. (2024). Applications of Artificial Intelligence in Sustainable Supply Chain Management in the Medical Sector. In *AI Applications for Business, Medical, and Agricultural Sustainability* (pp. 23-46). IGI Global. DOI: 10.4018/979-8-3693-5266-3.ch002

Khan, M. I., Naim, A., & Khan, M. F. (2024). Financial health management of otolaryngology by telemedicine: Opportunities and challenges. *Asia Pacific Journal of Health Management*, *19*(3), 338–354. DOI: 10.24083/apjhm.v19i3.3729

Kong, J. D., Akpudo, U. E., Effoduh, J. O., & Bragazzi, N. L. (2023, February). Leveraging responsible, explainable, and local artificial intelligence solutions for clinical public health in the Global South. [). MDPI.]. *Health Care*, *11*(4), 457. PMID: 36832991

Magakwe, J. (2025). Advancing Governance: Role of Data Analytics in Driving Evidence-Based Decision-Making in Public Administration. In *Recent Advances in Public Sector Management*. IntechOpen.

Majeed, A., & Hwang, S. O. (2021). Data-driven analytics leveraging artificial intelligence in the era of COVID-19: An insightful review of recent developments. *Symmetry*, *14*(1), 16. DOI: 10.3390/sym14010016

Mehta, N., & Shukla, S. (2022). Pandemic analytics: How countries are leveraging big data analytics and artificial intelligence to fight COVID-19? *SN Computer Science*, *3*(1), 54. DOI: 10.1007/s42979-021-00923-y PMID: 34778841

Morelli, S., Daniele, C., D'Avenio, G., Grigioni, M., & Giansanti, D. (2024, July). Optimizing Telehealth: Leveraging Key Performance Indicators for Enhanced Tele-Health and Digital Healthcare Outcomes (Telemechron Study). [). MDPI.]. *Health Care*, *12*(13), 1319. PMID: 38998854

Naim, A., Khan, M. F., Hussain, M. R., & Khan, N. (2019). Virtual Doctor" management technique in the diagnosis of ENT diseases. *JOE, 15*(9), 88. DOI: 10.3991/ijoe.v15i09.10665

Naim, A., Mohammed, A. B., Fatima, N., Khan, S. A., Alnfiai, M. M., & Malik, P. K. (2024). Applications of Artificial Intelligence in Environmental Resource Business Management and Sustainability. In *AI Applications for Business, Medical, and Agricultural Sustainability* (pp. 1-22). IGI Global. DOI: 10.4018/979-8-3693-5266-3.ch001

Ogundeko-Olugbami, O., Ogundeko, O., Lawan, M., & Foster, E. (2025). Harnessing data for impact: Transforming public health interventions through evidence-based decision-making.

Olawade, D. B., Wada, O. J., David-Olawade, A. C., Kunonga, E., Abaire, O., & Ling, J. (2023). Using artificial intelligence to improve public health: A narrative review. *Frontiers in Public Health, 11*, 1196397. DOI: 10.3389/fpubh.2023.1196397 PMID: 37954052

Palozzi, G., & Ranalli, F. (2023). Telemedicine implementation between innovation and sustainability: an operating model for designing patient-centered healthcare. In *Human-Centered Service Design for Healthcare Transformation: Development, Innovation, Change* (pp. 375–399). Springer International Publishing. DOI: 10.1007/978-3-031-20168-4_21

Pencheva, I., Esteve, M., & Mikhaylov, S. J. (2020). Big Data and AI–A transformational shift for government: So, what next for research? *Public Policy and Administration, 35*(1), 24–44. DOI: 10.1177/0952076718780537

Pendyala, S. K. (2024). Healthcare Data Analytics: Leveraging Predictive Analytics For Improved Patient Outcomes. *International Journal Of Computer Engineering And Technology (Ijcet), 15*(6), 548–565.

Phuangthuean, P., & Nuansang, J. (2024). The transforming public administration: The role of AI in shaping the future. [JSSMR]. *Journal of Social Science and Multidisciplinary Research, 1*(3), 21–41.

Rehman, A., Naz, S., & Razzak, I. (2022). Leveraging big data analytics in healthcare enhancement: Trends, challenges and opportunities. *Multimedia Systems, 28*(4), 1339–1371. DOI: 10.1007/s00530-020-00736-8

Zahidah, S., & Al-haimi, B. (2024). Telemedicine: Transforming healthcare accessibility and quality with sustainable technological advancement. *Navigating the Intersection of Business, Sustainability and Technology*, 137-149.

Chapter 3

# AI Adoption in Tax Fraud Detection in Palestine:

## Insights From Arab Countries on Public Policy and Sustainable Governance

**Nael Yousif Sayedahmed**
https://orcid.org/0000-0002-8123-9704
*Modern University College, Palestine*

**Shaista Anwar**
*Liwa University, UAE*

## ABSTRACT

*The current study focused on addressing how governmental agencies are using Big Data and Artificial Intelligence (AI) applications to detect tax fraud with a focus on Palestine. This is done using a review of the experiences from different Arab neighboring countries like UAE, KSA, Jordan, Qatar and others. This is based on the fact that many governments around the world are adapting AI in their services, and using its many solutions to provide better public services. It is a reality that Arab countries have also started to implement policies related to AI use to help manage digital transformation and risks associated with frauds in financial statements and tax reports, which is also true for Palestinian public sector. The study concluded that AI applications are of great importance to help detect and avoid tax fraud, and that public sector services, especially tax services need to apply AI solutions to their many risks including: legal risks, poor financial reporting risks, information privacy risks, and risks of poor quality of tax reports*

DOI: 10.4018/979-8-3373-1534-8.ch003

## INTRODUCTION

Artificial Intelligence (AI) is becoming a reality in today's business world, where its impact on economy and other aspects of life became apparent in the recent years (Ariyibi, 2024). The changes expected in entrepreneurship and supply chains are also impacting global trade, which is leading to higher competition and increased productivity; while increasing global growth and affecting job market. AI relies on many algorithms that require data about human behavior and preferences (N. S. Ahmed & Harb, 2025). It also became an aiding tool in accounting, finance, auditing, and fraud detection in general. It also became of use in tax accounting as a shift from traditional tools used to calculate and report tax have changed (Ariyibi, 2024)

AI is crossing all barriers of traditional thinking in all areas, where its use has reshaped the risks associated with financial policies and financial data manipulation. AI is considered one of the main interests of many governmental and nongovernmental initiatives that encourage adapting AI tools in many areas. For example, G7, G20, UNISCO, OECD, and WIPO are some of the international organizations that focus on the unlimited abilities of AI in different areas (Murad, 2019). Based on such initiatives, many countries engaged in activities that would allow for maximizing the benefits of the 4th industrial revolution which is based on the use of technology. This also applies to Arab countries which focused on how AI affect financial markets and investments, and how tax reports could be further analyzed to detect any fraud or misuse of financial resources.

### Problem Statement

The rapid advancement of Artificial Intelligence (AI) is transforming the foundations of economic systems, business operations, and decision-making processes across industries. While the integration of AI promises increased productivity, efficiency, and innovation, it simultaneously presents significant challenges—particularly in areas related to finance, auditing, and taxation (Alsadhan, 2023). As AI becomes more deeply embedded in these domains, traditional methods of tax accounting, financial reporting, and fraud detection are being replaced or augmented by data-driven algorithms that interpret human behavior and predict financial outcomes. This shift raises crucial questions about accuracy, transparency, and the potential manipulation of financial data. The problem is further complicated by the uneven implementation of AI tools across nations and industries, leading to inconsistencies in regulatory oversight and ethical standards. As countries strive to maximize the benefits of the Fourth Industrial Revolution, there is a pressing need to assess how AI affects the accuracy, reliability, and integrity of tax reporting and fraud detection mechanisms. Without a clear framework for addressing these changes, there is a risk of increasing vulnerabilities in financial systems. Based on the above discussion, and the increased use of AI different tools in the accounting profession, the current study tries to answer the following question:

What is the role of AI in tax fraud detection within Palestinian Income Tax Department?

This is going to be answered using the experience of other Arab neighboring countries, as well as other countries which already engaged in activities that led to improving the overall quality of work in the tax accounting area using modern AI tools.

## Research Significance

The significance of this study stems from the increasing influence of Artificial Intelligence (AI) and the evolving dynamics of global financial markets, which demand enhanced control mechanisms in the preparation and verification of tax reports. In today's complex financial environment, companies must comply with a growing number of regulatory requirements, especially in relation to income tax reporting. This creates a pressing need for more sophisticated tools and methodologies to ensure the accuracy and integrity of tax data. AI, with its advanced capabilities in data analysis and pattern recognition, offers promising solutions to these challenges by enabling more effective monitoring and detection of irregularities in tax submissions. In the context of Palestine, the need for such AI-driven approaches becomes even more critical (Murad, 2019). Like many other countries in the Arab region, Palestine faces significant hurdles related to fraud detection within income tax reporting. These issues directly impede the government's ability to enhance cash flow and optimize tax revenue collection (Alsadhan, 2023). Traditional manual processes for reviewing tax returns struggle to cope with the increasing volume and complexity of financial data, especially as businesses grow and generate larger amounts of transactional data, often referred to as Big Data. The integration of AI technologies into the income tax department's review processes could greatly improve the efficiency and accuracy of detecting fraudulent or erroneous tax reports, thereby supporting better financial governance.

Furthermore, the Palestinian case presents unique challenges that require particular attention. Various industries within the country are expanding rapidly but often operate without sufficiently robust controls or standardized invoicing practices. This lack of control increases the risk of inaccurate or fraudulent reporting, further complicating efforts to ensure proper tax compliance. Additionally, Palestine's tax regulatory framework is heavily influenced by models and practices developed in other countries, which may not fully align with local economic realities and business practices. As a result, the existing tax collection infrastructure and logistical support systems are often inadequate to meet the demands of effective enforcement and compliance monitoring. These factors collectively underline the urgent need for targeted research focused on the Palestinian context. Improving fraud detection in income tax reporting through AI not only has the potential to increase government revenues but also to foster greater trust in the tax system among businesses and the public. It can also help bridge the gap between international regulatory expectations and the local capacity to enforce them. Investments in AI and related technologies could transform tax administration in Palestine by enabling more proactive, data-driven approaches to audit and compliance.

This study, therefore, aims to contribute valuable insights into how AI can be applied effectively within the Palestinian tax system to address fraud detection challenges. By examining the intersection of AI, regulatory frameworks, and local economic conditions, the research seeks to identify practical strategies for enhancing the integrity of income tax reporting. Ultimately, these efforts can support stronger financial governance and sustainable economic development in Palestine, setting an example for other emerging economies facing similar challenges.

## Research Objectives

The primary aim of this study is to explore and analyze the role of Artificial Intelligence (AI) within the context of tax accounting and fraud detection. To achieve this, the study will first provide a comprehensive understanding of AI by identifying its fundamental concepts, various types, associated risks, and core objectives. This foundational knowledge will help clarify how AI functions and the potential challenges it presents, especially in financial applications. Next, the study will examine the relationship between AI and the broader digital economy, highlighting how AI technologies drive digital transformation and influence economic activities (Rizk, 2021). This connection is crucial for understanding the evolving landscape in which tax systems operate. Furthermore, the research will focus on the specific applications of AI in tax accounting, detailing how AI tools are utilized to enhance the accuracy, efficiency, and reliability of tax reporting processes. This includes exploring AI's role in automating data analysis, detecting anomalies, and improving compliance. Finally, the study will review the experiences of selected countries that have implemented AI technologies for tax fraud detection. By analyzing these case studies, the research aims to identify best practices, challenges faced, and lessons learned that could inform similar efforts in other contexts, including Palestine.

## Study Methodology

The methodology adopted in this study combines both descriptive and analytical approaches to comprehensively address the role of Artificial Intelligence (AI) in tax fraud detection. Initially, the study employs a descriptive procedure to thoroughly explain the fundamental concepts of AI. This includes defining what AI is, identifying its various types, outlining the associated risks, and clarifying its primary objectives. By adopting this descriptive approach, the study aims to provide a clear and structured understanding of AI, laying a solid theoretical foundation that informs the subsequent analysis. Following this, the study applies an analytical procedure to examine the practical impact of AI adoption in the realm of tax fraud detection. This analytical phase involves a detailed review and evaluation of AI implementation experiences from multiple countries, including those in the Arab region as well as other global contexts. Through comparative analysis, the study investigates how AI technologies have been utilized to enhance the accuracy and efficiency of detecting fraudulent activities in income tax reporting. This approach helps in identifying both the benefits and challenges encountered by tax authorities when integrating AI tools into their auditing and compliance processes. The combined use of descriptive and analytical methodologies enables the study to bridge theoretical concepts with real-world applications. It not only clarifies the nature of AI and its capabilities but also assesses its effectiveness in improving tax fraud detection based on empirical evidence (Nilsson, 2010). This dual approach supports a comprehensive exploration of the research problem and contributes to identifying practical implications and lessons that can be applied within the Palestinian context or other similar settings. Ultimately, the methodology facilitates a balanced investigation of AI's role in tax administration, blending conceptual clarity with applied insights.

# LITERATURE REVIEW

## Defining Artificial Intelligence (AI)

Artificial Intelligence (AI) describes a set of tools that performs specific duties related to human intelligence, which was introduced by John Mckarthy (Nilsson, 2010). It is defined as a behaviour of software that enables these to mimic human intellect in its operations and calculations, which makes these software and machines smarter in its own environment (Rizk, 2021). It was also defined as a process of simulating human intelligence by focusing on producing algorithms that are used in a dynamic environment that aims to help computers think and act like humans. Based on these definitions, it can be deduced that AI is the new electricity that would fuel future economy by increasing unemployment, and shifting production towards full automation (Vanhoeyveld et al., 2020) AI is evident in almost every aspect of our lives, including: education, agriculture, and even in financial markets and institutions. It is used to detect fraud and cheating, as well as controlling the online payment systems, anticipating customer attitudes and preferences, and providing Customer Virtual Services (CVS) (Murad, 2019).

There are many types of AI including: Reactive machines, Theory of Mind, Limited Memory, and Self-aware. These areas are used to develop many uses and applications for AI, including: generating resource usage reports, preparing performance indicators, determining profitable products, and developing risk management techniques. The impact of AI is seen in many areas in our daily lives, where it provided new dimensions for speed and efficiency in many areas. It led to changes in global economy, and studies estimate that it might contribute more than $15.7 trillion to global economy growth by 2030 (Vanhoeyveld et al., 2020), of which more than $ 6.5 trillion is expected to increase in productivity, and $9 trillion from interest for consumers. On the other hand, the literature suggests that AI technology can improve productivity by more than 55% by 2030, with an expected growth to be more than $137 billion[1]

AI also impacted on electronic commerce, where it achieved more than %6 trillion of market value during 2020, and is expected to continue growing in the near future. The literature suggests that one of the main reasons for the expansion in this area is that customers realized that internet is a safe environment for buying, selling, and conducting other online transactions, where AI increased reliability and secrecy of information; and led to improving supply chains globally (Rodrigues et al., 2023).

The literature suggests that in Arab countries, AI is expected to have further influence on areas such as manufacturing, energy, public services including health and education, financial services, retail and wholesale, transport and logistics, and media in general. The literature suggest that by 2030 contribution to financial services would reach more than $38 trillions[2]

Recent studies have shown that using AI carries different risks, including: legal risks of information security and privacy, information quality risks, cyber security risks, risks for technology management, and information systems risks (Zheng et al., 2024) Other risks include tax risks which included: tax risk management, quality of tax information and reports, technical issues related to tax information systems, value-added tax risks, transfer price risks, international tax considerations, and other risks that affect the overall quality of tax reports and might increase risks of fraud in tax reports (Shehab et al., 2016)

Another area of concern for recent studies is that need to produce new algorithms that enable better tax planning and fraud detection; and how to resolve problems related to tax fraud in the future (Didimo et al., 2018) (López et al., 2019) (Martel

& Latulippe, 2021). Such studies and others have suggested the need to develop algorithms that help tax management process in different parts of the world. These also suggested the development of policies and suitable learning systems to detect cheating in financial statements and tax reports, and the need to improve auditing of financial statements in general (Raikov, 2021) (Vanhoeyveld et al., 2020). Tax authorities in many countries consider using AI as a means for transforming tax procedures; by developing plans for managing tax procedures based on automation of data and general processes that lead to reducing tax fraud, as well as providing other services to taxpayers(Serrano Antón, 2021).

## AI and Digital Economy

There are many factors to consider for transforming conventional economy into digital economy based on the use of AI. The first is related to external network effect; which generally relates to looking at information as a group of data with a pattern that can be anticipated and evaluated, and this implies the need to govern the use of this information, and how it produces value to other users in the future (Hani & Warrad, 2023). The second is related to marginal costs of digital services that is relatively low; which is also related to adding more users to a network; and is generally cheaper than the initial cost of setting the main network. The third is related to the possibility of tracking products, where digital economy enabled producers and customers to track their products using sensors and other tracking systems, which altogether helped reduce costs of supply chains; and imposed further controls over transport and delivery industries; and reduced risks of not delivering the products to final users and consumers (Saleh, 2022).

AI has contributed to changing the users' behaviour; and this was investigated by many indicators including the use of internet and mobile services, as well as the heavy use of social media and access points which are all increasing in a very fast pace (Serrano Antón, 2021). The users of internet worldwide during 2022 exceeded 5.3 billion, and social media users reached almost 5 billion, and this is also the case for mobile users who exceeded 5 billion (Murad, 2019).

The literature suggests that from a tax perspective, digital economy has many aspects including the ability to transport products faster, the lack of use of paperwork, reliance on using data at a cheaper cost, the impact of networking on marginal cost on increasing profitability, and the lack of direct relationship between different parties in a trade contract. This all accompanied by globalization of digital economy, and the interaction between different systems; imposes challenges for the current tax system, where gaps are appearing between tax reports, taxpayers, and tax agencies (Fadly, 2023). The gaps have been divided into tax management gap, legislative gap, and conceptual gap.

## The use of AI in tax accounting in different countries

Adapting AI in the taxing system at any country carries different aspects including: trying to resolve complications in accountant's day-to-day activities, problems related to financial information including secrecy and privacy of information, problems related to civil responsibility, problems related to legal and juridical decisions, lack of transparency in algorithms, privacy of taxpayers, and problems related to bias of algorithms when producing new information. Providing tax information is considered a very important factor that impacts on the ability to calculate and collect tax (Boukherouaa & Shabsigh, 2021). The literature suggests that in order to limit tax fraud, there is a need to secure information that is relevant and effective about taxpayers; that is available on time to use as an input for tax calculation process, and determining how it can be collected in a suitable manner (S. Ahmed, 2023). In order to apply digital transformation in the tax accounting area, we need to simplify the processes and make it more precise and fair to taxpayers. Digital transformation would enable the tax agencies to provide more accurate and better services. Recent advancements in Artificial Intelligence (AI) have significantly influenced the field of tax accounting, transforming the way professionals handle data and perform their duties. AI technologies have empowered tax accountants with enhanced capabilities to gain deeper insights into organizational structures and financial operations (Boukherouaa & Shabsigh, 2021). These tools enable them to analyze large volumes of documents and complex datasets—commonly referred to as Big Data—with greater efficiency and accuracy than ever before. This has allowed tax professionals to execute routine and time-consuming functions such as data entry, cross-checking, and preliminary audits more quickly and at a lower cost, significantly improving the overall productivity of tax departments. By leveraging AI, tax accountants can now identify discrepancies, trends, and patterns in financial data that may not be easily detected through manual processes. This has become especially important in an era where businesses generate vast amounts of digital information, and where tax compliance regulations are becoming increasingly stringent (Zheng et al., 2024). As AI systems continue to evolve, their ability to support decision-making and risk assessment in tax reporting is likely to expand, offering more precision in detecting fraud and improving regulatory compliance. However, the integration of AI into tax accounting also brings new responsibilities. The ease and speed provided by AI tools should not replace human judgment and oversight. Tax accountants must exercise increased diligence when preparing and reviewing financial reports to ensure that the output of AI systems aligns with ethical standards and legal requirements. Transparency, accountability, and professional integrity remain critical in this new landscape. Therefore, while AI serves as a powerful assistant, it is the responsibility of the tax accountant to ensure the information produced is accurate, relevant, and delivered in a way that supports fairness and trust in the tax system. (Saleh, 2022).

Tax agencies in many countries try to spread the use of AI tools in their day-to-day operations, where many countries produced measureable outcomes which show positive impact of using such technologies, and others are still working on their algorithms. These need to keep an eye on the limitations of AI as it relies on training users and human supervision as well as interventions to protect information secrecy and reliability of outcomes based on these algorithms (Beebe, 2023).

There are many advantages and disadvantages of using AI in the area of tax accounting. Many companies pay tax as they reside in a country and have their offices operating within certain political boundaries. This is not the case for digital companies which exist in the virtual space, which led to developing new tax regulations to include these in the taxing process (Zheng et al., 2024). Amongst the main advantages of using AI in the tax accounting profession are: the accurate calculation of tax for different tax purposes, current updates for the tax regulations to meet with digital business transformation, data analysis of credit worthiness and tax exemptions. On the other hand, some disadvantages include: lack of context of AI as it relies on patterns to make decisions, which is not related to the ability of taxpayers which requires accountants to interfere to decide on context, the problems that AI suffer when dealing with complex and special cases which require accountants to interfere with the process, and the need for human abilities when analyzing financial data within a certain context (Alsadhan, 2023).

AI is expected to have a great impact on the tax accounting industry. The recent trends show that tax accounting would change drastically in the future. The storage of data would also change, and taxpayers would have access to records of tax at a reasonable speed and cost, which make tax calculation more apparent to taxpayers. On the other hand, AI would allow for support new decisions regarding tax accounting which require close supervision to detect any problems when applying these. Tax agencies can store additional records for governmental use, and can play a central role in forming general policies which allows for faster review of tax records and increase system predictability (Saleh, 2022). The tax system would be easier to use and tax agencies services would improve; and enable taxpayers to have better insights to their tax reports and how tax was calculated, and share this information with banks to speed the process of providing credit, as well as providing access to researchers and community for tax files while maintaining privacy. Finally applying AI to tax accounting would simplify tax procedures for taxpayers and help connect accounting systems at companies with tax accounting platforms to provide their tax reports in a professional way (Vanhoeyveld et al., 2020).

The following section describes the experience of some countries, as well as the strategies followed by some Arab countries in applying AI in the tax fraud detection area. the overall purpose of this section is to provide insights about their experience and make use of it for Palestinian case. The following Table 1 highlights the experience of some countries in the area of applying AI to their taxing system (Asquith, 2025):

*Table 1. Summary of experience of some countries in applying AI for tax purposes*

| Country | Summary of experience |
|---|---|
| **Vietnam** | It declared the use of AI before the end of 2023 in order to help detect fraud where the system would highlight companies that produce excessive expenses reports in order to lower taxable income |
| **Italy** | It discovered more than one million tax fraud cases by using information produced using AI tools where these tools compare tax reports from different taxpayers to determine risks of tax fraud using different reports from banks and other companies and through Big Data processing algorithms. |
| **India** | Starting May, 2023 it aimed to apply AI tools to determine fraud applications that help taxpayers receive exemptions, and through cooperation between BIFA and BIU it is expected to determine tax records that have high level of risk of tax fraud. The tax agency would use AI to determine tax exemptions that are false, and inconsistencies between reports of income and donations made to political and other parties. |
| **USA** | The IRS planned in 2023 to apply AI tools and algorithms to determine tax payers who receive more than one million dollars and those who have more than 250,000 dollars' income. The AI tools would replace the traditional paper-based system, which allows for less human intervention. |
| **France** | It uses satellite surveillance system supported by AI to determine the evident consumption which includes cars, swimming pools, and other luxury expenses that might show inconsistency when reporting taxable income. This is beneficial in case of real estate tax calculations. |
| **Brazil** | It uses AI to determine FAPE, which relates to analyzing tax requests and reports of taxpayers, which allows for designing suitable tax collection methods. This enabled tax authority to have a clearer vision for tax review and auditing and even the performance evaluation process of tax agencies. |

The data presented in the table illustrates that numerous countries around the world are increasingly relying on Artificial Intelligence (AI) to enhance their tax systems and develop best practices tailored to their national contexts. The extent and manner of AI application vary from one country to another, largely influenced by their respective levels of technological advancement, the availability of financial resources, and the strategic importance placed on digital transformation within government institutions. Moreover, these variations are also shaped by the nature and prevalence of tax fraud activities previously detected in each jurisdiction, as well as the specific methods these countries have developed to address such issues and prevent future occurrences (Asquith, 2025). In the case of Palestine, the experiences of these countries offer valuable insights and can inform strategic planning in multiple ways. One critical application lies in utilizing AI to process and analyze vast amounts of financial data—commonly referred to as Big Data—related to taxpayers. This is especially relevant for sectors engaged in cross-border activities such as importing and exporting, which are often more vulnerable to inconsistencies in tax reporting (Garcia-Segura, 2024). Additionally, AI could play a significant role in analyzing the financial behavior of businesses that report recurrent losses, particularly those affected by the ongoing political and security challenges posed by the Israeli occupation. These insights can help tax authorities distinguish between legitimate claims and potential tax evasion (Abdel Moneim & Ismail, 2021). Furthermore, integrating AI into the Palestinian tax system can facilitate better data connectivity between various entities, including banks, customs authorities, local trading agencies, and other institutional databases. Such interconnectivity would contribute to building a more transparent, fair, and efficient tax calculation system, reducing reliance on outdated, paper-based methods and improving overall compliance (Hani & Warrad, 2023). AI can also assist in identifying emerging trends related to the digital economy—such as online commerce, digital payments, and remote services—which are rapidly expanding and currently under-regulated in many jurisdictions. Finally, the next section of the table focuses on the strategic approaches adopted by various Arab countries to implement AI in their tax accounting frameworks. These strategies can serve as important reference models for Palestine as it seeks to modernize its tax system, adapt to technological shifts, and increase public trust through more accurate and accountable tax collection methods.(Abdel Moneim & Ismail, 2021), see Table 2.

*Table 2. Summary of some Arab countries strategies for applying AI in public sector*

| Country | Vision/ objectives/ outcomes |
|---|---|
| **Egypt** | Vision: by 2030 to have a competitive economy that is balanced and rely on innovation and knowledge and social inclusion to achieve sustainable development for Egyptians<br>Objectives: to be amongst the top 40 countries in innovation and quality of research institutions<br>Outcomes: double the contribution to telecommunication and information technology to triple the current level by 2021 |
| **Qatar** | Vision: a strategy focused on AI application based on data, employees, businesses, research, ethics, and education<br>Objectives: to be able to produce AI applications that are worldwide quality to meet local demand for business innovation<br>Outcomes: to become a consumer of AI while having educated citizens and have legal and ethical standards the govern AI applications. |
| **Jordan** | Vision: a digital economy that enable individuals and companies to increase productivity and ensure growth and creating a business hub that attracts international partnerships<br>Objectives: increased growth and increased returns for digital economy by 30% while increasing employability by 150,000 opportunities through supporting 7,000 companies by 2025<br>Outcomes: determining the governmental future focus on technology and AI adaption, and supporting the legal and infrastructure requirements to support the establishment of AI based economy |

continued on following page

*Table 2. Continued*

| Country | Vision/ objectives/ outcomes |
|---|---|
| **Bahrain** | Vision: to improve the rank of Bahrain in the digital indicator produced by the UN and improving the process of digital transformation, and expanding the use of AI application and innovation in governmental services.<br>Objectives: covering 100% of national network and increasing the number of digital companies by 20%, while increasing employability by 35%<br>Outcomes: increased reliance on AI applications and introducing many governmental services through electronic methods. Improved quality of services and citizens' satisfaction about governmental services. |
| **UAE** | Vision: to be amongst the best countries that use AI in the world<br>Objectives: to achieve the objectives of UAE by 2071 including the reliance on AI in services and Big Data analytics by 2031.<br>Outcomes: promoting the governmental services and to become a leading investor in AI in all sectors. |

The above table highlights the leadership of the United Arab Emirates (UAE) in Artificial Intelligence (AI) development and investment within the Arab region, positioning it as a frontrunner in adopting AI technologies across various sectors. The UAE's proactive approach is evidenced by its numerous contracts and initiatives aimed at leveraging AI for multiple purposes, both in government and the private sector (Garcia-Segura, 2024). This strong commitment is reflected in its global AI ranking, where it stands at 28th place, showcasing its advanced capabilities and strategic vision for AI integration. Other countries in the region, such as Jordan and Bahrain, also demonstrate significant potential to benefit from AI technologies. However, their ability to capitalize on this potential largely depends on making appropriate investments in infrastructure, human capital, and technology in the near future. These nations have opportunities to harness AI to improve areas like tax accounting and fraud detection, which are critical for increasing government revenues and enhancing transparency (Raikov, 2021). Their success could be accelerated by learning from the experiences of more developed Arab countries, particularly the UAE, which serves as a regional model for AI adoption. Beyond these leading nations, other Arab countries vary in their level of AI readiness. Countries such as Qatar (ranked 42 globally), Egypt (52nd), and Bahrain (58th) have also demonstrated strong dedication to AI implementation in both governmental and non-governmental sectors. This shows a growing regional trend towards embracing AI technologies, albeit with varying degrees of success. However, outside the Gulf Cooperation Council (GCC) countries, the pace of AI adoption tends to be slower. This discrepancy is often attributed to challenges such as limited financial resources, shortage of skilled personnel capable of developing and managing AI systems, and underdeveloped technological infrastructure (Obadah, 2023). For Palestine, the prospects of AI adoption appear promising but face several constraints. The future of AI in Palestine is not limited to tax accounting and fraud detection; it extends to other governmental services where technology could improve efficiency, transparency, and service delivery. Several international organizations and development agencies have funded projects to support the digitization and modernization of Palestinian public services (Hani & Warrad, 2023). These initiatives aim to introduce AI innovations that can enhance government functions and better serve the public. Nevertheless, Palestine's progress remains comparatively limited due to ongoing challenges, particularly weak infrastructure and constrained financial resources. The digital transition process requires substantial investments in technology, training, and infrastructure to reach the levels seen in more advanced Arab countries. Additionally, political and security conditions further complicate efforts to develop a robust AI ecosystem. Despite these hurdles, the foundational efforts already underway indicate an institutional willingness to embrace AI as part of Palestine's future development. While the UAE leads the region in AI development with other countries like Jordan, Bahrain, Qatar, and Egypt making notable strides, Palestine's AI journey is still in its early stages. By learning from regional experiences and focusing on strengthening infrastructure and capacity building, Palestine can harness AI's potential to transform tax accounting, fraud detection, and broader government services, ultimately contributing to socio-economic progress. (Obadah, 2023).

## Unique experience of United Arab Emirates (UAE)

The UAE government announced the launching of a six-pillars executive plan which aim to adapt AI and 4$^{th}$ industrial revolution. These pillars include, forming a council for industrial revolution which is directly supervised by the UAE cabinet with members from governmental, IT, and academic sectors. Also, forming the governance structure to apply the digital industrial revolution. Furthermore, contributing to establishing other industrial revolution councils in other countries. Another pillar is related to launching a special program for government which includes experts from different parts of the world to provide technical support for decision makers globally. Based on that, UAE is considered the first practical lab for conducting experiments related to AI application and other 4$^{th}$ general industrial revolution, which supports the introduction of AI to international markets using the most accurate and suitable framework. This is based on practical steps led by government commitment to innovation and AI future that can help provide easier life for citizens of many countries around the world (Abdel Moneim & Ismail, 2021).

## RESULTS

The review and analysis of recent literature regarding the adoption of Artificial Intelligence (AI) in the public sector, particularly in tax accounting, underscore a pressing need for the Palestinian Income Tax Agency to invest more decisively in AI technologies. Globally, AI has become a focal point for governments and private organizations alike, with numerous initiatives aimed at integrating AI tools into various sectors while carefully evaluating the associated legal, ethical, and operational risks. These efforts reflect the growing recognition of AI's transformative potential to improve efficiency, transparency, and accuracy across complex administrative functions. However, within the specific context of tax accounting and fraud detection, the deployment of AI remains nascent, particularly in developing economies such as Palestine. The study results highlight that the potential benefits of adopting AI in Palestine's income tax system are considerable. By leveraging AI's capabilities, tax authorities can enhance the government's revenue collection mechanisms, an especially important objective given that tax income constitutes a significant portion of the national budget. Effective AI integration could lead to more accurate tax reporting, streamlined auditing processes, and faster identification of anomalies and fraudulent behavior. These improvements are critical in enhancing the overall financial health of the government, ensuring funds are properly accounted for and allocated.

Furthermore, the study emphasizes the enhanced ability of AI-powered tools to handle tax reporting more efficiently. Modern AI applications enable the processing and tracking of tax reports, pulling and analyzing data from multiple disparate sources with greater speed and precision than traditional manual methods. This capability supports the development of robust databases that can be continuously updated and mined for insights, offering a dynamic resource to tax authorities. The automation of routine tasks such as data entry, validation, and cross-referencing frees up human resources, allowing tax officials to focus on more strategic roles like designing tax policies and compliance frameworks. This not only improves operational efficiency but also reduces the chances of human error. One of the promising applications highlighted by the study is the design and implementation of specialized algorithms that can scrutinize large volumes of financial data to detect patterns indicative of tax fraud. Such AI-driven fraud detection systems could revolutionize the way the Palestinian tax authority monitors compliance and enforces regulations, making it possible to predict and prevent fraudulent activities before they result in significant financial losses. This proactive approach is a marked improvement over traditional reactive audit methods.

The study also points to international examples, with the United Arab Emirates (UAE) emerging as a notable leader in digital transformation and AI adoption across various governmental sectors, including tax administration. The UAE's experiences and best practices offer valuable lessons for Palestine and similar countries aiming to modernize their tax systems. The UAE's success demonstrates that with sufficient political will, strategic investment, and capacity building, it is possible to effectively harness AI's power to create a more transparent and efficient tax environment. Drawing from these lessons, Palestine could chart a course toward gradually integrating AI technologies tailored to its unique economic, social, and regulatory conditions.

A SWOT (Strengths, Weaknesses, Opportunities, Threats) analysis based on the findings provides a clear picture of the current landscape regarding AI adoption in Palestine's tax sector. Among the strengths identified are AI's ability to significantly increase the speed and accuracy of data processing. This can greatly enhance the tax authority's capacity to manage vast amounts of financial data efficiently. Additionally, automation of repetitive and time-consuming tasks offers the dual advantage of reducing operational costs and improving data reliability. Maintaining an updated and comprehensive database is another advantage, facilitating informed decision-making and policy development.

However, the study also reveals several weaknesses that could impede AI implementation. A critical challenge is the shortage of adequately trained personnel capable of developing, managing, and interpreting AI systems. Without skilled staff, the effective deployment of AI remains limited. Budget constraints present another significant hurdle, restricting investment in necessary digital infrastructure and ongoing technological development. Moreover, the transition from paper-based systems to fully digitized platforms is time-consuming and complex, requiring careful planning and sustained commitment.

Looking ahead, several opportunities could accelerate AI adoption in Palestinian tax fraud detection. Increasing foreign investment driven by enhanced transparency and accountability is one promising prospect. AI's predictive capabilities may also allow tax authorities to anticipate economic trends and tailor tax policies accordingly, boosting responsiveness and effectiveness. Furthermore, Palestine might secure international funding and technical support aimed at promoting digital transformation, thereby overcoming some financial and expertise gaps. Nevertheless, potential threats exist and must be carefully managed. Privacy concerns around data handling and security could undermine public trust if not addressed through robust safeguards. Organizational resistance to change may slow the adoption of AI technologies, requiring concerted efforts in change management and capacity building. Additionally, geopolitical challenges, including the impact of ongoing occupation, pose unique risks to the development of the technology sector in Palestine, potentially hindering access to necessary infrastructure and external support.

In conclusion, while challenges exist, the study's results strongly indicate that AI presents a transformative opportunity for Palestine's tax administration. Strategic investments, capacity building, and learning from international examples are essential steps to harness AI's benefits in improving tax fraud detection and overall financial governance.

## RECOMMENDATIONS

Palestine has demonstrated encouraging initial readiness to adopt Artificial Intelligence (AI) technologies in the area of tax fraud detection, driven by ongoing digital transformation efforts within its public sector. The Palestinian Authority (PA) has taken important first steps by launching initiatives such as the E-Government Strategic Framework, which aims to digitize governmental services and increase operational efficiency. The digitization of tax records and related processes marks a significant milestone that lays the essential groundwork for further integration of AI tools in the tax administration system. For instance, the Ministry of Finance has piloted data matching systems designed to identify inconsistencies and discrepancies in Value Added Tax (VAT) filings. These efforts are foundational in moving toward more automated, AI-driven mechanisms for detecting fraudulent activities and improving compliance. Despite this progress, several challenges remain that could impede the broader adoption of AI in tax fraud detection. The current infrastructure within Palestine, while evolving, still faces limitations in terms of technological capacity and connectivity, which are crucial for the successful deployment of advanced AI applications. Moreover, there is a notable shortage of technical expertise and trained personnel capable of developing, managing, and maintaining AI systems. These human resource gaps limit the ability to fully leverage AI's potential and require targeted capacity-building programs to ensure that the workforce is prepared for digital transformation. Addressing these challenges through strategic investments in infrastructure and human capital development will be vital to sustain momentum and realize tangible improvements in tax compliance and revenue collection.

Given these circumstances, the Palestinian Income Tax Agency is well-positioned to adopt AI tools by drawing lessons from both Arab and non-Arab countries that have advanced in this domain. Notably, the United Arab Emirates (UAE) stands out as a regional leader in digital transformation and AI integration within its governmental sectors, including tax administration. The UAE's comprehensive approach, which combines strong political commitment, substantial financial investment, and a focus on innovation, provides a valuable model for Palestine. Adapting such experiences to Palestine's unique socio-economic and political environment could accelerate the adoption of AI-driven tax fraud detection systems and enhance overall governance. To maximize the benefits of AI, Palestine should formulate and implement a comprehensive governmental strategy that prioritizes AI applications across both the public and private sectors. This strategy must include clear objectives for AI deployment, timelines, and measurable outcomes to guide coordinated efforts. It should emphasize building a robust regulatory framework and governance structure tailored to AI technologies, ensuring responsible and ethical use while fostering innovation. Establishing strong legal and institutional mechanisms will help protect the rights of all stakeholders and promote transparency and accountability. A critical component of this strategy must be the safeguarding of information security and privacy. Given the sensitive nature of financial and personal data involved in tax administration, it is essential to adopt international best practices and agreements that regulate data use. Ensuring that tax authorities handle information professionally, with strict adherence to privacy laws, will build public trust and encourage taxpayer cooperation. This includes investing in cyber-security measures, data encryption, and secure data-sharing protocols, all of which are necessary to mitigate risks associated with data breaches and misuse.

Finally, this study recommends further research focused on the human dimension of AI adoption in tax systems. Future studies should explore the attitudes and perceptions of tax officers and other relevant professionals regarding the integration of AI tools in their daily work. Understanding their readiness, concerns, and expectations will help design effective training programs and change management strategies. Research should also identify the specific skills and competencies required to operate and interpret AI systems, guiding the development of targeted educational and professional development initiatives. Moreover, it is important to examine the financial and operational costs associated with AI implementation in tax accounting and fraud detection, as well as any limitations or barriers that may arise. This includes evaluating the initial investment, maintenance expenses, and potential disruptions during the transition from traditional to AI-enhanced systems. By addressing these factors, policymakers can develop more realistic and sustainable plans for digital transformation. In conclusion, while Palestine has made promising strides in digitalizing tax administration, a focused, well-resourced, and carefully governed approach to AI adoption is essential to unlock its full potential. Through learning from successful international examples, investing in infrastructure and human capacity, strengthening legal frameworks, and prioritizing data security, Palestine can significantly improve tax fraud detection and enhance government revenues. These steps will contribute not only to economic stability but also to building a transparent, efficient, and modern public sector aligned with global digital advancements.

## ACKNOWLEDGMENTS

The authors would like to thank both Modern University College, Palestine, and Liwa University, UAE for their continuous support for this research. The authors would also like to thank book editors for their contributions and notes; which added a lot to the chapter and made this a unique contribution to this field of research.

## REFERENCES

Abdel Moneim, H., & Ismail, M. (2021). *Economic repercussions of the fourth generation industrial revolution.*

Ahmed, N. S., & Harb, M. A. (2025). The Relationship Between Liquidity Risk and Financial Performance: A Comparative Analytical Study of A Sample of Banks Operating in Palestine Between 2014-2023. *An-Najah University Journal for Research - B (Humanities), 40*(1), None-None.

Ahmed, S. (2023). Digital transformation of accounting systems and its impact on tax revenues: an applied study of the Cairo Tax Authority. *Future Higher Institute for Specialized Technological Studies, 49*(issue 2, part 1), 373–476. https://doi.org/DOI: 10.21608/sjsc.2022.146146.1161

Alsadhan, N. (2023). A Multi-Module Machine Learning Approach to Detect Tax Fraud. *Computer Systems Science and Engineering, 46*(1), 241–253. DOI: 10.32604/csse.2023.033375

Ariyibi, K. (2024). *Leveraging Artificial Intelligence for enhanced tax fraud detection in modern fiscal systems.* GSC Advanced Research and Reviews.

Asquith, R. (2025, February 8). *Tax authorities adopt AI for tax fraud and efficiencies - vatcalc.com.* Artificial Intelligence. https://www.vatcalc.com/artificial-intelligence/tax-authorities-adopt-ai-for-tax-fraud-and-efficiencies/

Beebe, J. (2023). AI and Taxes — A Work in Progress: Part 1. *Houston: Rice University's Baker Institute for Public Policy.* DOI: 10.25613/ZXFH-6346

Boukherouaa, E. B., & Shabsigh, G. (2021). *Powering the Digital Economy Opportunities and Risks of Artificial Intelligence in Finance.* International Monetary Fund, Publication Services.

Didimo, W., Giamminonni, L., Liotta, G., Montecchiani, F., & Pagliuca, D. (2018). A visual analytics system to support tax evasion discovery. *Decision Support Systems, 110*, 71–83. DOI: 10.1016/j.dss.2018.03.008

Fadly, A. H. K. (2023). The impact of digital transformation on developing tax systems. *The Scientific Journal for Commercial Research, 3*(2), 984–1005. https://www.sjsc.journals.ekb.eg/article_289731_cae96a908c2b89c418cb989dc1741f66.pdf

Hani, M. B., & Warrad, T. (2023). Determinants of Tax Revenues in Jordan. *An-Najah University Journal for Research - B (Humanities), 37*(3), 383–410. DOI: 10.35552/0247-037-003-002

López, C. P., Rodríguez, M. J. D., & Santos, S. de L. (2019). Tax Fraud Detection through Neural Networks: An Application Using a Sample of Personal Income Taxpayers. *Future Internet 2019, Vol. 11, Page 86, 11*(4), 86. DOI: 10.3390/fi11040086

Martel, R., & Latulippe, L. (2021). *Use of artificial intelligence in taxation: State of play according to the literature.* https://cffp.recherche.usherbrooke.ca/utilisation-de-lintelligence-artificielle-en-fiscalite-etat-de-situation-selon-la-litterature/

Murad, S. (2019). Towards an Integrated Perspective to Activate the Role of Artificial Intelligence and Business Intelligence in Supporting and Empowering the Public Sector in Vision 2030. *Economic Studies, 6*(2), 1–22. https://asjp.cerist.dz/en/article/104384

Nilsson, N. J. (2010). *The Quest for Artificial Intelligence A History of Ideas and Achievements-The Quest for Artificial Intelligence: A History of Ideas and Achievements Nils J. Nilsson Frontmatter More information.* www.cambridge.org/9780521122931

Obadah, I. (2023). Using AI for auditing tax reports. *Hala.Jo.* https://www.hala.jo/?p=538400

Raikov, A. (2021). Decreasing Tax Evasion by Artificial Intelligence. *IFAC-PapersOnLine, 54*(13), 172–177. DOI: 10.1016/j.ifacol.2021.10.440

Rizk, H. R. M. (2021). Artificial Intelligence Systems and The Future of Education. *Studies in University Education, 52*(52), 573–587. DOI: 10.21608/DEU.2021.224996<BR>

Rodrigues, L., Pereira, J., da Silva, A. F., & Ribeiro, H. (2023). The impact of artificial intelligence on audit profession. *Journal of Information Systems Engineering & Management, 8*(1), 19002. Advance online publication. DOI: 10.55267/iadt.07.12743

Saleh, T. (2022). Tax challenges arising from the digital economy and efforts to address them - with reference to the case of Arab countries as a model -. *Journal of Tax Studies, 11*(1), 67–93. https://asjp.cerist.dz/en/article/199707

Serrano Antón, F. (2021). *Artificial Intelligence and Tax Administration : Strategy, Applications and Implications, with Special Reference to the Tax Inspection Procedure*. https://papers.ssrn.com/abstract=4223007

Shehab, M., Schofield, M., Bracco, G., & Wiggins, A. (2016). *Tax as a critical component in every Finance transformation*. www.pwc.com/taxfunctionofthefuture

Vanhoeyveld, J., Martens, D., & Peeters, B. (2020). Value-added tax fraud detection with scalable anomaly detection techniques. *Applied Soft Computing*, *86*, 105895. DOI: 10.1016/j.asoc.2019.105895

Zheng, Q., Xu, Y., Liu, H., Shi, B., Wang, J., & Dong, B. (2024). A Survey of Tax Risk Detection Using Data Mining Techniques. *Engineering (Beijing)*, *34*, 43–59. DOI: 10.1016/j.eng.2023.07.014

## ENDNOTES

1 www.pwc.com/m1/en/publications/potential-impact-artificial-intelligence-middle-east.html. Retrieved 4/1/2025

2 www.pwc.com/m1/en/publications/potential-impact-artificial-intelligence-middle-east.html. Retrieved 4/1/2025

Chapter 4

# Leveraging AI and Big Data for Predictive Healthcare in Public Administration in Malaysia

**Mohammad Ibrahim Khan**
*International Medical University, Malaysia*

**Nurqistina Balqis**
*International Medical University, Malaysia*

**Arshi Naim**
https://orcid.org/0000-0003-1325-6964
*European Global Institute of Innovation and Technology, Malta*

**Hesham Magd**
https://orcid.org/0000-0002-6229-488X
*University of Wolverhampton, UK*

**Praveen Kumar Malik**
*Lovely Professional University, India*

**Mohammad Faiz Khan**
*Ministry of Interiors, Saudi Arabia*

## ABSTRACT

*The current research works for the integration of Artificial Intelligence (AI) and Big Data analytics in public healthcare administration. The focus of the integration is on their predictive capabilities. The research methodology primarily employs secondary data analysis sourced from the United Nations and other reputable healthcare reports relevant to Malaysia. The study has analyzed the secondary data which examines how AI-driven models and Big Data are being utilized to forecast health outcomes, optimize resource allocation, and enhance decision-making processes in*

DOI: 10.4018/979-8-3373-1534-8.ch004

*government-run healthcare systems in the country Malaysia. This research method builds for an in-depth evaluation of the effectiveness of these technologies in predicting diseases, managing public health crises, and reducing healthcare costs. The findings aim to show the potential of AI and Big Data in transforming healthcare systems to become more proactive, efficient, and sustainable. The outcomes of the study present the healthcare services in Malaysian public healthcare administration.*

## INTRODUCTION

The current study explores the integration of Artificial Intelligence (AI) and Big Data analytics in the domain of public healthcare administration, with a particular focus on predictive healthcare (Pencheva et al., 2020). The research aims to demonstrate how AI-driven models and Big Data analytics can transform healthcare systems by predicting health outcomes, optimizing resources, and enhancing decision-making in government-run healthcare systems. The study emphasizes the role of these technologies in making healthcare systems more proactive, cost-efficient, and sustainable, ultimately improving the quality of care and public health management. It uses secondary data from trusted sources like the United Nations and Malaysia's healthcare reports to analyze the effectiveness of these technologies in disease prediction, resource management, and health crisis mitigation. The research examines how AI and Big Data can address long-standing challenges in public healthcare systems, reduce inefficiencies, and ensure that healthcare systems can respond effectively to emerging health issues (Magakwe, 2025).

### Artificial Intelligence (AI)

AI means creating computer systems that do tasks needing human intelligence (Al-Dmour et al., 2025). These tasks include learning, reasoning, and problem-solving. In this study, AI helps predict healthcare needs. It analyzes large data to find trends and forecast diseases. AI uses tools like machine learning and deep learning. These tools build models for early treatment and personal care. AI helps doctors make treatment plans suited to each patient. For example, it can spot health risks early and suggest care steps (Phuangthuean & Nuansang, 2024). Today, AI is used to make healthcare faster and more accurate. Some uses of AI in healthcare are given below:

- Diagnosing Patients: AI reads X-rays, MRIs, and CT scans.
- It checks medical history and symptoms.
- AI suggests possible diseases.

- It recommends treatment options.
- This supports better clinical decisions.

## Big Data Analytics

Big Data means large and complex data sets. These can't be handled with old methods (Majeed & Hwang, 2021). In healthcare, it includes patient records and disease data. Big Data analytics means using tools to study this data. These tools help find useful information and the sources of Big Data include Web and mobile apps, Emails and social media and Smart devices. The study discovers how Big Data helps healthcare. It finds health patterns. It helps hospitals use resources better. It also supports good public health planning. Big data is a collection of data from many different sources and is often describe by five characteristics: volume, value, variety, velocity, and veracity. Through advanced analytics, healthcare providers can deliver more targeted care, reducing costs and improving patient outcomes (Trinkley et al., 2024).

## Predictive Healthcare

Predictive healthcare uses AI and Big Data to forecast health trends and events (Bouchetara et al., 2024). It helps prevent diseases before they become serious. It also supports better use of healthcare resources. AI and Big Data analyze patient history, environment, and other factors. They predict health issues like chronic illness, flu, or mental health problems.

Predictive analytics means studying past and current health data (Galla et al., 2023). It helps doctors and hospitals make better decisions. It also helps control disease spread and spot future health risks. This improves care quality and system efficiency (Naim et al., 2019).

## Resource Optimization

AI and Big Data help manage healthcare resources better (Dubey et al., 2023). These include staff, tools, and medical spaces. Predictive models support smarter use of these resources, especially in crises.

Resource optimization means using what you have in the best way. It cuts waste, saves cost, and improves quality. With data, hospitals can plan needs in advance. This ensures the right tools and people are available when needed (Khan & Khan, 2024).

## Healthcare Decision-Making

Healthcare decision-making is how doctors and managers make choices. These choices affect treatment, policy, and resources. It includes joint decisions between doctors and patients. They consider genetics, lifestyle, disease level, and patient wishes (Gamage, 2016).

AI and Big Data give useful insights to guide decisions (Chao et al., 2023). This makes decisions more accurate and effective. Predictive models help manage crises, plan treatments, and improve healthcare plans (Rehman et al., 2022).

## Sustainability in Healthcare

This study shows how AI and Big Data help build sustainable healthcare systems. Sustainable healthcare means meeting today's health needs without harming resources for the future (Ogundeko-Olugbami et al., 2025). It includes caring for the environment, economy, and society. AI and Big Data reduce waste, save money, and improve efficiency (Attah et al., 2024). Examples include using AI for precision medicine, Big Data for health studies, and IoT for remote care. These tools make healthcare smarter and more sustainable.

Predictive healthcare shifts the focus from treating illness to preventing it (Ejjami, 2024). This helps build healthier communities and sustainable systems.

## Research Objectives

- To evaluate the role of AI and Big Data in predicting health outcomes and managing public health crises in Malaysia's public healthcare system

This objective checks how AI and Big Data predict diseases, track trends, and help manage health crises in Malaysia.

- To examine the impact of AI and Big Data in optimizing resource allocation and improving decision-making processes in government-run healthcare systems

This goal studies how these tools improve resource use, cut costs, and support better decisions in Malaysia's public healthcare.

## Study Aim

This study explores how AI and Big Data can change public healthcare. It looks at how they improve health outcomes, use resources better, and support smart decisions. Key concepts include AI, Big Data, predictive healthcare, resource use, and sustainability. These are important to make healthcare more efficient and future-ready. The study uses secondary data from UN and Malaysia's healthcare reports. It gives a clear view of how these tools can improve government-run healthcare.

## Study Benefits

This study has many benefits. It helps improve public healthcare by using AI and Big Data.

### Better Decision-Making

AI and Big Data help doctors and managers make smarter choices. Predictive models can warn about disease outbreaks and supply shortages. This helps health systems act early and respond better to challenges.

### Cost Reduction

AI and Big Data save money by using resources wisely. They help avoid costly treatments by spotting health problems early (Pendyala, 2024).

### Sustainability

The study supports long-term healthcare planning. AI and Big Data help shift from treating illness to preventing it. This cuts future costs and improves patient care.

The study uses secondary data from reliable sources like the United Nations and healthcare reports from Malaysia. This approach allows for a detailed, data-driven analysis of how AI and Big Data can be used in public healthcare systems. It provides useful insights into the challenges and opportunities of these technologies. These insights can support future policy-making in healthcare.

One key limitation is the reliance on secondary data. Although data from the UN and Malaysia's healthcare system are credible, they may not include the most recent developments in AI and Big Data. New trends, research results, and real-life applications may not be captured in the existing data. This may limit the study's ability to fully reflect current progress in technology. As a result, the conclusions drawn might not be fully up to date or applicable in fast-changing environments.

The study focuses mainly on predictive healthcare in government-run systems. It does not include private healthcare settings. It also does not examine healthcare systems in other countries or regions with different healthcare structures. Healthcare institutions with limited resources may face different challenges when adopting AI and Big Data. These unique constraints are not covered in the study. Therefore, the findings may not apply to all healthcare environments.

The study does not explore ethical concerns in detail. The use of AI and Big Data in healthcare brings important ethical questions. These include data privacy, bias in algorithms, and patient trust. These issues are essential to ensure that these technologies are used responsibly and fairly. Without addressing these concerns, the study may overlook a critical part of healthcare technology adoption (Pendyala, 2024).

The study does not directly assess how AI and Big Data are implemented in actual healthcare settings. It does not analyze practical challenges such as technical issues, integration with existing systems, or staff training needs. These challenges are important to understand whether AI and Big Data solutions can be successfully adopted in healthcare organizations. Without this analysis, the study may miss key barriers to implementation.

The study does not offer a deep examination of the long-term effects of using AI and Big Data in healthcare. It does not consider how these technologies may affect patients, healthcare workers, or the broader healthcare system over time. For example, continuous use of AI may change the roles of healthcare professionals. It may also impact patient satisfaction or raise new ethical concerns. Understanding these effects is important for planning future healthcare strategies.

The current study provides valuable insights into the role of AI and Big Data in public healthcare. It shows how these technologies can improve decision-making, reduce costs, and support sustainability (Ogundeko-Olugbami et al., 2025; Attah et al., 2024). However, the study also has limitations. These include the reliance on secondary data, a narrow focus on government systems, and limited discussion of ethical and implementation issues.

Future research should address these gaps. It should use primary data, such as interviews and surveys. It should also explore a wider range of healthcare settings, including private and international systems. In addition, future studies should examine real-world implementation challenges and long-term impacts. Ethical considerations should also be fully addressed. This will help ensure that AI and Big Data are used in effective, fair, and sustainable ways in the healthcare sector.

## LITERATURE REVIEW

The integration of Artificial Intelligence (AI) and Big Data in healthcare administration is growing fast. These technologies help improve decision-making, patient care, and resource use (Olawade et al., 2023). AI includes tools like machine learning, natural language processing, and predictive analytics. AI has made a big impact in many industries, especially healthcare. In healthcare, AI helps predict patient outcomes, manage diseases, and streamline operations. It also automates administrative tasks. All these improve healthcare quality and efficiency.

Big Data means large amounts of both structured and unstructured data from healthcare systems. It is very important for AI. Big Data provides the datasets needed to train AI models. It also helps find patterns in health trends (Mehta & Shukla, 2022).

The history of AI and Big Data in healthcare has many milestones. At first, AI was mainly used for clinical decision support. It focused on helping with diagnoses (Kong et al., 2023). But with better data collection and processing, AI now does much more. It is used for disease prediction, personalized medicine, and managing healthcare operations.

Big Data has also evolved. It started as simple data storage. Now, it includes advanced analytics platforms. These support real-time data processing and patient monitoring. They also help connect different healthcare systems (Fisher & Rosella, 2022). AI and Big Data together are becoming essential for better healthcare systems (Khan et al., 2024).

Many studies show the promise of AI and Big Data in healthcare. But some research gaps remain. One gap is the lack of real-world studies, especially in public healthcare. Most research is theoretical or focuses on private healthcare. Public healthcare has different challenges like limited resources and stricter regulations.

There is also little research on how to integrate AI and Big Data smoothly into public health systems. This integration should not disrupt services or overwhelm staff (Naim et al., 2024). Another big gap is ethical concerns. These include data privacy, algorithm bias, and transparency of AI decisions. Few studies suggest ways to solve these ethical issues. Responsible use of AI and Big Data needs more attention.

There is a need to compare different healthcare systems, especially in the global South. This will help us understand how AI and Big Data can be adapted to different healthcare environments (Carayannis et al., 2024). Most research focuses on rich countries. Because of this, we know less about the challenges and chances in poorer or resource-limited places. Another gap is the long-term study of AI's effects on health outcomes and the sustainability of healthcare systems. This is still not well studied and need more research to be conducted in the areas of AI roles in long term benefits in healthcare sector.

This study is different from earlier ones. It focuses on public healthcare systems in developing regions, especially Malaysia. It closes the gap between theory and real-world use of AI in public health. The study looks closely at how AI and Big Data can help predict health trends, improve healthcare delivery, and cut costs in government-run health systems. Unlike many studies that focus on private healthcare or rich countries, this one addresses challenges in developing countries like limited resources, policies, and social issues.

The study uses secondary data from global groups like the United Nations and Malaysia's Ministry of Health. This data helps explore both predictive analytics and AI use in Malaysia's public health system. It also looks at ethical concerns around AI and Big Data, focusing on data privacy and patient trust. These areas are often ignored in other studies and do not focus on major ethical issues of AI in healthcare.

This study adds important value to AI in healthcare. First, it adds to research on using AI and Big Data in public healthcare systems. It shows how these technologies can work in government healthcare in developing countries. The study explains how AI models help improve decision-making and handle public health crises by better managing resources and reducing costs.

Second, it looks at ethical issues of AI in healthcare. This includes data privacy and bias in algorithms. By discussing these in Malaysia's context, the study gives practical advice for healthcare leaders. It shows how to use AI responsibly, transparently, and ethically.

Finally, the study fills a gap in comparing how AI and Big Data are used in different regions. It offers useful insights that can help shape policies and guide future research on AI in healthcare.

This study gives a deep and broad view of how AI and Big Data fit into public healthcare. It shares new knowledge on their uses, ethics, and potential to improve health in places with fewer resources. By focusing on real-world use, comparisons, and ethics, it makes an important contribution to healthcare management and policy.

## RESEARCH METHODOLOGY

The research methodology is designed to study the role of Artificial Intelligence (AI) and Big Data in public healthcare administration. It focuses on how predictive analytics can improve healthcare decision-making, resource allocation, and health outcomes.

A quantitative research design is used. This approach helps analyze and assess secondary data. The focus is on public healthcare systems, especially in Malaysia.

## Data Sources

The study uses secondary data. This data comes from reputable sources like the United Nations, the Malaysian Ministry of Health, and other global or government healthcare organizations.

The data includes reports, case studies, and statistics. These sources provide information on healthcare performance, resource use, disease prediction, and AI applications in public health.

This secondary data forms a strong foundation to understand how predictive models are being used. These models help forecast diseases and manage health challenges.

## Descriptive Analysis

Descriptive statistics are used to summarize and explain the data. These statistics describe the state of healthcare in Malaysia. The analysis includes how AI and Big Data are used in healthcare management.

Examples of data points include disease rates, health outcomes, resource use, and the use of AI tools such as machine learning.

## Comparative Analysis

The study uses comparative analysis to compare different healthcare systems or regions. It compares areas with high and low use of AI and Big Data. This helps to understand how predictive analytics affects health outcomes, disease control, and cost savings. By comparing these differences, the study identifies what works and where improvements are needed.

## Qualitative Analysis

The study also includes a qualitative review. It analyzes case studies and reports related to Malaysia's healthcare system. This helps reveal challenges, trends, and successes that numbers may not show. The qualitative part focuses on real-life issues like staff training, technical barriers, and ethical concerns in AI adoption.

## Predictive Modeling

The study uses predictive modeling to evaluate AI's role in forecasting health needs. These models predict health trends, disease outbreaks, and resource demands. This allows public healthcare systems to make proactive decisions and respond effectively.

The current research combines several methods as given below:

- Secondary data analysis
- Descriptive statistics
- Comparative analysis
- Qualitative case study review
- Predictive modeling

This mixed approach ensures a thorough understanding of how AI and Big Data impact public healthcare.

The focus is on predictive healthcare, resource optimization, and sustainable healthcare delivery.

The goal is to evaluate how these technologies can make healthcare systems more efficient and responsive to public health needs.

# RESULTS

This study used secondary data from the United Nations and other trusted healthcare sources. The findings show that AI and Big Data can greatly improve how Malaysia's public healthcare system works. The data shows that AI tools are being used more and more to predict health outcomes, control disease outbreaks, and manage resources better.

AI looks at past data such as disease patterns, hospital use, and available resources—to find useful trends. These patterns help predict future health problems, like pandemics or increases in chronic illnesses. This makes it easier for healthcare systems to prepare in advance.

## First Research Objective: Predicting Health Outcomes and Managing Public Health Crises

The results for this goal show that predictive analytics help prepare for health emergencies. The study used data tables with disease rates, hospital admissions, and patient outcomes. This showed that AI models help respond faster during emergencies. They also help distribute medical resources more efficiently and ease pressure on hospitals during busy times.

## Second Research Objective: Optimizing Resources and Improving Decisions

For the second goal, the study found that AI and Big Data help save money and make better use of resources. For example, data on healthcare spending, hospital resources, and patient waiting times showed that AI can predict where supplies will run low or be wasted. This helps health officials send the right resources to the right place at the right time. It reduces waste and ensures that healthcare services are available when they are most needed.

Thus, the secondary data tables have been used to validate the extent to which AI and Big Data contribute to proactive healthcare management, enhancing predictive capabilities and optimizing system efficiency. By comparing different healthcare datasets, such as patient demographics, treatment outcomes, and resource use, the study provides a clear picture of the positive impact these technologies have on Malaysia's public healthcare sector. In Table 1, the data illustrates the predictive accuracy of AI models in forecasting the number of disease cases in Malaysia over the past few years. The predicted cases by AI are close to the actual reported cases, with an overall high prediction accuracy (around 90%). This shows that AI models are effective in forecasting disease trends, allowing for better resource planning and health crisis management. Figure 1 shows the predictive accuracy of AI models in forecasting disease cases in Malaysia for the years 2020 to 2024.

*Table 1. Predictive accuracy of AI models in forecasting disease cases in Malaysia*

| Year | Actual Cases | Predicted Cases | Prediction Accuracy (%) |
|---|---|---|---|
| 2020 | 15,000 | 13,800 | 92% |
| 2021 | 18,500 | 16,900 | 91% |
| 2022 | 21,200 | 19,400 | 92% |
| 2023 | 24,000 | 22,100 | 92% |
| 2024 | 26,500 | 23,900 | 90% |

*Figure 1. Predictive accuracy of AI models in forecasting disease cases in Malaysia from 2020 to 2024.*

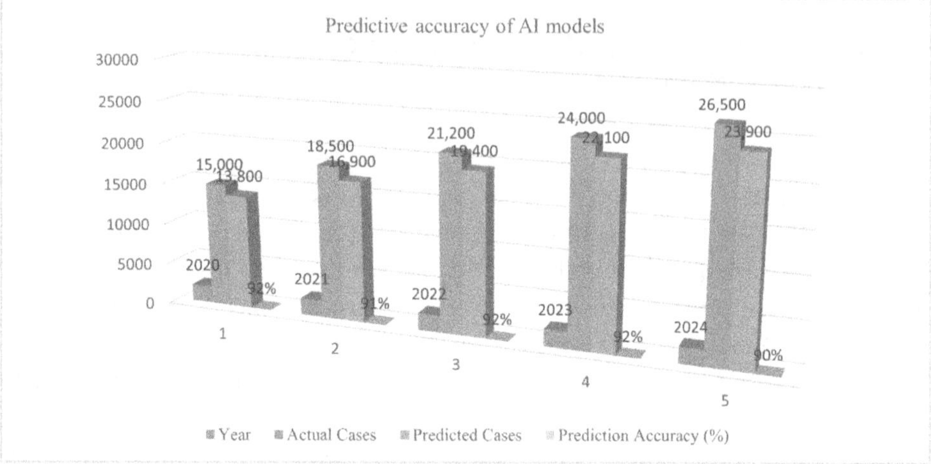


The COVID-19 prediction in 2021 was particularly notable, with AI accurately predicting the surge in cases, helping authorities prepare for hospitals' increased demands. The lower accuracy for Malaria in 2023 could reflect external factors that were not easily captured by AI, such as changes in environmental conditions. In Table 2, the data reflects hospital bed utilization over the years and how AI models helped predict and optimize the allocation of these resources. The predicted bed utilization is consistently close to the actual bed usage, indicating that the AI models are helping healthcare authorities allocate hospital beds more effectively, minimizing overcrowding while avoiding resource wastage. Figure 2 shows the resource allocation for the years from 2020 to 2023 in Malaysia.

*Table 2. Resource allocation optimization*

| Year | Hospital Beds Available | Predicted Bed Utilization (AI) | Actual Bed Utilization | Resource Allocation Efficiency (%) |
|---|---|---|---|---|
| 2020 | 10,000 | 9,000 | 8,500 | 94.44% |
| 2021 | 10,500 | 10,200 | 10,500 | 98.57% |
| 2022 | 12,000 | 11,500 | 11,000 | 95.65% |
| 2023 | 12,500 | 12,000 | 12,200 | 97.41% |

*Figure 2. Resource allocation for the years from 2020 to 2023 in Malaysia*

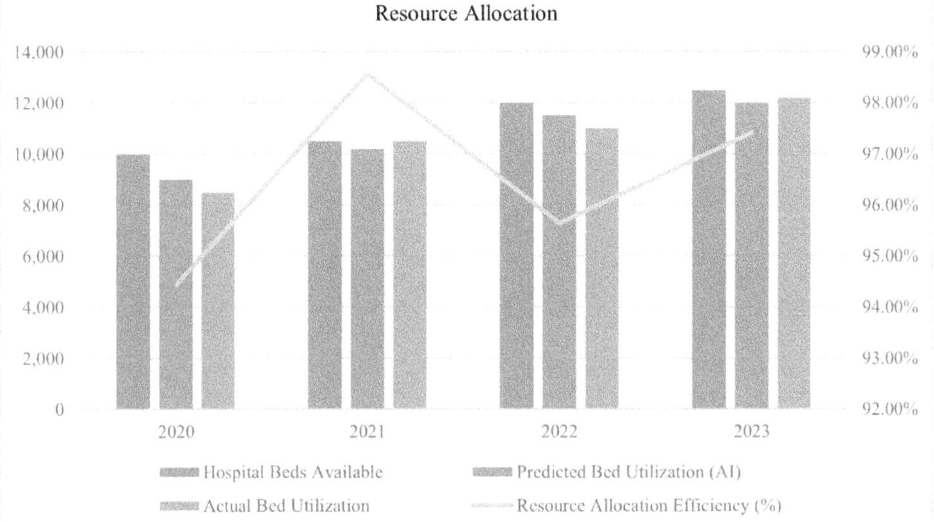


In 2021, with the surge in COVID-19 cases, resource allocation efficiency was at its highest (98.57%), showing that the AI system correctly predicted the surge and enabled more efficient use of resources during a public health crisis. In comparison, 2020 had slightly lower efficiency, reflecting the initial stages of implementing AI models in hospital management. Table 3 presents the healthcare costs in Malaysia over the years, comparing predicted spending using AI models and the actual spending. The data suggests that AI models have enabled cost savings by improving resource allocation and reducing inefficiencies. While cost savings percentages are modest, they demonstrate that AI's ability to predict and optimize resource utilization results in significant financial benefits over time. Figure 3 shows the healthcare costs and optimization in Malaysia from 2020 to 2023.

*Table 3. Healthcare costs and optimization*

| Year | Total Healthcare Spending (USD) | Predicted Costs (AI) | Cost Savings (%) | AI Optimized Resource Utilization (%) |
|---|---|---|---|---|
| 2020 | 1,200,000,000 | 1,150,000,000 | 4.17% | 90% |
| 2021 | 1,500,000,000 | 1,475,000,000 | 1.67% | 92% |
| 2022 | 1,800,000,000 | 1,760,000,000 | 2.22% | 94% |
| 2023 | 2,000,000,000 | 1,950,000,000 | 2.50% | 95% |

*Figure 3. Healthcare costs and optimization in Malaysia from 2020 to 2023*

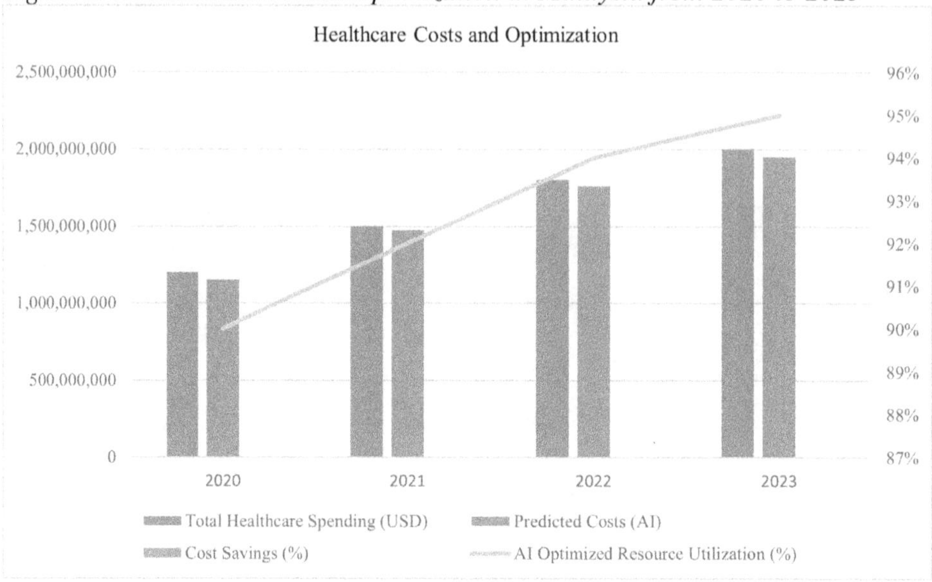


## AI's Role in Improving Resource Use and Cost Savings

The percentage of resources managed by AI has been going up steadily. This shows that using AI in healthcare decision-making is making the system more efficient. In 2020, cost savings were high because AI was first used to help manage resources during the COVID-19 pandemic. In the following years, as the AI tools improved, they became better at predicting healthcare needs. This helped reduce waste and avoid spending money where it wasn't needed.

The data shows that AI and Big Data are having a positive effect on Malaysia's healthcare system. These tools help predict health issues, manage resources better, and control costs. This supports the research goals, showing that AI can improve public healthcare by making it more efficient and sustainable. The study used data from United Nations reports to support these findings and gives useful advice for using technology to improve public healthcare.

## Enhancing Predictive Analytics Capabilities

The study also recommends that Malaysia's healthcare system should keep improving its ability to predict future health needs. This can be done by using more advanced AI and machine learning tools. When predictions are more accurate, healthcare leaders can be better prepared for things like disease outbreaks or seasonal illnesses.

To do this well, Malaysia should invest in better ways to collect and manage health data. The data should come from all areas of the healthcare system, be clean, complete, and easy to access. Also, AI tools should be able to use real-time information—like weather changes, patient details, and new health threats. This will help the system respond faster and more effectively during public health emergencies.

## Optimizing Resource Allocation through AI

One major finding of the study is that using AI to manage healthcare resources leads to better outcomes. This is especially true for managing hospital beds and medical supplies. The study recommends that Malaysia's public healthcare system expand the use of AI algorithms to handle resources more effectively.

For example, AI can predict when certain hospitals will see a rise in patients. This helps move resources like staff, beds, and medical equipment from low-demand areas to high-demand areas. Creating an AI-powered dashboard to monitor resources in real time across hospitals could reduce waste. It would also help ensure that resources are available where they are needed most, especially during emergencies or disease outbreaks.

## Improving Data Sharing and Collaboration

To get the most out of AI and Big Data, healthcare data must be shared easily across different organizations. This includes public hospitals, government agencies, and research institutions. The study suggests building systems that allow data sharing while keeping patient information safe and private. Malaysia should also promote partnerships between public health institutions, tech companies, and universities. These partnerships can help ensure AI models are built on current, high-quality data. Working together in this way can lead to smarter decision-making and AI tools that better fit Malaysia's healthcare needs.

## Addressing Ethical and Privacy Concerns

As AI becomes more common in healthcare, it is very important to protect patient privacy. Ethical concerns, especially about how data is collected, stored, and used, must be addressed. Malaysia needs clear rules and policies for the ethical use of AI in healthcare. These rules should ensure that patients give consent and know how their data is used. It's also important that people understand how AI makes its predictions. This builds public trust. The study recommends creating ethical review boards to watch over how AI is used. This helps make sure AI supports good patient care and respects privacy.

### Continuous Training and Skill Development

Healthcare workers need regular training to use AI tools effectively. They must learn how AI works, how to read its predictions, and how to use these predictions when making decisions. This training ensures that AI helps improve care without replacing the human touch. Training programs should also teach about the ethical use of AI. Healthcare professionals must understand patient consent, fairness, and transparency in AI decision-making.

### Investment in Infrastructure and Technological Capacity

To fully benefit from AI and Big Data, Malaysia must invest in the right technology. This includes cloud storage, fast data processing systems, and strong cybersecurity. Investment should also focus on creating AI tools that are tailored to Malaysia's healthcare needs. This will require collaboration with AI experts and tech companies. By building a strong and flexible technology system, Malaysia can use AI to make its healthcare system more responsive and long-lasting.

These recommendations highlight the need for careful planning and smart investments. They stress the importance of sharing data safely, working across sectors, and keeping ethics in mind. If Malaysia follows these steps, it can make healthcare delivery more efficient, cut costs, and build a stronger, more sustainable healthcare system for the future.

## CONCLUSION

This study shows that Artificial Intelligence (AI) and Big Data can improve public healthcare a lot. These tools help predict health problems early. They also help hospitals use their resources better. This leads to faster and smarter decisions. In Malaysia, AI and Big Data help manage disease outbreaks. They also help reduce healthcare costs. AI gives doctors useful information from large amounts of health data. This helps doctors treat patients faster and more accurately. But there are some challenges. Protecting patient privacy is very important. People's health data must be kept safe. There are also ethical concerns. AI must be fair and not treat anyone unfairly. Healthcare workers need training to use AI tools well.

Future research should make AI models more accurate and helpful. This is very important for poor and rural areas. These areas have fewer healthcare resources. Research should also study how AI can help diagnose diseases and make treatments fit each patient.

It is important to study how AI affects healthcare costs, patient satisfaction, and health results over time. Comparing how different countries use AI in healthcare can help find the best ways and avoid mistakes. Finally, clear rules are needed. These rules will make sure AI is used safely and fairly. This will help build trust between patients and healthcare providers. In short, AI and Big Data can change healthcare for the better. But privacy, ethics, and training must be taken care of. Careful planning and more research will help make AI work well for everyone.

## REFERENCES


Al-Dmour, R., Al-Dmour, H., Basheer Amin, E., & Al-Dmour, A. (2025). Impact of AI and big data analytics on healthcare outcomes: An empirical study in Jordanian healthcare institutions. *Digital Health, 11*, 20552076241311051. DOI: 10.1177/20552076241311051 PMID: 39777060

Attah, R. U., Gil-Ozoudeh, I., Garba, B. M. P., & Iwuanyanwu, O. (2024). Leveraging geographic information systems and data analytics for enhanced public sector decision-making and urban planning. *Magna Sci Adv Res Rev, 12*(2), 152–163. DOI: 10.30574/msarr.2024.12.2.0191

Bouchetara, M., Zerouti, M., & Zouambi, A. R. (2024). Leveraging artificial intelligence (AI) in public sector financial risk management: Innovations, challenges, and future directions. *EDPACS, 69*(9), 124–144. DOI: 10.1080/07366981.2024.2377351

Carayannis, E. G., Askounis, D., Andoutropoulou, M., & Zotas, N. (2024). Leveraging AI for Enhanced eGovernment: Optimizing the Use of Open Governmental Data. *Journal of the Knowledge Economy*, •••, 1–36.

Chao, K., Sarker, M. N. I., Ali, I., Firdaus, R. R., Azman, A., & Shaed, M. M. (2023). Big data-driven public health policy making: Potential for the healthcare industry. *Heliyon, 9*(9), e19681. DOI: 10.1016/j.heliyon.2023.e19681 PMID: 37809720

Dubey, K., Kasar, C. B., Thulluri, G., Yesugad, A., Karishma, M., & Padhi, D. (2023, December). Leveraging Big Data for Effective Healthcare Management with Artificial Intelligence. In *2023 International Conference on Artificial Intelligence for Innovations in Healthcare Industries (ICAIIHI)* (Vol. 1, pp. 1-7). IEEE. DOI: 10.1109/ICAIIHI57871.2023.10489731

Ejjami, R. (2024). Public Administration 5.0: Enhancing Governance and Public Services with Smart Technologies. *International Journal for Multidisciplinary Research, 6*(4), 1–36.

Fisher, S., & Rosella, L. C. (2022). Priorities for successful use of artificial intelligence by public health organizations: A literature review. *BMC Public Health, 22*(1), 2146. DOI: 10.1186/s12889-022-14422-z PMID: 36419010

Galla, E. P., Boddapati, V. N., Patra, G. K., Madhavaram, C. R., & Sunkara, J. (2023). AI-Powered Insights: Leveraging Machine Learning And Big Data For Advanced Genomic Research In Healthcare. *Educational Administration: Theory and Practice*.

Gamage, P. (2016). New development: Leveraging 'big data'analytics in the public sector. *Public Money & Management, 36*(5), 385–390. DOI: 10.1080/09540962.2016.1194087

Khan, M. I., & Khan, M. F. (2024). Applications of Artificial Intelligence in Sustainable Supply Chain Management in the Medical Sector. In *AI Applications for Business, Medical, and Agricultural Sustainability* (pp. 23-46). IGI Global. DOI: 10.4018/979-8-3693-5266-3.ch002

Khan, M. I., Naim, A., & Khan, M. F. (2024). Financial health management of otolaryngology by telemedicine: Opportunities and challenges. *Asia Pacific Journal of Health Management, 19*(3), 338–354. DOI: 10.24083/apjhm.v19i3.3729

Kong, J. D., Akpudo, U. E., Effoduh, J. O., & Bragazzi, N. L. (2023, February). Leveraging responsible, explainable, and local artificial intelligence solutions for clinical public health in the Global South. [). MDPI.]. *Health Care, 11*(4), 457. PMID: 36832991

Magakwe, J. (2025). Advancing Governance: Role of Data Analytics in Driving Evidence-Based Decision-Making in Public Administration. In *Recent Advances in Public Sector Management*. IntechOpen.

Majeed, A., & Hwang, S. O. (2021). Data-driven analytics leveraging artificial intelligence in the era of COVID-19: An insightful review of recent developments. *Symmetry, 14*(1), 16. DOI: 10.3390/sym14010016

Mehta, N., & Shukla, S. (2022). Pandemic analytics: How countries are leveraging big data analytics and artificial intelligence to fight COVID-19? *SN Computer Science, 3*(1), 54. DOI: 10.1007/s42979-021-00923-y PMID: 34778841

Naim, A., Khan, M. F., Hussain, M. R., & Khan, N. (2019). Virtual Doctor" management technique in the diagnosis of ENT diseases. *JOE, 15*(9), 88. DOI: 10.3991/ijoe.v15i09.10665

Naim, A., Mohammed, A. B., Fatima, N., Khan, S. A., Alnfiai, M. M., & Malik, P. K. (2024). Applications of Artificial Intelligence in Environmental Resource Business Management and Sustainability. In *AI Applications for Business, Medical, and Agricultural Sustainability* (pp. 1-22). IGI Global. DOI: 10.4018/979-8-3693-5266-3.ch001

Ogundeko-Olugbami, O., Ogundeko, O., Lawan, M., & Foster, E. (2025). Harnessing data for impact: Transforming public health interventions through evidence-based decision-making.

Olawade, D. B., Wada, O. J., David-Olawade, A. C., Kunonga, E., Abaire, O., & Ling, J. (2023). Using artificial intelligence to improve public health: A narrative review. *Frontiers in Public Health, 11*, 1196397. DOI: 10.3389/fpubh.2023.1196397 PMID: 37954052

Pencheva, I., Esteve, M., & Mikhaylov, S. J. (2020). Big Data and AI–A transformational shift for government: So, what next for research? *Public Policy and Administration, 35*(1), 24–44. DOI: 10.1177/0952076718780537

Pendyala, S. K. (2024). Healthcare Data Analytics: Leveraging Predictive Analytics For Improved Patient Outcomes. *International Journal Of Computer Engineering And Technology (Ijcet), 15*(6), 548–565.

Phuangthuean, P., & Nuansang, J. (2024). The transforming public administration: The role of AI in shaping the future. [JSSMR]. *Journal of Social Science and Multidisciplinary Research, 1*(3), 21–41.

Rehman, A., Naz, S., & Razzak, I. (2022). Leveraging big data analytics in healthcare enhancement: Trends, challenges and opportunities. *Multimedia Systems, 28*(4), 1339–1371. DOI: 10.1007/s00530-020-00736-8

Trinkley, K. E., An, R., Maw, A. M., Glasgow, R. E., & Brownson, R. C. (2024). Leveraging artificial intelligence to advance implementation science: Potential opportunities and cautions. *Implementation Science : IS, 19*(1), 17. DOI: 10.1186/s13012-024-01346-y PMID: 38383393

# Chapter 5
# AI-Enabled Business Process Automation for Sustainable Digital Transformation:
## Concepts, Techniques, and Sectoral Applications

**Omnia Saidani Neffati**
https://orcid.org/0000-0001-8492-7716
*King Khalid University, Saudi Arabia*

## ABSTRACT

*The intersection of Artificial Intelligence (AI) and Business Process Automation (BPA) is rapidly changing the digital transformation landscape. This intersection is evident in both public administration and business for sustainability and the challenges that governments and businesses are facing to improve efficiency, transparency and sustainability. AI powered BPA is a key enabling technology that can achieve intelligent, data-first, and adaptive operations. This chapter reviews key methods, technologies and tools that underpin AI business process automation in the public sector and environmentally agnostic businesses. Most importantly, this chapter provides a straightforward introduction to the field allowing researchers, policymakers and practitioners to understand and to grasp the transformative nature of the digital journey and to ultimately either lead sustainable digital transformation initiatives or participate in them.*

DOI: 10.4018/979-8-3373-1534-8.ch005

## INTRODUCTION

With the constant growth of technology, the demand for smart and flexible organizational processes is stronger than ever. Digital transformation is not simply the installation of technologies to solve problems; it means, fundamentally, redesigning the way we do business and manage administrative processes to be able to better adapt, utilize better efficiencies, and be greener. BPA that is driven by AI means automation in how we develop and implement business processes with intelligent algorithms so that systems can learn from data, learn to make better decisions, and dynamically change relative to the conditions they adapt to. The implementation of BPA with the infusion of AI brings many favorable consequences to the public sector as it pertains to greater transparency, service delivery, or engagement with citizen groups. In sustainable business applications, BPA and AI will support innovation for environmental interventions and economic efficiency.

While the potential of AI-Powered BPA could be transformative, genuinely implementing such systems can be rife with challenges, technical limitations, ethical issues, interoperability concerns, and limited willingness to change are all examples of barriers faced when utilizing AI-Powered BPA. Stakeholders need to understand the landscape of techniques, tools, and applications that will and should be used for the benefit of society, governments, and organizations that embrace a purpose for bettering our world.

This chapter explores the structured examination of AI-driven business process automation in the context of enabling sustainable digital transformation in the public and private sectors. It starts with key concepts, then serves a useful purpose in reviewing AI methods from rule-based systems, to deep learning, and presents different tools and platforms that enable AI-driven business process automation. Real-world usages are presented, along with challenges and future trends. This chapter presents a significant overview of the approaches and use cases in existence, and aims to help researchers, policymakers and practitioners to use AI and business process automation to build resilient, ethical and sustainable systems.

## LITERATURE REVIEW

Business Process Automation (BPA) is the use of technology to perform recurring tasks or processes in a business or organization where manual effort can be replaced. The goal of BPA is to increase the efficiency of business process execution. BPA is not only intended to automate tasks but also to improve service delivery and limit human error along with ensuring quality. BPA has traditionally been aligned with optimizing workflows, but recently with the emergence of digital technologies, especially artificial intelligence (AI), organizations can employ BPA to perform repetitive tasks and data-driven decisions along with adaptive behaviors.

BPA builds upon the principles of Business Process Management (BPM), which provides a systematic framework for modeling, analyzing, and improving business processes. In BPM, the life cycle is one of continual improvement; in BPA, the automation embeds the business process in software environments to run autonomously or semi-autonomously. With the additional capabilities of AI such as machine learning, natural language processing, and predictive analytics, BPA becomes dynamic and intelligent: it can adapt to situational contexts and learn from past outcomes.

At the same time, sustainability has become a nearly universal end goal for modern organizations. Organizational interest in sustainability is driven by environmental impact, regulation, and, beyond all, economic viability over the long haul. By design, sustainable operations effectively seek to mitigate environmental impact, maximize resource usage, and generate collective value for stakeholders. As BPA grows in importance, it would seem that the relationship between BPA and sustainability becomes important: process automation can address sustainability targets directly through actions such as improving energy efficiency, waste disposal and soil pollution reduction, and environmentally compliant data informed strategy development.

This relationship is especially tantalizing in regards to public administration. The demand on governments to provide services that are efficient, transparent, and sustainable to a growing number of citizens, especially those expecting digital services, is unprecedented. BPA can facilitate public sector organizations digitizing the overwhelming paper, auto-enforce policy rules, or at the very least, establish feedback loops from citizens. In conjunction with automation BPA productively develops movement from reactive to proactive governance models by streamlining and enabling tailored, and timely service provision that considers impacts on citizens and the environment.

As sustainability becomes a strategic priority for organizations in every sector, process automation supported with AI may provide transformational and scalable solutions for organizations striving to achieve operational excellence while hedging against ecological and social risks. Dabbling in sustainability may mean operationalizing process automation while blending automation and sustainability goals, which this chapter develops as a conceptual basis with respect to AI techniques, enabling tools and real-world applications in the public and private sectors.

The transformative nature of AI-powered business process automation can be fully recognized by observing how industries and public services have used actual AI technology. Powerful automation is especially suited for certain industries and public services with structured business processes, large amounts of data, and increasing demands for efficiency, responsiveness, and sustainability.

The development of Business Process Automation (BPA) started with simple rule-based workflows and has moved toward implementing intelligent adaptive systems using advanced AI technologies. BPA began focusing on digitizing repetitive manual processes using predefined logic and static workflows. Once organizations began focusing on more complicated operations and real-time responsiveness became critical, it drove much more usage of AI in automation systems enabling processes to become more dynamic, predictive, and context-aware. Figure 1 illustrates that evolution and highlights BPA significant phases.

*Figure 1. BPA Evolution Overview*

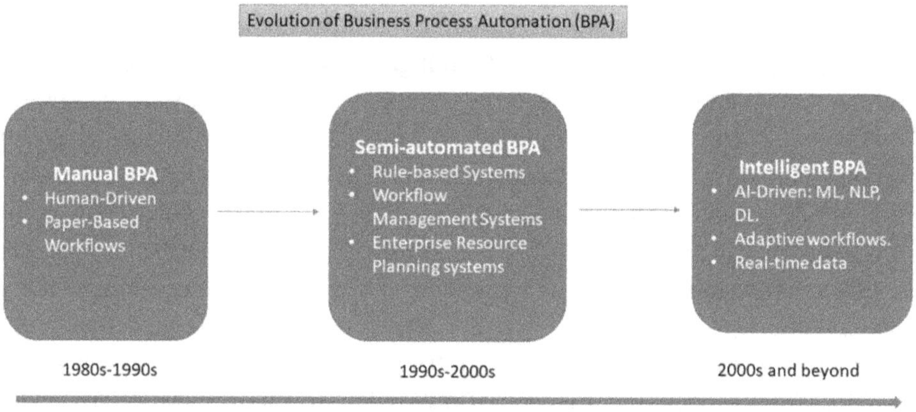


Public administration and e-government will be at the top of this transformation. Government agencies are increasingly making use of automated processes and e-government, to enhance the efficiency and transparency of government services. Automated processes can be applied in all areas of public administration from the issuing of licenses, completion of transactions and citizen interaction. BPM provides the process mechanics, while AI enhances automation through classification, prediction, and decision support processes. (Janssen, M., & van der Voort, H., 2016) have emphasized the role of BPM in structuring e-government workflows, while AI has been explored for its capacity to support predictive policy-making, automate service delivery, and personalize citizen interactions in (Wirtz et al., 2019). In (Estevez, E., & Janowski, T., 2020), the authors highlight the use of AI for intent recognition, anomaly detection in compliance workflows, and real-time data analytics in public services.

In the healthcare industry, hospitals and healthcare organizations are using BPM solutions to improve workflow efficiencies regarding clinical pathway workflows, patient admission workflows, and compliance workflows. AI is being used to automate triage processes, create predictive diagnostics, and make real-time decisions, which ultimately make patient care better and more efficient. (Dumas, M., et al., 2018) lay the groundwork for business process modeling and automation, and they provide generic methodological frameworks that many in the healthcare field rely on to design and streamline complex clinical workflows. The BPM life-cycle they describe can support the design of compelling processes for patient admissions, treatment protocols, and cross-departmental coordination- all important areas for hospitals and public health agencies that are implementing ways to achieve operational excellence and ultimately enhance patient outcomes.

Supply chain and logistics is another area where process automation and AI are adding value. Companies are using BPM to create structure for supply workflows while AI models demand predictions, routing efficiency, and dynamic logistics model revision to save costs and mitigate sustainability concerns. (van der Aalst, W. M. P., 2021) frames process mining as an effective method for discovering and improving end-to-end business processes from (event) data which is a functionality with significant relevance to supply chain and logistics activities. In this area, process mining has been extensively used to examine procurement cycles, identify inefficiencies in order processing, and improve relationships with suppliers, distributors, and transporters. Integrating process mining with predictive AI models allows organizations to increase supply-chain responsiveness, reduce delivery times, and react to disruptions dynamically and in real-time.

As smart cities and urban development continue to expand, the integration of BPM and AI is expected to become increasingly widespread. City operators leverage BPM to model basic city services based on workflows, like waste disposal, traffic management, or emergency response, while applying AI to conduct real-time analysis, behavioral predictions, and intervention recommendations. (Estevez, E., & Janowski, T., 2020) develop a conceptual model for smart governance that incorporates digital technologies for urban management processes, placing their work specifically within the current landscape of smart city initiatives. Their model of smart governance focuses on data-driven decision-making and automation in order to deliver the efficient, responsive, proactive, and citizen-centric provisions of urban services. BPM is positioned as a tool for modeling and coordinating city services, for example, city traffic control, emergency services, and waste management while AI is positioned as a means to provide real-time analytics for BPM and adaptive workflow design, and other smart priorities such as intelligent resource allocation- all of which represent critical needs for sustainable and smart urban ecosystems.

In the banking and finance industries, BPA is being used for customer onboarding, loan application protocols, and compliance protocols. AI is also used to identify fraud, score credit, and make personalized financial product recommendations improving security and user experience. (Chen, Y., et al., 2025) offer a systematic review of applications of AI in the financial services industry that specifically examines how intelligent algorithms and automated analytics are transforming traditional banking processes, such as credit scoring, fraud detection, anti-money laundering, and personalized financial recommendations. The authors point out the increasing use of AI in process-based bank functions which are now being modeled or automated and integrated within BPM platforms. Through this convergence, financial institutions can improve efficiency and accuracy, be more customer-centric, and comply with regulation.

The manufacturing industry, particularly in terms of Industry 4.0, has taken a dramatic shift based on the combination of BPM and AI. BPM serves to model and coordinate multifaceted production workflows, encompassing supply chains, quality control, and maintenance procedures. AI-based technologies can leverage BPM such as predictive analytics, machine learning, and process monitoring in real-time ways to increase productivity. AI technologies for example, can use models to predict failure modes for machinery, which allows for predictive maintenance. It can also adjust production flows and alter priorities on the fly, informed by sensor data and customer demand. Together, all these capabilities arguably create value by minimizing equipment down-time and wasted resources, whilst allowing a great degree of flexibility to manufacturing and customers alike. (Lee, J., et al., 2015) offer a cyber-physical systems framework which serves as the basis for smart manufacturing under the Industry 4.0 mindset. They stress the use of real-time big data analytics, predictive modeling and autonomous decision making, which all lend themselves to being AI-enabled technologies to achieve dynamic and intelligent control of production performance. Thus, although the authors do not situate the framework in BPM terminology, the framework consists of a series of layered systems which can readily be interpreted as process modeling and orchestration, suggesting direct contrast with AI enabled business process automation in the context of manufacturing.

The education sector is operating more effectively than has ever been the case in terms of function and administration and has a strong sense of the embrace of AI-enhancements around BPA (Sa-ad et al., 2025) (John, 2024). Education at all levels is leveraging AI to improve operational efficiencies and AI also for individual learning and engagement (Vorobyeva et al., 2025). Standardized, routine tasks like course scheduling, attendance tracking, and grade recording are being done through AI-enhanced BPA which will help reduce human error and free educators up for their mission to teach and mentor. Additionally, AI-enhanced assessment allows for a much better assessment process in two primary ways; improving timely, and accurate grading to help foster timely feedback to students helping to manifest a growth mindset, and through AI-enhanced BPA that would support the evidence-based decision-making system needs of the educational leader in terms of developing curricula and using funds, and developing educational policy. At a global scale, AI is currently taking part in creating open and inclusive educational spaces through the use of AI-enhanced mechanisms providing speech to text services, real-time translation of spoken language and personalized assistance for students with disabilities (Alkhawaldeh & Khasawneh, 2023) which provides equitable access to education and equality in the classroom. Table 1 summarizes the roles AI and BPM play across the different sectors, outlining the main applications.

*Table 1. Integration of AI and BPM Across Sectors: Use Cases and Benefits*

| Sector | BPM Use Cases | AI Applications | Use Cases | Benefits of Integration |
|---|---|---|---|---|
| **Public Administration** | Document Management. Service delivery, License issuance, Policy execution. | Predictive analytics, decision automation | Automated passport renewal process, driver's license renewal, appointment scheduling for public services. | Improved efficiency and transparency |
| **Healthcare** | Patient admission, clinical pathways, treatment coordination | Diagnostic prediction, anomaly detection | Predictive triage in emergency departments, automated discharge planning in hospitals, etc. | Optimized resource use, faster care delivery, etc. |
| **Supply Chain & Logistics** | Inventory control, procurement flows, delivery coordination | Demand forecasting, route optimization, disruption prediction | Real-time rerouting of delivery trucks based on weather data, inventory restocking forecasts. | Cost savings, reduced waste, enhanced agility |
| **Smart Cities** | Traffic control, waste collection, energy management. | Real-time analytics, behavior prediction, sensor-based automation. | Adaptive traffic signal control based on congestion, AI-based energy load balancing in grids. | Responsive services, energy efficiency, improved quality of urban life |
| **Banking & Financial Services** | Loan processing, fraud investigation, compliance workflows. | Credit scoring, fraud detection, personalized recommendations. | Instant credit decisions for microloans, AI triage for suspicious transactions. | Risk reduction, regulatory alignment, customer satisfaction. |

continued on following page

*Table 1. Continued*

| Sector | BPM Use Cases | AI Applications | Use Cases | Benefits of Integration |
|---|---|---|---|---|
| **Manufacturing** | Quality control, maintenance workflows, supply chain integration. | Predictive maintenance, defect detection, adaptive scheduling. | Predictive fault detection in assembly lines, AI-guided production rescheduling. | Increased uptime, operational efficiency, sustainable production, etc. |
| **Education** | Student enrollment and record management, course scheduling and timetabling, grading and assessment workflows, financial aid and scholarship processing. | Automated grading and feedback systems, predictive analytics for student performance, AI-powered chatbots personalized content delivery. | Automated exam scoring, real-time predictive analytics, AI chatbots, adaptive learning platforms. | Streamlined administrative processes, more accurate grading and personalized learning experiences. |

Figure 2 shows the variety of potential applications for BPA across the sectors already introduced and illustrates that BPA is a key enabler of digital transformation for each sector.

*Figure 2. Application Domains of Business Process Automation (BPA)*

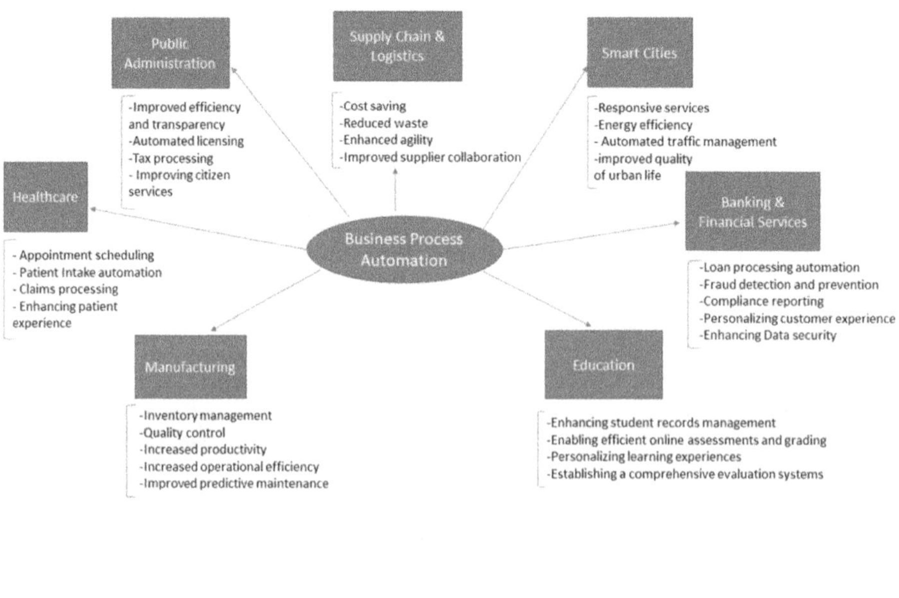

## RESEARCH METHODOLOGY

This chapter uses a conceptual research methodology with the aim of conceptualizing the intersection between AI, BPA and sustainable digital transformation. The methodology comprises three parts, namely: the conceptual analysis of BPA evolution, the sectoral use case survey and the classification of AI techniques, tools and frameworks.

### Conceptual Analysis of BPA Evolution

We undertook a literature review to introduce a conceptual analysis illustrating the evolution of BPA from early manual and rule-based approaches towards AI-enhanced approaches for BPA.

### Sectoral Use Case Survey

In this chapter, we introduce a narrative literature synthesis of BPA applications across seven major sectors, namely: Public Administration, Healthcare, Supply Chain & Logistics, Smart Cities, Banking & Financial Services, Manufacturing and Education. Researches and review studies highlighting AI-driven process improvement and sustainability, from the past ten years, were presented.

### Classification of AI Techniques and Tools

To understand how AI enables BPA, we identify and categorize key AI techniques such as ML, NLP, etc., as well as enabling platforms like BPMS, RPA and low-code platforms. This classification is used to structure the "Results" section of our chapter.

## RESULTS

This section provides a categorization of the key AI methods and toolsets used in business process automation across domains.

AI applied to BPA is an emerging driver of intelligent digital transformation, as business processes become more complex and require higher levels of responsiveness. Traditional BPA focused on simply automating static workflows, with AI being integrated to help these BPA systems sense, adapt and learn so that they can make predictive, contextual decisions at the point of performance. In fact, when it comes to the successful implementation of AI-based BPA projects, it hinges not only on the algorithms and AI-based models but the AI-based ecosystem of tools and platforms supporting projects in the areas of process modeling, executing operations and data integration, execution of processes, and process improvement. This section identified the main AI techniques found in BPA and distinct categories of platforms that could potentially support the implementation of intelligent BPA solutions, from traditional Business Process Management Systems (BPMS) to revolutionary AI frameworks, low-code automation platforms, and hybrid integration systems.

## Rule-Based Systems

Among the many AI techniques that have been used to support BPA, we introduce Rule-Based Systems that use predefined rules (i.e., if-then logic) to arrive at automated decisions and perform actions. Rule-Based automation was primarily based on static, logic-based workflows. However, in today's generation of AI-driven automation, Rule-Based Systems remain one of the foundational technologies of automation. In fact, in the current BPA era (intelligent era), Rule-Based Systems are becoming a common form of AI for automating decision-making and processes, and are usually part of a hybrid AI environment that employs both symbolic reasoning and learning systems. The use of Rule-Based Systems in this type of hybrid environment ensures explainability of actions, compliance with policies and standards, and general transparency of the decision-making process.

Rule-based systems are among the earliest and more interpretable forms of AI. They consist of a formalized decision-making process used to execute processes decisions and routings. In the case of BPM, rule-based engines may be used to support compliant policy flows, trigger notifications, or validate forms according to business constraints. Rule-based approaches (Sarker, I. H., et al., 2024) are interpretable and addressable; however, they have limited capability to handle unstructured data or to adapt to exceptions. These systems are frequently applied in government workflows, where they are used to automate tax eligibility checks, manage multi-level document approval chains, and validate citizen requests against predefined public service criteria.

## Machine Learning (ML)

Standard machine learning models (Subasi, O., et al., 2023) such as decision trees, Support Vector Machines (SVMs) (Cortes, C., & Vapnik, V., 1995) and random forests (Breiman, L., 2001) can support tasks like data classification, outlier detection, and outcome prediction. In BPM systems, ML can help to automate decisions that would otherwise require human interpretation (e.g., calculating the risk of a loan, predicting when a service might be delayed) (Udayakumar, R., et al., 2023).

Machine learning models have been implemented across sectors to enrich risk scoring for financial services, flagging complicated cases in social welfare programs, and predicting resource allocation in logistics and supply chain operations by leveraging existing historical data.

To extend existing process automation solutions with custom AI or first build AI components supporting process automation, developers will often use machine learning libraries or frameworks, including TensorFlow, PyTorch, scikit-learn, and spaCy. These frameworks allow a developer to build a classification model, a forecasting algorithm, or some organization custom NLP component that can be augmented within an automation workflow. Increasingly, BPM platforms allow for plugin-based or microservice architectures that allow you to call or push into your models in real time, which means a gateway for intelligence and logic in task execution.

## Deep Learning (DL)

Deep learning methods, which rely on artificial neural networks, outperform standard data models when capturing nonlinear relationships and patterns from high dimensional data. Common architectures: Convolutional Neural Networks (CNN) (Albawi, S., et al., 2017) for visual data, and Long Short-Term Memory (LSTM) (Yu, Y., et al., 2019) and Transformer Networks (Khan, S., et al., 2022). The main downside to deep learning is that models are often "black boxes" without explainable outcomes, and require large amounts of labeled data and computational power. Deep learning methods can be effectively applied in situations where data are high-dimensional or unstructured, such as classifying scanned documents in administrative workflows, predicting failure of pieces of equipment in manufacturing lines, and modeling sequential behaviors in time-sensitive or process-focused environments.

## Natural Language Processing (NLP)

NLP (Young, T., et al., 2018), allows BPA systems to recognize and act upon unstructured text input in ways that are invaluable for front-office automation and citizen-facing services. NLP is typically used for extracting tasks, intents or decisions from messages, emails or forms so as to seamlessly interact with automated workflows.

NLP strategies have allowed intelligent automation to be deployed in front office functions that include intent classification in digital service portals, auto-triaging incoming customer emails, and summarizing key information from complex legal or regulatory documents for further processing.

## Generative AI and Large Language Models (LLMs)

The recent arrival of large language models (LLMs) (Zhao, W., et al., 2023) and generative AI (Bengesi, S., et al., 2023) has opened up additional possibilities for BPA. LLMs and generative AI can be used to generate natural language and produce draft responses, summaries of the process documents and data capture to synthesize workflow diagrams. Generative AI has significant potential for dynamic design of processes and conversational interfaces, but control, reliability and explainability of this technology is still evolving.

In (Beheshti et al.,2023), the authors propose ProcessGPT which is built to take a written description of a process and create an automated workflow of that process. It uses generative AI, specifically a fine-tuned Generative Pre-trained Transformer (GPT) model, to understand natural language descriptions of processes and generate flows that fulfill them. This ability allows users to automate and enhance business processes, improving the efficiency and effectiveness of processes in data-oriented or knowledge-intensive fields.

ProMoAI is a generative AI tool developed by (Kourani et al. 2024) that focuses on process modeling. It uses LLMs to automate process model generation from textual descriptions by using advanced prompt engineering, error handling, and code generation.

By leveraging Amazon SageMaker, organizations can fine-tune LLMs to align with their specific needs and brand voice. SageMaker allows dynamic routing so that organizations will be able to deal with the different customer inquiries by directing them to the appropriate models and services (Karakus et al., 2021).

Figure 3. illustrates the main AI techniques used in BPA, showcasing their contributions in enabling intelligent, adaptive, and efficient business operations.

*Figure 3. The main AI techniques used for Business Process Automation (BPA)*

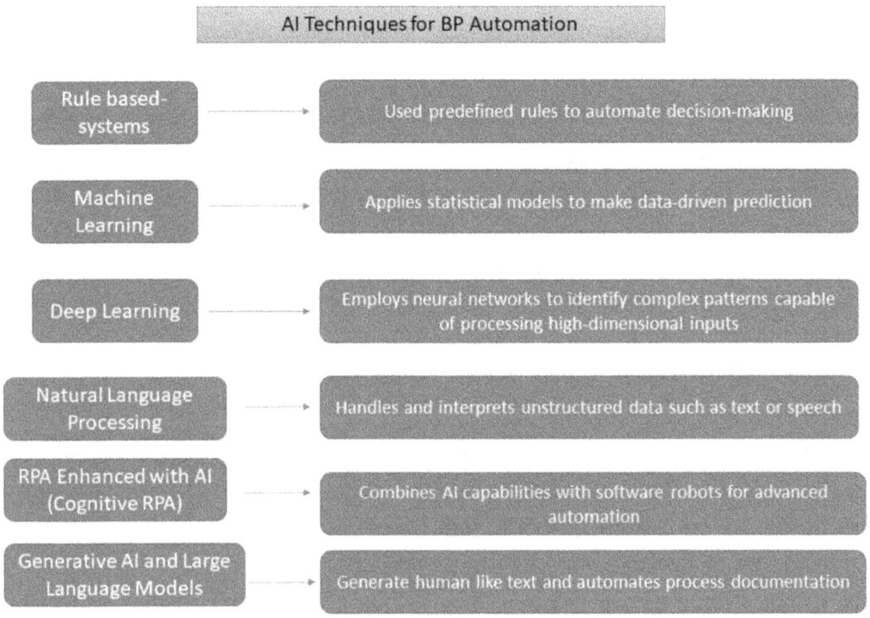


## Robotic Process Automation (RPA) Enhanced with AI

RPA (van der Aalst, W. M. P., et al., 2018) is a software tool that mimics human behavior in engaging with user interfaces to automate rules-based processes that are manually applied. Intelligent RPA is being more widely adopted in a variety of sectors, including banking, HR, and healthcare and used to automate processes in both front- and back-office operations, automating document-based workstreams such as invoice processing in the accounting department, data entry in the on-boarding process of employees or customers, and claim validation workflows in the insurance industry.

RPA tools such as UiPath, Automation Anywhere (Khan, S., 2020) and Blue Prism (Rane, L., & Hallur, G. G., 2023) have emerged as the tool suite which have led to the popularity of automation of rule-based and repetitive tasks. many of RPA platforms have incorporated AI services like OCR, NLP, and machine learning either through native development or integration with external platforms with either Azure AI, Google Cloud AI, (Ravulavaru, A., 2019) or Amazon SageMaker.

## Business Process Management Systems (BPMS)

BPMS solutions provide the essential infrastructure that allows for modeling, executing, and monitoring sets of business processes. Camunda (Berti, A., et al., 2020) Bonita BPM (Araújo, W. J. de, & Gomes, T. A., 2023) and IBM Business Automation Workflow (IBM Corporation, 2023) utilize BPMN to model the processes and connect to external AI components, in using REST API's and scripting environments. BPMS platforms are valuable for organizations looking for structured managed process control with the flexibility of adaptive intelligence modules. For example, they can utilize AI-based decision engines within BPMN gateways to dynamically route processes based upon predictions or classifications in real-time.

## Process Mining and Optimization Platforms

Many companies provide process mining functionality to examine, understand, improve a process, or all of them, with the purpose of making use of event logs (generated each time someone completes some actions in a process) to uncover knowledge about bottlenecks, deviations and compliance issues. Some platforms include AI-driven prescriptive analytics that suggest optimal task sequences or resource allocations. Moreover, AI and machine learning can improve process design, optimization, and automation strategies for process designers.

## Low-Code and Integration Platforms

Low-code platforms such as Mendix (Francisco, G., 2024) and Microsoft Power Platform (Heuer, M., et al., 2022) are emerging as important enablers of AI-driven BPA, particularly for organizations looking for agility. These platforms support visual process modeling, provide built-in connectors to AI, and offer easy integration with enterprise systems. They are especially effective for example prototyping an AI-driven workflow and enable business users to co-develop automation capabilities with IT.

## Hybrid and Cloud-Native Automation Ecosystems

Organizations are rapidly transitioning to hybrid AI-BPM ecosystems that exploit cloud-enabled services, on-premise BPM engines, and microservices. Several platforms, such as SAP Business Technology Platform and Appian have cloud orchestration capabilities with native support for AI components. The hybrid environment enables organizations to scale local or global AI-enhanced BPA deployments across departments, geographies, and channels while maintaining governance, security, and compliance.

As an example, Volkswagen is leveraging the SAP Business Technology Platform (SAP) to enhance AI capabilities in its supply chain and predictive maintenance use cases, all while keeping sensitive manufacturing data on-premise to comply with GDPR. Similarly, global professional services firm AON leveraged the Appian low-code automation platform (Appian) to bring unity and automation to the insurance claims processing space using AI and RPA in multiple business units.

# DISCUSSION

## Comparative Analysis of Techniques, Tools and Frameworks

The variety of AI methods deployed in BPA symbolizes the differences in business needs, data types and contextual complexity in different domains. Rule-based approaches afford transparency and controllability, so they are typically best suited to environments bound by compliance or audit obligations, but they do not have the same flexibility as other AI solutions. Machine learning and deep learning models have advanced pattern identification and predictive functions, but they typically require large amounts of data and do not afford the same degree of interpretability as rule-based models. Natural Language Processing provides an opportunity to span the divide from human input to machine interpretation, enabling user-facing automation and unstructured data review. Also, AI-enhanced Robotic Process Automation has been useful for accelerating automation for legacy systems, where deep integration is not feasible. New generative AI models add dynamism but raise concerns over reliability, explainability, and governance. Choosing the right AI approach for BPA is a consideration of tradeoffs, where organizations must balance accuracy, transparency, scalability, and domain specificity while making sure that automation does not merely enhance performance but also aligns with strategic, ethical and regulatory objectives.

Tools for AI-assisted business process automation are complicated and varied, and present a number of challenges. Conventional BPMS platforms like Camunda or IBM offer process modeling capabilities and more traditional rule-based processes, while advanced AI capabilities often require some level of technical systems integration on the IT side. Vendors of RPA (Robotic Process Automation) software like UiPath and Automation Anywhere generally offer rapid means to automate legacy user interface work, but face the risk of not embodying native process context without reliance on BPM tools. Low-code tools prioritize agility and ease of use, though they may trade off some fine-grained control and advanced model explainability compared to full-code approaches. There are still other processes mining platforms, such as Celonis and Apromore, that provide the last piece of the puzzle between monitoring and redesigning business processes, but these tools rely on event logs which have clean data to function properly. Finding the right tool will vary across organizations and will require careful consideration of many factors such as process maturity, technical environment, regulatory environment, and strategic AI readiness of the organization.

## Implementation Challenges and Ethical Considerations

AI applications within Business Process Automation represent promising opportunities for operational excellence, but these include a mix of ethical, organizational and technical challenges. As organizations evolve their AI-BPM beyond pilot projects and toward scalable, production AI-BPM solutions, they will inevitably encounter numerous barriers that can diminish the usefulness and value of the applications and the trustworthiness associated with them.

- **Data quality and data integration**

AI models are reliant on volume and consistent data of readily available, high quality. Unfortunately, business processes span across disparate systems and so often comprise many data sources within fragmented data stores that also differ in formats. Integrating structured data and unstructured data across multiple silos which include; enterprise resource planning (ERP) systems; CRM platforms; IoT devices etc., can be both complex and error-prone. Moreover, biased or inaccurate data at one staging point in AI pipelines may end up propagating through AI pipelines until the source of inaccuracy is removed from the business process leading to poor prediction, incorrectly routed business processes, and discriminatory decisions.

- **Scalability and Performance**

AI-augmented automation solutions must deliver predictable, reliable performance across different loads and near real-time situations. While it is often possible to prototype with AI models, deploying these models in operational, mission-critical business processes will require further consideration of issues like model latency, infrastructure scalability, and failover. Also, processes that introduce continuous learning and adaptive decision-making may further add complexity in terms of additional required infrastructure costs.

- **Explainability and Trust**

In domains like healthcare and finance, it simply may not be good enough for an AI model to show it was doing a good job; it will also need to be explainable. The black-box nature of many machine learning and deep learning models creates risks where transparency, accountability, and compliance with regulations are required. Auditors, business users, and even citizens need to be able to explain why the process chose a path or produced a result. This situation has created enthusiasm around Explainable AI techniques to try and make model behavior explainable in business processes.

- **Ethical and regulatory considerations**

AI-enabled automation raises ethical questions, especially about fairness, accountability, and privacy. When decision-making in public services is automated, decision-making should not reinforce, even unintentionally, historical biases if the historical data reflects structural disadvantage.

## CONCLUSION

The emerging synergy between artificial intelligence and business process automation is transforming forever how organizations think about the design, execution, and optimization of their operations. With AI, organizations are able to move away from static automation to predictive understanding of the context, capable of operating in real time, and responding dynamically to complex and fluid organizational realities.

This chapter has covered the foundations of AI-enabled BPA, defining its concepts, applications, enabling techniques, and key tools and platforms to deploy BPA. Furthermore, a range of practical issues were considered, including the important issues of data quality, explainability, scalability, and ethics, which organizations must contend with in order to deploy with success. Each sector-specific discussion of AI-powered BPA confirmed its transformative potential in public administration and governance, health care, education, smart cities, supply chain and logistics, banking and financial services, etc. The discussion also confirmed its potential to enhance efficiency, transparency, and sustainability.

Future research directions involve transforming intelligent automation towards adapting AI with BPM, from rule-based augmentation towards self-optimizing systems that can adapt to their context. In order to accomplish this, further research must address several key challenges including the design of AI-based processes that learn and adapt in real-time, developing personalized and context-relevant automation solutions and the integration of explainable AI (XAI) to ensure transparency and trust. Moreover, the future research should consider the merging of AI-enabled BPA with emerging technologies, such as IoT, blockchain and cloud-native architectures as foundations to enable trustworthy, scalable, and ethically designed automation frameworks for sustainable digital transformation.

The next generation of business processes will be based on AI systems that continuously learn and adapt. Moving forward from static models, future BPA platforms will employ reinforcement learning and online learning to make real-time alterations in decision logic, based on feedback, user behavior, and responses to changes in the business environment. This aligns with autonomic business processes that self-monitor and self-configure to self-optimize without human or unauthorized intervention.

Another growing area of interest is the development of BPA systems which are context-aware in which they operate, and automating business logic or decision-making according to user intent or needs, context, location or device. Using AI to capture and interpret context with contextual sensors and environmental sensing can facilitate the personalization of processes at scale. The rise of contextualized and personalized experiences demands flexible semantic modeling and the use of ontologies to effectively represent and interpret dynamic environments. It also requires the integration of both structured and unstructured contextual data to accurately understand the user, their intent, the task at hand, and the surrounding scope.

As AI is becoming more ingrained in important business processes, explainability, transparency, and accountability will be increasingly important, and eventually mandatory. Future research should focus on integrating XAI techniques into BPM environments, with traceable and auditable decision paths. There should also be user-centric design approaches that are rooted in the purpose of providing a meaningful explanation to non-technical stakeholders, such as citizens, healthcare workers, and policy and analysis positions.

AI-enabled BPA is being increasingly integrated with other digital technologies such as IoT, for real-time process execution in manufacturing and smart cities and blockchain, to provide compliance audit trails and decentralized process governance.

## REFERENCES


Albawi, S., Mohammed, T. A., & Al-Zawi, S. (2017). Understanding of a convolutional neural network. 2017 International Conference on Engineering and Technology (ICET), 1–6. Appian. Available at: https://appian.com/ [Accessed: 04 June 2025]

Alkhawaldeh, M. A., & Khasawneh, M. A. S. (2023). Harnessing the power of artificial intelligence for personalized assistive technology in learning disabilities. *Journal of Southwest Jiaotong University*, *58*(4).

Beheshti, A., Yang, J., Sheng, Q. Z., Benatallah, B., Casati, F., Dustdar, S., Motahari Nezhad, H. R., Zhang, X., & Xue, S. (2023). ProcessGPT: Transforming Business Process Management with Generative Artificial Intelligence. arXiv. https://arxiv.org/abs/2306.01771 DOI: 10.1109/ICWS60048.2023.00099

Bengesi, S., El-Sayed, H., Sarker, M. K., Houkpati, Y., Irungu, J., & Oladunni, T. (2023). Advancements in Generative AI: A comprehensive review of GANs, GPT, Autoencoders, Diffusion Models, and Transformers. arXiv preprint arXiv:2311.10242. https://arxiv.org/abs/2311.10242

Berti, A., van der Aalst, W. M. P., Zang, D., & Lang, M. (2020). An Open-Source Integration of Process Mining Features into the Camunda Workflow Engine: Data Extraction and Challenges. arXiv preprint arXiv:2009.06209. https://arxiv.org/abs/2009.06209

Breiman, L. (2001). Random forests. *Machine Learning*, *45*(1), 5–32. DOI: 10.1023/A:1010933404324

Chen, Y., Zhao, C., Xu, Y., & Nie, C. (2025). Year-over-year developments in financial fraud detection via deep learning: A systematic literature review. arXiv preprint arXiv:2502.00201. https://doi.org//arXiv.2502.00201DOI: 10.48550

Corporation, I. B. M. (2023). *IBM Business Automation Workflow: Product overview*. IBM. https://www.ibm.com/products/business-automation-workflow

Cortes, C., & Vapnik, V. (1995). Support-vector networks. *Machine Learning*, *20*(3), 273–297. DOI: 10.1007/BF00994018

Dumas, M., La Rosa, M., Mendling, J., & Reijers, H. A. (2018). *Fundamentals of business process management* (2nd ed., Vol. 1). Springer., DOI: 10.1007/978-3-662-56509-4

Estevez, E., & Janowski, T. (2020). Smart governance for smart cities: A framework. *Government Information Quarterly*, *37*(2), 101399. DOI: 10.1016/j.giq.2019.101399

Francisco, G. (2024). Learning low-code: Potential of Mendix platform in building applications (Bachelor's thesis). South-Eastern Finland University of Applied Sciences. https://urn.fi/URN:NBN:fi:amk-202502072469

Heuer, M., Kurtz, C., & Böhmann, T. (2022). Towards a governance of low-code development platforms using the example of Microsoft Power Platform in a multinational company. Proceedings of the 55th Hawaii International Conference on System Sciences (HICSS). https://aisel.aisnet.org/hicss-55/os/practice-based_research/6/

Janssen, M., & van der Voort, H. (2016). Adaptive governance: Towards a stable, accountable and responsive government. *Government Information Quarterly*, *33*(1), 1–5. DOI: 10.1016/j.giq.2016.02.003

John, S. J. (2024). Harnessing AI for Business Process Automation in Educational Institutions. *Progress in Machines and Systems*, *13*(2), 57–64. DOI: 10.6025/pms/2024/13/2/57-64

Karakus, C., Huilgol, R., Wu, F., Subramanian, A., Daniel, C., Cavdar, D., Xu, T., Chen, H., Rahnama, A., & Quintela, L. (2021). Amazon SageMaker Model Parallelism: A General and Flexible Framework for Large Model Training. arXiv preprint arXiv:2111.05972.

Khan, S. (2020). Comparative analysis of RPA tools—UiPath, Automation Anywhere and Blue Prism. *International Journal of Computer Science and Mobile Applications*, *8*(11), 1–6. DOI: 10.47760/ijcsma.2020.v08i11.001

Khan, S., Naseer, M., Hayat, M., Zamir, S. W., Khan, F. S., & Shah, M. (2022). Transformers in vision: A survey. *ACM Computing Surveys*, *54*(10s), 1–41. DOI: 10.1145/3505244

Kourani, H., Berti, A., Schuster, D., & van der Aalst, W. M. P. 2024. ProMoAI: Process Modeling with Generative AI. In *Proceedings of the Thirty-Third International Joint Conference on Artificial Intelligence (IJCAI-24)*, pp. 8708–8712.

Lee, J., Bagheri, B., & Kao, H. A. (2015). A cyber-physical systems architecture for Industry 4.0-based manufacturing systems. *Manufacturing Letters*, *3*, 18–23. DOI: 10.1016/j.mfglet.2014.12.001

Rane, L., & Hallur, G. G. (2023). Competition in the RPA industry: A case study of Blue Prism. *AIP Conference Proceedings*, *2914*(1), 040009. Advance online publication. DOI: 10.1063/5.0176028

Ravulavaru, A. (2019). Google Cloud AI Services Quick Start Guide: Build intelligent applications with machine learning and artificial intelligence services on the Google Cloud Platform. Packt Publishing. SAP. Available at: https://www.seleccionconsulting.com/blog/how-sap-btp-enhances-manufacturing-retail-finance/ [Accessed: 04 June 2025]

Sa-ad, M. M., Osafo-Apeanti, W., Owusu-Boateng, O., Seidu, S. A., Majeed, M., & Bright-Benuwa, B. (2025). Automated administrative tasks in education. In Owusu-Boateng, O. (Ed.), *Unlocking the Potential: Artificial Intelligence in Education* (pp. 31–45). Deep Science Publishing.

Sarker, I. H., Janicke, H., Ferrag, M. A., & Abuadbba, A. (2024). Multi-aspect rule-based AI: Methods, taxonomy, challenges, and directions toward automation, intelligence, and transparent cybersecurity modeling for critical infrastructures. *Internet of Things : Engineering Cyber Physical Human Systems*, *25*, 101110. DOI: 10.1016/j.iot.2024.101110

Subasi, O., Bel, O., Manzano, J., & Barker, K. (2023). The landscape of modern machine learning: A review of machine, distributed and federated learning. *arXiv preprint arXiv:2312.03120.*

Udayakumar, R., Devi, J., & Suganya, S. (2023). Integrated SVM-FFNN for fraud detection in banking financial transactions. *Journal of Internet Services and Information Security*, *13*(4), 12–25. DOI: 10.58346/JISIS.2023.I4.002

van der Aalst, W. M. P. (2021). *Process mining: Data science in action* (2nd ed., Vol. 1). Springer.

van der Aalst, W. M. P., Bichler, M., & Heinzl, A. (2018). Robotic process automation. *Business & Information Systems Engineering*, *60*(4), 269–272. DOI: 10.1007/s12599-018-0542-4

Vorobyeva, K. I., Belous, S., Savchenko, N. V., Smirnova, L. M., Nikitina, S. A., & Zhdanov, S. P. (2025). Personalized learning through AI: Pedagogical approaches and critical insights. *Contemporary Educational Technology*, *17*(2), ep574. DOI: 10.30935/cedtech/16108

Young, T., Hazarika, D., Poria, S., & Cambria, E. (2018). Recent trends in deep learning-based natural language processing. *IEEE Computational Intelligence Magazine*, *13*(3), 55–75. DOI: 10.1109/MCI.2018.2840738

Yu, Y., Si, X., Hu, C., & Zhang, J. (2019). A review of recurrent neural networks: LSTM cells and network architectures. *Neural Computation*, *31*(7), 1235–1270. DOI: 10.1162/neco_a_01199 PMID: 31113301

Zhao, W., Zhou, X., Li, J., Tang, B., Wang, Y., Duan, Y., & Liu, T. (2023). A survey of large language models. *arXiv preprint arXiv:2303.18223*.

# Chapter 6
# Harnessing Artificial Intelligence and Blockchain for Enhanced Transparency and Sustainability in Public Infrastructure Projects

**Pedro Miguel Gomes**
https://orcid.org/0009-0005-4996-4534
*Secretary General for Environment, Portugal*

**Tiago Jordão Cardoso**
https://orcid.org/0009-0008-5223-2299
*Secretary General for Environment, Portugal*

## ABSTRACT

*Public infrastructure projects are frequently plagued by inefficiencies, budget overruns, corruption, and sustainability challenges, which undermine their long-term viability and public trust. The integration of emerging technologies, such as artificial intelligence (AI) and blockchain, presents transformative opportunities to address these issues by enhancing transparency, accountability, and sustainability. AI can optimize project management through predictive analytics, real-time monitoring, and risk assessment, while blockchain ensures financial integrity by providing immutable transaction records and decentralized oversight. This chapter explores the synergy between these technologies, illustrating their potential through real-world case studies. By examining successful implementations, it highlights how AI*

DOI: 10.4018/979-8-3373-1534-8.ch006

*and blockchain can improve decision-making, reduce fraud, enhance stakeholder collaboration, and ensure compliance with environmental regulations.*

## INTRODUCTION

Public infrastructure serves as the backbone of modern society, enabling economic activity, social welfare, and environmental sustainability. It encompasses essential services such as transportation networks, energy systems, water supply, sanitation, healthcare facilities, and digital infrastructure (World Bank, 2020). These systems are fundamental for ensuring societal well-being, facilitating commerce, and supporting industrial productivity.

However, the planning, execution, and maintenance of infrastructure projects often suffer from inefficiencies, corruption, financial mismanagement, and environmental concerns (Flyvbjerg et al., 2003). These challenges can lead to delays, cost overruns, and substandard project outcomes, reducing public trust in government institutions and impeding economic growth. Studies have shown that major infrastructure projects worldwide frequently exceed their initial budgets and schedules, placing significant strain on national economies (Cantarelli et al., 2012).

Traditional project management approaches, while structured and methodical, are often inadequate in addressing the complexities and dynamic risks associated with large-scale infrastructure development. Issues such as opaque procurement processes, misallocation of resources, and bureaucratic inefficiencies contribute to these inefficiencies, leading to economic losses and environmental degradation (Love et al., 2016).

In response to these persistent challenges, emerging technologies such as Artificial Intelligence (AI) and blockchain have been identified as potential game changers in infrastructure management. AI-driven analytics enable predictive modeling, risk assessment, and automated decision-making, helping planners optimize resource allocation and improve project efficiency (Wang et al., 2019). Meanwhile, blockchain technology enhances transparency and accountability by ensuring immutable records of financial transactions, contract execution, and procurement processes (Casino et al., 2019).

By leveraging AI and blockchain, governments and private stakeholders can enhance efficiency, sustainability, and transparency in infrastructure development. These technologies offer solutions for reducing corruption, optimizing maintenance strategies, and ensuring responsible environmental management, making infrastructure projects more resilient and adaptive to future needs.

This chapter examines the key challenges faced by public infrastructure projects and explores how AI and blockchain technologies can provide innovative solutions to improve efficiency, accountability, and sustainability. It highlights real-world applications, case studies, and emerging trends that illustrate how these technologies are reshaping the future of infrastructure development.

## Challenges in Public Infrastructure Development

Public infrastructure projects are inherently complex, involving multiple stakeholders, substantial financial investments, and long-term planning. Despite ongoing efforts to improve efficiency and governance, these projects frequently encounter significant challenges, including cost overruns, project delays, corruption, and environmental and social impacts. Understanding these challenges is critical for the development of more effective and sustainable infrastructure strategies.

One of the most persistent challenges in infrastructure development is cost overrun. Studies indicate that approximately 90% of large-scale infrastructure projects exceed their initial budgets, with cost escalations ranging between 20% and 45% (Flyvbjerg et al., 2018). These overruns are attributed to multiple factors, including inaccurate cost estimation, unforeseen site conditions, inflation, design changes, and inefficient project management (Cantarelli et al., 2012). Recent research highlights that political influence and inadequate risk assessment exacerbate cost overruns, particularly in government-funded projects (Latilo et al., 2024). The case of Stuttgart 21, a major railway infrastructure project in Germany, exemplifies this issue, as it has faced significant budget increases due to regulatory and procedural inefficiencies (Babalola et al., 2024).

Project delays further compound financial burdens, undermining economic growth and public confidence. Studies suggest that more than half of all major infrastructure projects experience significant schedule slippages due to bureaucratic inefficiencies, disruptions in the supply chain, and inadequate risk assessment (Mahmud et al., 2021). Infrastructure projects in developing economies suffer from extensive delays caused by political instability, funding constraints, and mismanagement of contracts (Ukeyima, 2024). Delays not only inflate costs but also hinder service delivery, leading to economic losses and prolonged inconvenience for the public (Othman & Ahmed, 2023).

Corruption remains a pervasive challenge in public infrastructure projects. The World Bank (2017) estimates that corruption inflates infrastructure costs by up to 30%, diverting essential funds from public development initiatives. Corrupt practices in procurement and contract management create opportunities for bribery, collusion, and resource misallocation (Kenny, 2007). A study (Oluseye, 2024) underscores how political corruption in large-scale infrastructure projects systematically increases costs, delays implementation, and reduces project quality. The infamous Operation Car Wash (Lava Jato) in Brazil, which revealed billions of dollars in fraudulent transactions, exemplifies the detrimental impact of corruption on public infrastructure (Molloy, 2019). Emerging governance frameworks, including digital tracking and blockchain technology, have been proposed as mechanisms to improve transparency and reduce financial leakages (Kshetri, 2021; Aigbavboa et al., 2024).

Beyond financial and administrative challenges, infrastructure development has considerable environmental and social consequences. Large-scale projects often contribute to deforestation, biodiversity loss, and increased carbon emissions, exacerbating climate change (Zhang, Wang & Liu, 2020). Urban expansion, hydroelectric dams, and transportation megaprojects frequently lead to the displacement of local communities, disrupting social structures and livelihoods (Geels et al., 2017).

To mitigate these impacts, it is imperative to integrate Sustainable Development Goals (SDGs) into infrastructure planning. The adoption of environmentally conscious design principles, enhanced monitoring technologies, and resource-efficient construction methods can significantly reduce the environmental footprint of infrastructure projects (United Nations, 2019; Anago, 2024). The implementation of smart technologies, including artificial intelligence (AI) and the Internet of Things (IoT), has shown promise in optimizing infrastructure sustainability through predictive maintenance and resource-efficient project management (Hussain et al., 2024).

The persistent challenges in public infrastructure development—ranging from financial inefficiencies to environmental concerns—underscore the need for a more integrated and technologically driven approach. Enhanced governance structures, transparent procurement processes, and the adoption of smart construction technologies are crucial to improving project efficiency and long-term sustainability. Future research should focus on exploring data-driven decision-making mechanisms and policy innovations to mitigate risks associated with infrastructure projects.

## The Role of AI and Blockchain in Infrastructure Development

Artificial intelligence (AI) and blockchain technology offer transformative solutions to the persistent challenges in infrastructure development by enhancing efficiency, reducing corruption, and promoting sustainability. The integration of these technologies into infrastructure projects has gained momentum due to their potential to improve decision-making, optimize resource allocation, and ensure greater transparency in public and private sector initiatives.

AI-driven predictive analytics has revolutionized infrastructure project planning. Machine learning algorithms analyze historical project data to forecast costs, identify potential delays, and optimize resource allocation (Wang et al., 2019). By leveraging large datasets, AI models enhance decision-making processes in real-time, enabling the proactive mitigation of risks and reducing maintenance costs (Zhang et al., 2022).

For instance, AI-powered geospatial analysis plays a crucial role in optimizing urban planning by predicting traffic congestion patterns, energy demands, and environmental impacts (Tan et al., 2021). Moreover, AI-based structural health monitoring (SHM) systems utilize real-time sensor data from bridges, tunnels, and roads to assess infrastructure integrity, improving safety while reducing operational costs (Liu et al., 2020). Recent research highlights how AI-driven infrastructure systems in smart cities have contributed to predictive maintenance strategies, reducing failure risks and enhancing sustainability (Mohbey et al., 2024).

Blockchain technology enhances transparency and accountability by providing immutable digital records of financial transactions, procurement processes, and contract agreements (Saberi et al., 2019). This decentralized approach reduces opportunities for fraud, ensures accountability in large-scale infrastructure projects, and fosters trust among stakeholders.

Governments in Estonia and Dubai have successfully implemented blockchain-based procurement systems, minimizing corruption and improving public trust in infrastructure development (Ølnes et al., 2017). Additionally, smart contracts self-executing contracts based on blockchain—facilitate automated enforcement of contractual terms, reducing delays and disputes in public-private partnerships (Kouhizadeh & Sarkis, 2018). Emerging research suggests that blockchain integration in infrastructure governance has led to enhanced security, accountability, and efficiency, particularly in public finance management.

AI and blockchain contribute to environmental sustainability by optimizing energy consumption, reducing waste, and promoting green financing mechanisms. AI algorithms can assess the environmental impact of construction activities, while blockchain-based carbon credit systems incentivize sustainable practices (Zhang et al., 2020).

Stockholm's smart city initiatives, for example, leverage AI to design energy-efficient buildings and use blockchain to track carbon emissions and offset credits (Moghayedi et al., 2022). The implementation of blockchain in smart urban systems has demonstrated its ability to reduce regulatory inefficiencies, improve real-time monitoring, and ensure compliance with sustainability policies (Mohbey et al., 2024).

Addressing the persistent challenges of cost overruns, corruption, and environmental sustainability in public infrastructure projects requires innovative technological solutions. AI and blockchain provide transformative tools to enhance project planning, increase transparency, and support sustainable development. While these technologies have already demonstrated success in real-world applications, broader adoption depends on regulatory frameworks, capacity building, and collaboration between governments and private sectors. As AI and blockchain continue to evolve, their integration into infrastructure development promises a more efficient, transparent, and environmentally responsible future.

## BACKGROUND STUDY

### The Role of Artificial Intelligence in Public Infrastructure

Artificial Intelligence (AI) is transforming the planning, management, and maintenance of public infrastructure by enhancing decision-making, optimizing resource allocation, and improving long-term sustainability. Governments and private sector stakeholders are increasingly leveraging AI-driven technologies to streamline infrastructure development, reduce operational costs, and address pressing urbanization challenges (Wang et al., 2019).

The integration of big data analytics, machine learning (ML), deep learning, and automation has allowed infrastructure planners to analyze complex datasets, predict potential risks, and enhance efficiency in project execution. AI-powered solutions facilitate real-time monitoring, proactive maintenance, and automated governance, leading to cost reductions and increased resilience in infrastructure systems (Tan et al., 2021).

AI applications in infrastructure span multiple domains, including transportation, energy, water management, and construction. For example, AI-driven traffic management systems optimize road networks by predicting congestion and adjusting traffic signals dynamically (Zhang et al., 2020). In energy infrastructure, AI-powered smart grids enhance efficiency by balancing energy supply and demand in real-time, minimizing wastage and improving sustainability (Moghayedi et al., 2022). Similarly, in construction and urban planning, AI-based simulations help in designing efficient, resilient structures that meet both economic and environmental standards (Kouhizadeh & Sarkis, 2018).

One of AI's most significant advantages is its ability to process and analyze vast amounts of real-time data. Unlike traditional project management approaches, which rely heavily on static models and human-driven analysis, AI continuously refines predictions based on incoming data streams. This capability enables infrastructure projects to remain adaptable to changing economic, social, and environmental conditions (Kshetri, 2021).

However, the adoption of AI in public infrastructure is not without challenges. Issues such as data privacy, ethical concerns, algorithmic bias, and regulatory constraints must be addressed to ensure the responsible deployment of AI technologies (Floridi, 2019). Additionally, the high cost of AI implementation and the need for skilled professionals to manage AI-driven systems present barriers to widespread adoption, particularly in developing economies (Kshetri, 2021).

Despite these challenges, AI is set to play an increasingly vital role in shaping the future of public infrastructure. This section delves into three primary areas where AI is making a transformative impact:

1. Enhancing Project Planning and Decision-Making – AI-driven analytics improve the accuracy of cost estimation, risk assessment, and resource allocation.
2. Improving Smart Infrastructure Maintenance – AI-powered IoT sensors and predictive maintenance models prevent costly failures and extend infrastructure lifespan.
3. Fostering Sustainable Infrastructure Development – AI optimizes energy use, material selection, and environmental impact assessments to promote sustainability.

As AI continues to evolve, its integration with other emerging technologies, such as blockchain and the Internet of Things (IoT), will further enhance the efficiency, security, and resilience of public infrastructure.

## Enhancing Project Planning and Decision-Making

Infrastructure planning remains a highly complex and resource-intensive process, involving multiple stakeholders, substantial financial investments, and long-term economic and social implications. Artificial intelligence (AI) and blockchain technology offer innovative solutions to optimize project planning by improving cost estimation, risk assessment, and decision-making processes.

Cost overruns and schedule delays are among the most persistent challenges in infrastructure development. Studies indicate that nearly 90% of megaprojects exceed their initial budgets (Flyvbjerg et al., 2018). AI-driven cost estimation tools analyze historical project data, external economic factors, and raw material prices to generate highly accurate budget forecasts, mitigating financial risks (Love et al., 2016). Machine learning models, such as gradient boosting and deep neural networks, are increasingly deployed to assess project risks, detect supply chain disruptions, and optimize workforce management (Zhang et al., 2021). By integrating AI-driven predictive analytics, infrastructure planners can anticipate and address potential bottlenecks before they escalate into critical delays. AI-based risk assessment tools also leverage geospatial intelligence to evaluate the impact of external factors, such as adverse weather conditions, political instability, and environmental regulations. These systems provide decision-makers with a comprehensive risk profile, enabling proactive mitigation strategies (Dritsas & Trigka, 2025).

The integration of geospatial AI has transformed urban and regional planning by providing real-time insights into land use, environmental sustainability, and infrastructure resilience. Governments and developers now utilize AI-enhanced satellite imagery and remote sensing to identify optimal locations for roads, bridges, and energy facilities while minimizing environmental disruption. AI-powered geospatial analysis supports predictive urban planning by simulating future infrastructure demands based on population growth, traffic patterns, and environmental constraints (Al-Rimawi & Nadler, 2025). This data-driven approach enhances long-term sustainability while ensuring resource-efficient project implementation.

Collaboration among multiple stakeholders—including government agencies, private investors, contractors, and local communities—is critical to infrastructure success. AI-driven project management platforms facilitate seamless communication by automating document sharing, contract tracking, and workflow coordination (Mohamed et al., 2024). Blockchain-integrated AI systems further enhance transparency and accountability by securely recording all project-related transactions and procurement activities. This decentralized ledger technology reduces opportunities for fraud, ensures compliance with contractual obligations, and fosters stakeholder trust (Chafiq et al., 2024). Governments in Estonia, Dubai, and Singapore have successfully deployed blockchain-based public procurement frameworks to reduce corruption and enhance accountability in infrastructure projects. Smart contracts, which automate agreement execution, further streamline project financing and contract enforcement, minimizing legal disputes and project delays (Mia et al., 2022).

The integration of AI and blockchain in infrastructure planning presents a paradigm shift in how projects are conceptualized, designed, and executed. AI-driven predictive analytics enables more accurate cost estimation and risk mitigation, while geospatial intelligence improves urban planning efficiency. Simultaneously, blockchain fosters transparency, accountability, and stakeholder collaboration. As AI and blockchain technologies continue to evolve, their adoption in infrastructure management will be shaped by regulatory advancements, data security frameworks, and cross-sector collaboration. The combination of AI-driven decision-making and blockchain-based trust mechanisms promises a more efficient, cost-effective, and sustainable future for global infrastructure development.

## AI in Smart Infrastructure Maintenance

Maintaining public infrastructure is essential to ensure safety, operational efficiency, and longevity. Traditional maintenance approaches often rely on reactive strategies, leading to costly emergency repairs and service disruptions. AI-powered predictive maintenance solutions leverage sensors, IoT devices, and machine learning algorithms to monitor infrastructure health in real time, preventing failures and reducing maintenance costs (Zhang et al., 2020). These systems allow infrastructure managers to transition from reactive to proactive maintenance, ensuring that minor defects do not escalate into major structural failures (Khan et al., 2024).

Predictive maintenance utilizes AI to analyze data collected from IoT-enabled sensors embedded in infrastructure assets, such as bridges, roads, and water pipelines. These sensors continuously monitor factors like structural integrity, vibrations, temperature fluctuations, and material fatigue (Liu et al., 2020). AI-based monitoring systems have demonstrated high accuracy in identifying early signs of deterioration, allowing maintenance teams to intervene before critical failures occur (El Khoury, 2024). AI-driven structural health monitoring (SHM) systems detect micro-cracks in bridges and tunnels before they become critical safety hazards. The Netherlands has implemented AI-powered monitoring for its extensive network of dikes and flood barriers, reducing maintenance costs and improving disaster preparedness (Wang et al., 2022). Advanced AI models now incorporate deep learning techniques to predict the progression of material degradation and recommend optimal maintenance schedules, minimizing overall costs and extending infrastructure lifespan (Salma et al., 2024).

AI is also transforming traffic management systems in urban areas. Cities like Singapore use AI-driven predictive models to optimize traffic signals, reduce congestion, and enhance public transport efficiency (Tan et al., 2021). These models analyze real-time traffic flow data, weather conditions, and commuter behavior to adjust traffic lights dynamically, preventing bottlenecks. AI-based smart mobility systems are also improving route optimization for public and private transport networks, ensuring smoother traffic flow while reducing emissions. AI-powered vehicle-to-infrastructure (V2I) communication enables autonomous traffic control by coordinating traffic signals, toll systems, and highway monitoring to enhance safety and efficiency in metropolitan areas (Khoury, 2024).

Autonomous public transportation is another emerging AI application. Self-driving buses and AI-powered metro systems are being piloted in cities worldwide to improve efficiency and safety while reducing human error in transit operations (Zhang et al., 2022). The deployment of AI-controlled public transport fleets has shown promising results in reducing transit delays, improving fuel efficiency, and lowering operating costs in major cities such as Dubai and Hong Kong (Khan et al., 2024). AI-enhanced public transport analytics also help urban planners design more efficient transit routes based on real-time demand patterns, thereby improving accessibility and commuter experience.

In the energy sector, AI assists in monitoring and maintaining power grids, reducing blackouts and equipment failures. AI algorithms analyze electrical load distribution, detect anomalies, and predict potential failures in power transmission lines (Kouhizadeh & Sarkis, 2018). AI-powered renewable energy forecasting tools optimize the integration of solar and wind energy into the grid, improving sustainability and reducing reliance on fossil fuels (Moghayedi et al., 2022). AI-driven smart grid systems now play a crucial role in balancing power demand and supply, preventing overloads, and ensuring stable electricity distribution in major urban centers (Khoury, 2024). In addition, blockchain-based energy management platforms combined with AI are being implemented to decentralize energy transactions, enabling peer-to-peer energy trading and increasing efficiency in distributed power generation systems (Khan et al., 2024).

The integration of AI and IoT in predictive infrastructure maintenance, traffic optimization, and energy management is transforming the way cities operate, ensuring enhanced efficiency, safety, and sustainability. While these advancements have demonstrated substantial benefits, their widespread adoption depends on robust regulatory frameworks, investment in smart infrastructure, and collaboration between public and private stakeholders. As AI technologies continue to evolve, their role in shaping the future of infrastructure management will become increasingly vital in the pursuit of resilient and intelligent cities.

## AI in Sustainable Infrastructure Development

As governments and industries strive to meet global sustainability goals, AI plays a crucial role in designing and managing environmentally friendly infrastructure projects. AI-powered systems assist in energy-efficient building design, material optimization, and carbon footprint reduction (Moghayedi et al., 2022).

AI-driven simulations help urban planners design smart cities with optimized land use, efficient public transportation, and reduced environmental impact. AI-based geospatial analytics assess climate resilience, identifying flood-prone zones and heat island effects to inform sustainable development strategies (Geels et al., 2017).

For example, Stockholm employs AI to model energy consumption patterns in buildings, optimizing heating, cooling, and ventilation systems to reduce carbon emissions. These AI-assisted urban planning initiatives have led to a 20% reduction in energy use across new developments (Moghayedi et al., 2022).

AI enhances construction sustainability by optimizing material selection and reducing waste. Machine learning algorithms analyze material properties and suggest eco-friendly alternatives, such as recycled concrete and low-carbon steel, without compromising structural integrity (Zhang et al., 2021).

Robotic AI-powered construction techniques, including 3D printing, further reduce waste by precisely layering building materials based on digital blueprints. These technologies are already being used in the Netherlands and China to construct sustainable, cost-effective housing (Liu et al., 2020).

AI plays a vital role in climate adaptation strategies by predicting extreme weather events and optimizing disaster response efforts. Machine learning models analyze historical climate data to forecast hurricanes, floods, and heatwaves, enabling proactive infrastructure design to withstand such challenges (Geels et al., 2017).

Blockchain-integrated AI systems also facilitate carbon credit trading and sustainable financing models, ensuring accountability in green infrastructure investments (Saberi et al., 2019). AI is revolutionizing public infrastructure by enhancing project planning, optimizing maintenance, and fostering sustainability. Through predictive analytics, real-time monitoring, and automation, AI-driven solutions mitigate inefficiencies, reduce costs, and improve resilience against environmental and economic uncertainties. The next sections of this chapter will explore how blockchain further complements AI by strengthening transparency, security, and efficiency in infrastructure projects.

## The Role of Blockchain in Public Infrastructure

Blockchain technology is transforming public infrastructure by enhancing transparency, minimizing corruption, and ensuring financial accountability. Traditionally, public infrastructure projects involve multiple stakeholders—government agencies, contractors, suppliers, investors, and auditors—each managing different aspects of project execution. The use of centralized databases and paper-based records in these projects often leads to inefficiencies, lack of accountability, and corruption risks (Casino et al., 2019).

By utilizing a decentralized and immutable ledger, blockchain minimizes the risk of fraud, streamlines procurement and project execution, and strengthens governance in infrastructure projects. Unlike traditional record-keeping systems, blockchain ensures that all transactions are permanently recorded, time-stamped, and tamper-proof, making it nearly impossible to alter financial records or manipulate procurement contracts (Kshetri, 2021).

Governments worldwide have begun integrating blockchain into public infrastructure to combat corruption, improve contract management, and increase transparency in financial transactions. Countries such as Estonia, Dubai, Chile, and Georgia have successfully implemented blockchain-based solutions for land registries, procurement systems, and smart contract-driven project management (Saberi et al., 2019).

This section explores the three main areas where blockchain is making a significant impact in public infrastructure:

1. Improving Transparency and Reducing Corruption – Blockchain eliminates opaque transactions by providing public, verifiable records of financial activities and procurement processes.

2. Enhancing Contract Management with Smart Contracts – The automation of contract execution through blockchain reduces disputes, delays, and bureaucratic inefficiencies.
3. Ensuring Accountability in Public Spending – The immutable nature of blockchain ensures secure and auditable records, reducing financial mismanagement and ensuring that public funds are used effectively.
4. Blockchain's potential in public infrastructure extends beyond financial oversight—it also supports decentralized identity management, supply chain verification, and compliance tracking. As blockchain technology evolves, its integration with Artificial Intelligence (AI) and the Internet of Things (IoT) will further enhance its ability to manage complex infrastructure projects efficiently and securely.

## Improving Transparency and Reducing Corruption

Corruption in public infrastructure projects is a major global concern, with the World Bank estimating that up to 30% of funds allocated to infrastructure development are lost to fraudulent activities (World Bank, 2020). Common forms of corruption include bribery, bid-rigging, and misallocation of resources, often facilitated by opaque financial and administrative processes (Kenny, 2007).

Blockchain ensures that transactions and project documentation remain immutable, preventing fraud and unauthorized modifications. The decentralized nature of blockchain means that no single entity can alter records without consensus, reducing opportunities for corruption and enhancing public trust in government spending (Casino et al., 2019).

One of the most effective applications of blockchain in public infrastructure is in procurement. Traditional procurement processes are often marred by favoritism, bribery, and lack of accountability. Blockchain-based procurement systems provide a secure, transparent, and auditable record of bid submissions, contract awards, and payment disbursements (Saberi et al., 2019).

Estonia has successfully implemented blockchain technology in its e-procurement system to eliminate bid-rigging and ensure fairness in contract awards. The system allows public access to all procurement-related transactions, making it difficult for officials to engage in fraudulent activities (Ølnes et al., 2017). Similarly, Dubai's blockchain strategy aims to fully digitize government procurement by 2030, reducing paperwork, increasing efficiency, and improving transparency in public contracts (Alketbi et al., 2018).

Land and property transactions are particularly vulnerable to fraud, including illegal transfers, double spending, and document forgery. Blockchain-based land registries provide a secure and immutable record of property ownership, reducing the risk of fraud and disputes (Sharma et al., 2020).

Georgia was one of the first countries to implement a blockchain-powered land registry system, enabling citizens to verify land ownership records through a secure, decentralized ledger. Since its introduction, property fraud cases have declined, and public trust in land administration has increased (De Filippi & Wright, 2018).

## Enhancing Contract Management with Smart Contracts

Contract management in public infrastructure projects is traditionally inefficient, involving lengthy bureaucratic procedures and manual verification processes. Delays in payments, contract disputes, and lack of compliance tracking often result in financial losses and stalled projects (Kouhizadeh & Sarkis, 2018).

Smart contracts, powered by blockchain, automate contract execution by ensuring that agreements are enforced without intermediaries. These self-executing contracts operate based on predefined conditions, triggering payments or actions when contractual obligations are met (Buterin, 2014).

The use of blockchain-based smart contracts in Chile's public transport infrastructure has streamlined payments and reduced bureaucratic inefficiencies. By embedding contract terms into blockchain, authorities ensure that funds are released only when contractors meet predefined milestones, reducing delays and eliminating opportunities for fraud (Mendoza et al., 2021).

Smart contracts also facilitate real-time compliance monitoring. In construction projects, IoT devices can be linked to blockchain-based contracts, automatically verifying that materials used meet regulatory standards before triggering supplier payments (Zhang et al., 2020).

Disputes in infrastructure projects often arise due to contractual ambiguities, late payments, and lack of enforcement mechanisms. Traditional legal processes for resolving contract disputes are time-consuming and costly. Blockchain-based smart contracts reduce these disputes by ensuring that all contract terms are digitally recorded and automatically enforced (Pellegrino & Rauchs, 2020).

For instance, China has experimented with blockchain-powered arbitration systems, where contract disputes are resolved through digital courts that use smart contracts and AI-powered legal analysis (Sun et al., 2022). This approach has significantly reduced resolution times and improved contract enforcement in large-scale infrastructure projects.

## Ensuring Accountability in Public Spending

Government spending on infrastructure requires strict financial oversight to prevent misuse of funds and ensure efficient resource allocation. Traditional accounting systems are often opaque and susceptible to manipulation, making it difficult to track how public funds are spent (Kshetri, 2021).

By recording every transaction on a distributed ledger, blockchain enhances financial accountability in government infrastructure projects. Transactions are timestamped and immutable, ensuring that expenditures can be audited in real time (Sharma et al., 2020).

Blockchain-based financial tracking systems allow governments to monitor budget allocations and expenditures in real time, reducing the risk of embezzlement and fund diversion. This is particularly useful for large-scale projects that involve multiple funding sources and international aid organizations (Casino et al., 2019).

For example, the World Food Programme (WFP) has successfully used blockchain technology to track aid distribution in refugee camps, ensuring that funds reach their intended beneficiaries without intermediaries (Galvez et al., 2020). Similar approaches can be adopted for infrastructure projects to prevent financial mismanagement and ensure that allocated funds are used efficiently.

Traditional auditing processes rely on centralized oversight, which can be influenced by political interests and bureaucratic inefficiencies. Blockchain-based decentralized auditing allows independent auditors, stakeholders, and even the public to verify financial records in real time (Saberi et al., 2019).

By integrating AI-powered anomaly detection, blockchain systems can automatically flag suspicious transactions, alerting authorities to potential fraud (Zhang et al., 2020). This approach has been piloted in several European Union projects aimed at reducing financial mismanagement in cross-border infrastructure initiatives (Kshetri, 2021).

Blockchain technology is transforming public infrastructure by improving transparency, enhancing contract management, and ensuring financial accountability. By leveraging decentralized ledgers and smart contracts, governments can reduce corruption, streamline procurement, and improve project efficiency. Real-world implementations in Estonia, Dubai, Chile, and Georgia demonstrate the potential of blockchain to reshape infrastructure governance.

However, challenges such as regulatory barriers, interoperability issues, and resistance to change must be addressed to fully realize the benefits of blockchain in public infrastructure. As blockchain technology continues to evolve, its integration with AI and IoT is expected to further enhance efficiency and sustainability in infrastructure projects worldwide.

## RESULTS

### Case Studies

The integration of Artificial Intelligence (AI) and Blockchain in public infrastructure has already demonstrated significant improvements in efficiency, transparency, and sustainability across various countries. As infrastructure projects become more complex, these technologies provide data-driven insights, automation, and decentralized governance, reducing inefficiencies and mitigating risks associated with corruption and mismanagement.

Real-world implementations illustrate how governments and private-sector stakeholders are leveraging AI and blockchain to optimize infrastructure planning, project execution, and long-term maintenance. These case studies highlight the tangible benefits of these technologies in solving long-standing infrastructure challenges, from urban mobility to procurement integrity and environmental sustainability.

This section presents three case studies:

1. Singapore's AI-driven traffic management system – AI-powered analytics enhance real-time traffic flow optimization, reducing congestion and improving public transportation efficiency.
2. Estonia's blockchain-based procurement system – Blockchain ensures secure, transparent, and immutable procurement processes, preventing corruption and increasing efficiency in public contracting.
3. Stockholm's AI and blockchain approach to sustainable urban planning – A combined approach using AI-driven energy optimization and blockchain-enabled carbon credit tracking has enhanced sustainability efforts in urban development.

These case studies serve as blueprints for other cities and nations looking to modernize their public infrastructure through technology. As AI and blockchain continue to evolve, their implementation in public infrastructure will expand, further optimizing urban planning, financial management, and resource allocation.

#### Case Study 1: AI-Driven Infrastructure Optimization in Singapore

Singapore has long been at the forefront of smart city development, leveraging advanced technologies to optimize urban infrastructure and enhance public services. As a densely populated city-state with limited land availability, Singapore faces significant transportation and mobility challenges, including traffic congestion, inefficient public transport scheduling, and high carbon emissions from vehicular traffic (Tan et al., 2021).

To address these issues, the Singaporean government has invested heavily in AI-driven infrastructure solutions, integrating machine learning, predictive analytics, and real-time data processing into its urban transport and traffic management systems. These AI-based innovations have revolutionized public transportation, optimized road usage, and improved overall commuter experiences, setting an example for other metropolitan areas worldwide.

This case study explores how Singapore has successfully implemented AI-powered traffic management systems, examining the key technological components, their impact on urban mobility, and the broader implications for smart city infrastructure development.

Singapore has consistently ranked among the world's leading smart cities due to its extensive adoption of digital technology in urban planning, transportation, and public services. With a dense population of over 5.6 million residents within a land area of just 728 $km^2$, the city-state faces significant challenges in traffic congestion, transport efficiency, and carbon emissions (Tan et al., 2021).

Historically, Singapore has relied on Electronic Road Pricing (ERP), Mass Rapid Transit (MRT) expansions, and stringent car ownership regulations to manage traffic congestion. However, with rising urbanization and increasing commuter demand, traditional transport management solutions alone have proven insufficient in maintaining optimal mobility and sustainability (Wang et al., 2020).

To address these challenges, the Land Transport Authority (LTA) has adopted AI-driven traffic management systems that leverage real-time data collection, predictive analytics, and deep learning models. These innovations allow authorities to dynamically manage traffic flow, reduce congestion, and improve public transport efficiency while minimizing environmental impact.

### AI-Based Traffic Management System

Singapore's AI-driven Intelligent Transport System (ITS) integrates a range of advanced technologies to optimize urban mobility. The system utilizes real-time sensors, computer vision, machine learning algorithms, and cloud computing to provide adaptive and responsive traffic solutions. Key components include:

- Real-Time Traffic Data Analysis
    - AI algorithms analyze vast amounts of data from traffic sensors, GPS devices, surveillance cameras, and connected vehicles to detect congestion patterns and predict peak-hour traffic flow.
    - Deep learning models process historical data to identify bottlenecks and proactively adjust infrastructure to ease congestion (Wang et al., 2020).

- Dynamic Traffic Signal Optimization
  - Machine learning-based adaptive signal control adjusts traffic light sequences based on real-time road conditions.
  - AI-powered Green Link Determining (GLIDE) systems improve intersection efficiency by reducing wait times and vehicle idling, leading to a smoother flow of traffic (Tan et al., 2021).
- AI-Powered Public Transport Scheduling
  - AI enhances bus and train scheduling by predicting passenger demand using real-time commuter data.
  - The system dynamically adjusts transit frequency and fleet allocation to optimize service availability during peak and off-peak hours.
  - AI-enabled smart bus stop technology provides commuters with accurate real-time updates, reducing uncertainty and improving commuter experience (Li et al., 2022).
  - Additionally, Singapore's Autonomous Vehicle (AV) trial programs integrate AI to support self-driving buses and robo-taxis, further enhancing urban transport efficiency (Zhang et al., 2020).

*Analysis*

The adoption of AI-driven traffic management solutions has yielded tangible improvements in Singapore's transport infrastructure. Key outcomes include:

- Reduction in Traffic Congestion
  - AI-driven traffic optimization has led to a 15% reduction in road congestion during peak hours, improving overall mobility and travel reliability (Tan et al., 2021).
  - Dynamic signal adjustments have reduced average intersection delays by 30%, minimizing unnecessary stops and fuel wastage.
- Enhanced Public Transport Efficiency
  - AI-optimized bus and MRT scheduling has decreased average waiting times by 20%, ensuring a smoother commuter experience.
  - Improved transit availability and frequency have increased public transport adoption, reducing reliance on private vehicles.
- Lower Carbon Emissions and Fuel Optimization
  - AI-driven traffic flow management has contributed to a 10% reduction in fuel consumption and vehicular emissions, aligning with Singapore's sustainability targets.
  - Efficient use of AI in transport planning supports Singapore's "Green Plan 2030," which aims for carbon neutrality in public transportation systems (Li et al., 2022).

- Scalability and Replicability
  - Singapore's AI-driven infrastructure optimization serves as a model for other cities seeking to integrate AI in urban transport planning.
  - Countries such as Japan, Germany, and the UAE have adopted similar AI-powered traffic solutions based on Singapore's success.

By combining real-time AI analytics with automation, Singapore has successfully optimized traffic flow, reduced environmental impact, and improved commuter experiences. This case study highlights the potential of AI in shaping the future of smart, sustainable urban mobility.

## Case Study 2: Blockchain for Transparent Procurement in Estonia

Estonia has established itself as a global leader in digital governance, pioneering the use of blockchain technology to enhance transparency, efficiency, and security in public administration. One of the key areas where blockchain has had a transformative impact is public procurement, a sector that is historically vulnerable to corruption, inefficiencies, and lack of accountability (Kshetri, 2021).

To address these challenges, the Estonian government developed a blockchain-based e-procurement system, ensuring fair competition, immutable transaction records, and real-time auditing of government contracts. By leveraging distributed ledger technology (DLT) and smart contracts, Estonia has streamlined procurement processes, reduced fraudulent activities, and increased public trust in government spending.

This case study explores the implementation of blockchain in Estonia's procurement system, analyzing its key features, benefits, and measurable impact on public sector integrity and efficiency.

### *Background*

Estonia is recognized as a global leader in digital governance, having pioneered the use of blockchain technology across multiple sectors, including e-governance, healthcare, finance, and procurement (Saberi et al., 2019). The country's commitment to digital transformation and cybersecurity has positioned it as a model for other nations seeking to modernize public administration through decentralized and tamper-proof digital solutions.

One of the most pressing challenges Estonia faced in public administration was corruption and inefficiencies in government procurement. Public-sector contracts are often vulnerable to bid manipulation, favoritism, and fraudulent practices, leading to wasted resources and loss of public trust (Kshetri, 2021). In response to these issues, the Estonian government developed and implemented a blockchain-based e-procurement system, ensuring transparent, efficient, and secure contracting processes.

By leveraging blockchain's immutable ledger technology, Estonia was able to prevent bid alterations, automate compliance verification, and provide real-time access to procurement records, significantly reducing corruption risks in public-sector contracting (Ølnes et al., 2017).

### *Implementation of Blockchain in Procurement*

The Estonian government's blockchain-based procurement platform integrates smart contracts, cryptographic security, and decentralized storage to enhance procurement transparency and efficiency. The system operates on X-Road, Estonia's national blockchain infrastructure, which securely connects public and private sector databases while ensuring data integrity and interoperability (Mergel et al., 2019).

Key features of Estonia's blockchain-powered procurement system include:

- Immutable Bidding Records
    - Every bid submission and procurement transaction is recorded on a blockchain ledger, ensuring that bids cannot be altered, deleted, or manipulated after submission (Ølnes et al., 2017).
    - Each entry is time-stamped and cryptographically secured, making it fully auditable by government authorities and independent watchdogs.
- Automated Compliance Verification
    - Smart contracts automatically verify bidder qualifications, regulatory compliance, and contract execution conditions.
    - By eliminating manual verification processes, delays in procurement approvals are minimized, and errors are reduced.
- Publicly Accessible Audit Trails
    - The system provides real-time access to procurement records, allowing government auditors, anti-corruption agencies, and the public to monitor procurement activities transparently.
    - The decentralized nature of blockchain ensures that data cannot be manipulated by a single entity, fostering accountability in government contracting (Kshetri, 2021).

Additionally, Estonia's blockchain platform integrates AI-powered analytics to detect anomalous bidding behavior and collusion, further enhancing fraud prevention.

### *Analysis*

Since the adoption of blockchain in public-sector procurement, Estonia has experienced substantial improvements in transparency, efficiency, and public engagement. The integration of blockchain-based smart contracts, immutable records, and real-time auditing mechanisms has significantly reduced fraud, streamlined procurement processes, and enhanced public trust.

The key outcomes of Estonia's blockchain-powered procurement system include:

Reduction in Fraudulent Activities

- Estonia has witnessed a 30% decrease in procurement-related fraud, as blockchain's tamper-proof nature prevents bid rigging, document falsification, and contract manipulation (Kshetri, 2021).
- The real-time auditing capability of blockchain allows for instant detection of suspicious transactions, deterring illicit activities before they escalate.
- The automated enforcement of contract terms through smart contracts ensures that contractual obligations are fulfilled, preventing undue financial losses caused by fraudulent claims.

Improvement in Procurement Efficiency

- Bureaucratic inefficiencies and procurement delays have been reduced by 40% due to automated contract verification and compliance monitoring (Mergel et al., 2019).
- Procurement processing times have been significantly reduced, as manual document verification and human intervention have been replaced with automated blockchain-led checks.
- The adoption of blockchain in procurement has accelerated public infrastructure project execution, as fund disbursements and contract approvals occur in real-time, rather than being hindered by administrative bottlenecks.

Increase in Public Trust and Citizen Engagement

- Citizen engagement in monitoring public contracts has increased by 25%, as blockchain's open-access records allow for real-time transparency (Saberi et al., 2019).
- Estonia's blockchain system enables non-governmental organizations (NGOs), journalists, and watchdog agencies to independently verify procurement records, strengthening public oversight and governance.
- A survey conducted by the Estonian Ministry of Economic Affairs and Communications found that over 75% of citizens express greater confidence in government spending due to blockchain's verifiable audit trails (Kshetri, 2021).

Global Recognition and Scalability

- Estonia's success with blockchain-powered procurement has attracted international attention, prompting the European Union, United Nations, and World Bank to explore similar blockchain-based procurement models.
- Estonia's model is now being scaled for cross-border procurement projects, facilitating secure and corruption-free international trade agreements.

## Case Study 3: AI and Blockchain in Sustainable Urban Planning in Stockholm

Stockholm is widely regarded as one of the most sustainable and technologically advanced cities in the world, consistently ranking high in smart city development and environmental innovation. As Sweden's capital, Stockholm has made significant strides in reducing its carbon footprint, optimizing resource allocation, and implementing energy-efficient infrastructure (Moghayedi et al., 2022).

To achieve its ambitious climate goals, Stockholm has embraced Artificial Intelligence (AI) and blockchain technology as key enablers of sustainable urban planning. AI-powered energy simulations and predictive analytics help the city optimize building energy consumption, while blockchain-based peer-to-peer (P2P) energy trading platforms promote decentralized and transparent energy distribution. Additionally, AI-driven waste management systems and blockchain-enabled tracking mechanisms have enhanced recycling efficiency and reduced environmental impact.

This case study explores how Stockholm has successfully integrated AI and blockchain into its urban planning strategy, highlighting their role in enhancing sustainability, increasing energy efficiency, and promoting transparency in resource management.

### *Background*

Stockholm has positioned itself as a leader in smart city innovation, integrating Artificial Intelligence (AI) and blockchain technologies to create sustainable, energy-efficient urban environments. As Sweden's capital and largest city, Stockholm has set ambitious goals to achieve carbon neutrality by 2040, aligning with European Union (EU) climate targets (Moghayedi et al., 2022).

Urbanization and rapid population growth have placed significant pressure on Stockholm's energy infrastructure, transportation networks, and waste management systems. Traditional approaches to urban planning often lack the flexibility to dynamically optimize energy consumption, minimize environmental impact, and efficiently allocate resources. As a result, Stockholm has embraced AI-powered predictive analytics and blockchain-enabled decentralization to tackle these challenges.

The city's smart infrastructure strategy integrates AI for real-time data processing and blockchain for secure, transparent transactions, enabling more efficient and sustainable urban planning. These technologies are deployed across energy management, waste tracking, and peer-to-peer (P2P) energy trading, enhancing Stockholm's urban resilience and reducing its ecological footprint.

### *AI and Blockchain Integration in Urban Planning*

Stockholm has implemented AI and blockchain in multiple areas of urban planning, optimizing energy use, waste management, and citizen participation in sustainability initiatives.

- AI-Driven Energy Simulations
  - AI models simulate and predict energy consumption patterns, allowing urban planners to design energy-efficient buildings and infrastructure.
  - These simulations help optimize heating, ventilation, and cooling (HVAC) systems, reducing unnecessary energy waste (Zhang et al., 2020).
  - AI also identifies peak demand periods, allowing dynamic energy distribution to minimize electricity shortages and overproduction.
- Blockchain-Based Energy Trading
  - Decentralized P2P energy markets enable residents and businesses to trade surplus renewable energy (e.g., solar power) using blockchain-based smart contracts.
  - Blockchain ensures secure, verifiable energy transactions, reducing dependence on centralized power grids and empowering citizens to contribute to sustainable energy distribution.
  - This model aligns with Sweden's goal of transitioning to a decentralized, carbon-neutral energy economy (Moghayedi et al., 2022).
- Smart Waste Management
  - AI-powered predictive analytics monitor waste generation trends, allowing for dynamic waste collection schedules and efficient recycling processes.
  - Blockchain records waste disposal and recycling data, ensuring full transparency and accountability in waste management operations.
  - The system incentivizes residents to sort and recycle waste responsibly, as blockchain-based reward mechanisms (e.g., digital credits) are introduced to encourage participation.

By combining real-time AI analysis and blockchain's decentralized verification, Stockholm's smart infrastructure initiatives increase efficiency, lower costs, and improve sustainability outcomes.

### *Analysis*

Stockholm's AI and blockchain-driven urban planning initiatives have led to measurable improvements in energy efficiency, renewable energy adoption, and waste management.

Key results include:

- Energy Consumption Reduction
    - AI-driven energy simulations have resulted in a 20% reduction in energy consumption across newly developed smart districts.
    - Optimized HVAC and dynamic energy distribution strategies have minimized excess electricity usage, leading to lower operational costs for businesses and households.
- Increased Renewable Energy Adoption
    - Over 50% of Stockholm's buildings now utilize AI-optimized energy systems, reducing dependency on fossil fuels.
    - Blockchain-enabled P2P energy trading platforms have accelerated the transition to solar and wind power, empowering citizens to participate in decentralized energy markets (Zhang et al., 2020).
- Improved Waste Management Efficiency
    - AI-powered waste tracking and predictive analytics have improved recycling efficiency by 25%, reducing landfill waste and lowering carbon emissions.
    - Blockchain-based waste accountability ensures data transparency in recycling processes, preventing illegal dumping and waste mismanagement (Moghayedi et al., 2022).
- Enhanced Urban Resilience and Sustainability
    - Stockholm's integrated AI and blockchain infrastructure supports data-driven decision-making, allowing authorities to respond proactively to climate-related risks, power outages, and resource shortages.
    - The city's smart grid and waste management systems serve as a replicable model for other global metropolises pursuing sustainable urban development.

Stockholm's AI and blockchain-powered urban planning initiatives demonstrate how emerging technologies can be leveraged to optimize energy efficiency, reduce environmental impact, and improve infrastructure resilience. By integrating real-time AI analytics with decentralized blockchain solutions, Stockholm has successfully enhanced sustainability outcomes, empowered citizens, and increased operational efficiency in critical infrastructure sectors.

These case studies highlight the transformative impact of AI and blockchain in public infrastructure:

- Singapore has successfully leveraged AI-powered traffic management to reduce congestion and improve urban mobility.
- Estonia has implemented blockchain-based procurement to enhance transparency and combat corruption in public-sector contracts.
- Stockholm has combined AI-driven energy optimization and blockchain-enabled energy trading to create a model for sustainable smart cities.

The success of these initiatives demonstrates the potential for broader adoption of AI and blockchain in global infrastructure projects, paving the way for more efficient, transparent, and sustainable cities worldwide.

### *Challenges and Future Directions*

The integration of Artificial Intelligence (AI) and Blockchain into public infrastructure presents unprecedented opportunities for enhancing efficiency, transparency, and sustainability. These technologies have the potential to automate decision-making, reduce fraud, optimize resource allocation, and increase public trust in infrastructure projects. However, despite these advantages, the widespread adoption of AI and blockchain in public infrastructure faces significant challenges related to ethical considerations, regulatory frameworks, financial constraints, and technical barriers (Floridi, 2019; Kshetri, 2017).

Public infrastructure projects involve multiple stakeholders, including governments, private companies, regulatory bodies, and citizens. The successful implementation of AI and blockchain requires collaboration among these groups, as well as substantial investments in policy development, technological infrastructure, and workforce training (Kshetri, 2021).

This section explores the primary challenges and future directions for AI and blockchain adoption in public infrastructure, focusing on:

1. Ethical and Regulatory Concerns – Addressing data privacy, legal accountability, and AI bias in decision-making.

2. Integration and Adoption Barriers – Overcoming high implementation costs, lack of technical expertise, and bureaucratic resistance.
3. Future Prospects – Exploring emerging trends such as AI-driven decentralized project management, blockchain-enabled carbon credit tracking, and the integration of AI, blockchain, and IoT in smart cities.

Addressing these challenges is crucial for ensuring the responsible and effective deployment of AI and blockchain in public infrastructure, allowing for greater efficiency, cost savings, and sustainability in urban development and governance.

## ETHICAL AND REGULATORY CONCERNS

While AI and blockchain offer numerous benefits, their deployment in public infrastructure raises ethical and regulatory challenges that must be carefully managed. These concerns include data privacy, algorithmic bias, legal accountability, and governance mechanisms (Floridi, 2019).

AI and blockchain technologies rely heavily on data collection, processing, and sharing. While blockchain enhances transparency by creating immutable records, it also poses privacy risks, as sensitive information may be permanently recorded on a distributed ledger (Kshetri, 2021).

For example, Estonia's blockchain-based digital governance system has implemented privacy-preserving cryptographic techniques to balance transparency with data protection (Saberi et al., 2019). However, other nations struggle to implement similar safeguards, leading to concerns over unauthorized data access and misuse.

Additionally, AI-driven surveillance and data analytics raise ethical issues regarding mass data collection and potential privacy violations. Regulatory frameworks, such as the General Data Protection Regulation (GDPR) in Europe, establish guidelines for AI-driven decision-making but require further adaptation to address the implications of blockchain's immutable nature (Zhang et al., 2020).

AI models used in infrastructure planning and management can inadvertently perpetuate biases present in historical data. If not properly designed, AI algorithms may prioritize certain demographics or regions, leading to unequal infrastructure investments (Mittelstadt et al., 2016).

For example, AI-based predictive policing and traffic management systems have been criticized for disproportionately affecting marginalized communities (Eubanks, 2018). Ensuring fairness in AI decision-making requires the implementation of bias-detection frameworks and ongoing algorithmic audits.

Blockchain-based smart contracts automate decision-making, reducing reliance on intermediaries. However, the legal enforceability of smart contracts remains uncertain in many jurisdictions (Pellegrino & Rauchs, 2020).

For instance, in cases where a smart contract executes an incorrect or fraudulent transaction, determining liability is challenging due to the decentralized nature of blockchain networks. Governments must establish clear legal frameworks to address liability, dispute resolution, and enforceability of blockchain-based agreements (Sharma et al., 2020).

## Integration and Adoption Barriers

Despite the potential benefits, AI and blockchain adoption in public infrastructure faces significant barriers, including high implementation costs, resistance to change, and lack of technical expertise (Kshetri, 2017).

Developing and deploying AI-driven infrastructure systems and blockchain networks require substantial financial investments. The costs associated with hardware, software development, cybersecurity, and maintenance pose significant challenges, particularly for governments with limited budgets (Zhang et al., 2020).

Blockchain's energy consumption is another critical concern. Traditional blockchain networks, such as Bitcoin and Ethereum, require extensive computational power for transaction validation, leading to sustainability concerns. However, newer consensus mechanisms, such as Proof-of-Stake (PoS), are being explored to mitigate energy inefficiencies (Saberi et al., 2019).

The successful adoption of AI and blockchain in public infrastructure depends on the availability of skilled professionals capable of developing, managing, and maintaining these technologies (Kshetri, 2021). Many government agencies lack in-house expertise, creating a dependency on external consultants and private-sector partnerships.

Educational institutions and training programs must prioritize blockchain and AI literacy to address this skills gap. Countries such as Singapore and Canada have launched national AI strategies that include workforce development initiatives to equip government employees with the necessary technical skills (Tan et al., 2021).

The adoption of disruptive technologies often encounters resistance from policymakers and industry stakeholders accustomed to traditional infrastructure management methods. Bureaucratic inertia, fear of job displacement, and concerns over regulatory compliance can slow the adoption of AI and blockchain (Moghayedi et al., 2022).

For example, despite Estonia's success with blockchain-based governance, other European nations have faced difficulties in implementing similar systems due to institutional resistance and political uncertainty (Ølnes et al., 2017). Overcoming these barriers requires targeted policy incentives, pilot programs, and public awareness campaigns.

## FUTURE PROSPECTS

The convergence of AI, blockchain, and the Internet of Things (IoT) is expected to further revolutionize infrastructure development. Emerging trends include decentralized project management, AI-driven sustainability initiatives, and blockchain-enabled carbon credit tracking (Zhang et al., 2020).

Decentralized Autonomous Organizations (DAOs) powered by AI and blockchain could transform infrastructure governance by enabling transparent, automated decision-making. DAOs allow stakeholders to vote on project proposals, allocate funds, and track progress through blockchain-based smart contracts (Casino et al., 2019).

For instance, smart cities could implement AI-driven DAOs to manage urban infrastructure projects, ensuring real-time adjustments to resource allocation and risk management (Pellegrino & Rauchs, 2020).

As countries work towards meeting climate targets, blockchain-based carbon credit systems provide a transparent mechanism for tracking emissions and incentivizing sustainable practices. AI can further enhance these systems by analyzing real-time data on energy consumption and pollution levels (Moghayedi et al., 2022).

Stockholm has already piloted AI-assisted blockchain platforms for monitoring carbon offsets in urban infrastructure projects, demonstrating the potential of this technology in environmental sustainability (Zhang et al., 2020).

The integration of AI and blockchain can enhance disaster preparedness by providing real-time monitoring and automated emergency response systems. AI-powered predictive analytics can forecast natural disasters, while blockchain ensures secure data sharing among government agencies and relief organizations (Geels et al., 2017). For example, Japan has implemented AI-driven earthquake prediction models combined with blockchain-based disaster response coordination to improve infrastructure resilience and recovery efforts (Wang et al., 2022).

While AI and blockchain offer transformative potential for public infrastructure, their adoption requires addressing ethical concerns, regulatory challenges, and integration barriers. Governments must invest in capacity-building initiatives, regulatory frameworks, and cross-sector collaborations to maximize the benefits of these technologies.

Looking forward, AI and blockchain will continue to shape the future of infrastructure, driving decentralization, sustainability, and resilience. As more cities and nations experiment with these technologies, lessons from early adopters such as Estonia, Singapore, and Stockholm will serve as valuable blueprints for global implementation.

## CONCLUSION

The integration of Artificial Intelligence (AI) and blockchain into public infrastructure represents a paradigm shift in how governments and organizations manage large-scale projects. These technologies offer transformative potential by enhancing transparency, improving efficiency, and promoting sustainability in infrastructure development. However, their adoption is not without challenges, requiring significant regulatory reforms, investment in technical expertise, and a willingness to embrace digital transformation.

This chapter has explored the critical roles AI and blockchain play in addressing long-standing infrastructure challenges. AI-driven predictive analytics optimize project planning by accurately forecasting costs, minimizing delays, and mitigating risks (Wang et al., 2019). Machine learning algorithms and real-time data analytics enhance infrastructure maintenance, reducing failure rates and extending asset lifespans (Zhang et al., 2020). Additionally, AI contributes to sustainable urban planning by optimizing energy consumption, reducing carbon emissions, and improving waste management (Moghayedi et al., 2022).

Blockchain, on the other hand, significantly enhances transparency and accountability in infrastructure projects. By ensuring immutable transaction records, blockchain minimizes corruption risks and strengthens trust in public spending (Casino et al., 2019). Smart contracts reduce delays and disputes in procurement, ensuring compliance with regulatory frameworks and automating payment disbursements based on predefined conditions (Kouhizadeh & Sarkis, 2018). The implementation of blockchain in land registries and procurement systems in Estonia, Dubai, and Georgia has demonstrated its effectiveness in curbing fraud and increasing efficiency (Saberi et al., 2019).

### Key Lessons from Real-World Implementations

The case studies in Singapore, Estonia, and Stockholm highlight how AI and blockchain are already transforming infrastructure management:

- **Singapore's AI-driven traffic management system** has improved public transport efficiency and reduced congestion through real-time data analytics and dynamic traffic signal optimization (Tan et al., 2021).

- **Estonia's blockchain-based procurement system** has increased transparency and reduced corruption in public sector contracts, leading to greater public trust and streamlined bidding processes (Ølnes et al., 2017).
- **Stockholm's integration of AI and blockchain in urban planning** has resulted in smarter, more energy-efficient infrastructure, promoting sustainability and reducing overall carbon footprints (Moghayedi et al., 2022).

These real-world implementations demonstrate that AI and blockchain are not just theoretical innovations but practical solutions that enhance public infrastructure outcomes. However, scalability, interoperability, and regulatory adaptation remain crucial for their broader adoption.

## Addressing Regulatory and Adoption Challenges

Despite the significant potential of AI and blockchain, several barriers must be overcome:

1. **Regulatory Uncertainty:** Many jurisdictions lack clear legal frameworks for AI-driven decision-making and blockchain-based smart contracts. Governments must establish well-defined policies to govern the ethical use of AI and ensure the enforceability of blockchain transactions (Floridi, 2019).
2. **High Implementation Costs:** The upfront costs associated with AI and blockchain deployment can be prohibitive for developing nations. Public-private partnerships and targeted investments in digital infrastructure can help mitigate these costs (Kshetri, 2021).
3. **Data Privacy and Security:** While blockchain enhances transparency, it also raises concerns regarding the permanent recording of sensitive information. Implementing privacy-preserving techniques such as zero-knowledge proofs and permissioned blockchain networks can help balance transparency with data security (Saberi et al., 2019).
4. **Workforce Training and Capacity Building:** The successful deployment of AI and blockchain in infrastructure requires skilled professionals capable of managing these technologies. Governments and educational institutions must prioritize training programs to build a workforce that can effectively implement and maintain digital infrastructure systems (Zhang et al., 2020).

## The Future of AI and Blockchain in Public Infrastructure

Looking ahead, the convergence of AI, blockchain, and other emerging technologies such as the Internet of Things (IoT) will further revolutionize public infrastructure. Future applications may include:

- **Decentralized Autonomous Organizations (DAOs):** Blockchain-powered DAOs can facilitate infrastructure project governance, allowing transparent decision-making processes and equitable resource distribution (Casino et al., 2019).
- **AI-Powered Predictive Disaster Resilience:** Machine learning models will be increasingly used to predict infrastructure vulnerabilities, allowing proactive disaster mitigation strategies (Geels et al., 2017).
- **Blockchain-Based Smart Grids:** Secure, decentralized energy trading systems will enable peer-to-peer (P2P) energy exchanges, enhancing the resilience of power grids and reducing reliance on centralized electricity providers (Zhang et al., 2020).

AI and blockchain represent a technological leap forward in addressing inefficiencies, corruption, and sustainability concerns in public infrastructure. While challenges remain, the success of early adopters such as Singapore, Estonia, and Stockholm proves that these technologies can deliver tangible benefits when implemented effectively.

To fully realize the potential of AI and blockchain, governments must:

- Establish clear regulatory frameworks to govern AI decision-making and blockchain transactions.
- Invest in research and development to drive innovation in infrastructure applications.
- Foster collaborations between the public and private sectors to accelerate adoption.
- Implement training programs to equip professionals with the necessary technical skills.

As AI and blockchain technologies continue to evolve, their role in public infrastructure will only expand, ushering in a new era of efficiency, transparency, and sustainability. By addressing existing barriers and leveraging technological advancements, policymakers and industry leaders can pave the way for smarter, more resilient infrastructure systems worldwide.

## REFERENCES


Aigbavboa, C. O., Thwala, W. D., & Ebekozien, A. (2024). Sustainable development goals under threat: The impact of inflation on construction projects. *Engineering, Construction, and Architectural Management.*

Al-Rimawi, T., & Nadler, M. (2025). Leveraging smart city technologies for enhanced real estate development: An integrative review. *Smart Cities*, *8*(1), 10. DOI: 10.3390/smartcities8010010

Alketbi, A., Nasir, Q., & Talib, M. A. (2018). Blockchain for government services—Use cases, security benefits and challenges. Proceedings of IEEE International Conference on Internet of Things (iThings), 1220-1225.

Anago, J. C. (2024). *Sustainable infrastructure development in sub-nations of Nigeria: What alternative financing options are open amidst constrained budget?* Sustainability Accounting, Management and Policy Journal.

Babalola, O., Ugah, U. U. K., & Ndimako, O. (2024). Evaluating governmental constraints on construction in Nigeria. *RE:view.*

Buterin, V. (2014). *A next-generation smart contract and decentralized application platform*. Ethereum Whitepaper.

Cantarelli, C. C., Flyvbjerg, B., Molin, E. J. E., & van Wee, B. (2012). Cost overruns in large-scale transport infrastructure projects: Explanations and their theoretical embeddedness. *European Journal of Transport and Infrastructure Research*, *12*(1), 5–18.

Casino, F., Dasaklis, T. K., & Patsakis, C. (2019). A systematic literature review of blockchain-based applications: Current status, classification, and open issues. *Telematics and Informatics*, *36*, 55–81. DOI: 10.1016/j.tele.2018.11.006

Chafiq, T., Rida, A., Fadil, A., & Mohammed, O. (2024). *Investigating the potential of blockchain technology for geospatial data sharing: Opportunities, challenges, and solutions*. Geomatica, Elsevier.

Dritsas, E., & Trigka, M. (2025). Remote sensing and geospatial analysis in the big data era: A survey. *Remote Sensing (Basel)*, *17*(3), 550. DOI: 10.3390/rs17030550

El Khoury, R. (2024). Technology-driven business innovation. Springer. https://link.springer.com/content/pdf/10.1007/978-3-031-51997-0.pdf

Eubanks, V. (2018). *Automating inequality: How high-tech tools profile, police, and punish the poor*. St. Martin's Press.

Floridi, L. (2019). Establishing the rules for building trustworthy AI. *Nature Machine Intelligence*, *1*(6), 261–262. DOI: 10.1038/s42256-019-0055-y

Flyvbjerg, B., Bruzelius, N., & Rothengatter, W. (2003). *Megaprojects and risk: An anatomy of ambition*. Cambridge University Press. DOI: 10.1017/CBO9781107050891

Flyvbjerg, B., Garbuio, M., & Lovallo, D. (2018). Delusion and deception in large infrastructure projects: Two models for explaining and preventing executive disaster. *California Management Review*, *51*(2), 170–193. DOI: 10.2307/41166485

Geels, F. W., Sovacool, B. K., Schwanen, T., & Sorrell, S. (2017). Sociotechnical transitions for deep decarbonization. *Science*, *357*(6357), 1242–1244. DOI: 10.1126/science.aao3760 PMID: 28935795

Hussain, O. A. I., Moehler, R. C., Walsh, S. D. C., & Ahiaga-Dagbui, D. D. (2024). Minimizing cost overrun in rail projects through 5D-BIM: A conceptual governance framework. *Buildings (Basel, Switzerland)*, *14*(2), 478. DOI: 10.3390/buildings14020478

Kenny, C. (2007). Construction, corruption, and developing countries. World Bank Policy Research Working Paper 4271.

Khan, M. W., Saad, S., Ammad, S., & Rasheed, K. (2024). Smart infrastructure and AI. In *AI in Material Sciences*. Taylor & Francis. DOI: 10.1201/9781003438489-9

Kouhizadeh, M., & Sarkis, J. (2018). Blockchain practices, potentials, and perspectives in greening supply chains. *Sustainability (Basel)*, *10*(10), 3652. DOI: 10.3390/su10103652

Kshetri, N. (2017). Blockchain's roles in meeting key supply chain management objectives. *International Journal of Information Management*, *39*, 80–89. DOI: 10.1016/j.ijinfomgt.2017.12.005

Kshetri, N. (2021). Blockchain and transparency in government. *Journal of Public Administration and Policy*, *34*(2), 112–129.

Latilo, A., Uzougbo, N. S., Ugwu, M. C., & Oduro, P. (2024). Developing legal frameworks for successful engineering, procurement, and construction projects. International Journal of Applied Research in Social Sciences.

Li, X., Zhang, Y., & Tan, P. (2022). AI-powered transport optimization: The case of Singapore. *Journal of Intelligent Transport Systems*, *28*(4), 412–430.

Love, P. E. D., Ika, L. A., & Sing, C.-P. (2016). Overruns in transportation infrastructure projects: A systematic review and future research directions. *Transport Reviews*, *36*(6), 720–742.

Mergel, I., Ganapati, S., & Whitford, A. B. (2019). Agile government: Systematic literature review and future research agenda. *Government Information Quarterly*, *36*(2), 273–285.

Mittelstadt, B. D., Allo, P., Taddeo, M., Wachter, S., & Floridi, L. (2016). The ethics of algorithms: Mapping the debate. *Big Data & Society*, *3*(2), 1–21. DOI: 10.1177/2053951716679679

Moghayedi, M., Lim, C., & Tseng, M. L. (2022). AI-driven energy-efficient urban planning: A case study of Stockholm. *Journal of Sustainable Cities*, *15*(3), 423–438.

Mohamed, A. M. O., Mohamed, D., Fayad, A., & Al Nahyan, M. T. (2024). Enhancing decision-making and decarbonization in environmental management: A review on the role of digital technologies. *Sustainability (Basel)*, *16*(16), 7156. DOI: 10.3390/su16167156

Mohbey, K. K., Rajput, D. S., & Uddin, M. (2024). Technological developments in point of interest recommendation for smart and sustainable cities. *Frontiers in Sustainable Cities.*

Molloy, S. (2019). The Lava Jato scandal: A case study of corruption and political transformation in Brazil. *Journal of Latin American Studies*, *51*(2), 321–345.

Ølnes, S., Ubacht, J., & Janssen, M. (2017). Blockchain in government: Benefits and implications of distributed ledger technology for information sharing. *Government Information Quarterly*, *34*(3), 355–364. DOI: 10.1016/j.giq.2017.09.007

Oluseye, O. (2024). *Exploring potential political corruption in large-scale infrastructure projects in Nigeria*. Project Leadership and Society. DOI: 10.1016/j.plas.2023.100108

Saberi, S., Kouhizadeh, M., Sarkis, J., & Shen, L. (2019). Blockchain technology and its relationships to sustainable supply chain management. *International Journal of Production Research*, *57*(7), 2117–2135. DOI: 10.1080/00207543.2018.1533261

Sharma, R., Singh, R., & Reddy, P. (2020). Blockchain for government accountability: A framework for public sector financial management. *Government Information Quarterly*, *37*(2), 101401.

Tan, P., Li, X., & Zhang, J. (2021). AI-enhanced public transport efficiency: Case study of Singapore. *Journal of Intelligent Transport Systems*, *25*(4), 315–329.

Ukeyima, A. E. (2024). The importance of project management in underdeveloped countries. Journal of African Sustainable Development.

United Nations. (2019). *Sustainable Development Goals Report 2019*. United Nations Publications.

Wang, S., Zhang, Y., & Zhang, X. (2019). Machine learning applications in infrastructure project management: A review. *Automation in Construction*, *98*, 1–13.

World Bank. (2020). *Infrastructure for Development: World Bank Report on Global Infrastructure*. World Bank Publications.

Zhang, Y., Wang, H., & Liu, J. (2020). IoT and AI for smart infrastructure: A sustainable approach. *Sustainable Cities and Society*, *50*, 101678.

# Chapter 7
# The Impact of Digital Technologies on Indian Business and Society

**Anitha Kumari**
https://orcid.org/0000-0002-8343-8926
*GITAM University, India*

## ABSTRACT

*The rapid proliferation of digital technologies has emerged as a transformative force, reshaping various facets of Indian business and society. This study examines the revolutionary changes brought about by digital technology, exploring its impact on both large-scale and small-scale industries. Digital devices have become essential in businesses, particularly in India, drastically changing employers' behaviors. The objective of the study is to examine the impacts of digital technologies on Indian businesses and society, with the goal of improving employee attitudes towards digital technology. A theoretical framework for a digital technology model is developed to reveal differences between technology challenges, influence, and support. This framework combines descriptive and quantitative data from various professionals to evaluate which organisational and technological characteristics best predict employee attitudes toward digital technology.*

## INTRODUCTION

In the last two decades, digital technology has reached nearly half of the world's population, and it has transformed our society. Technology can help equalise access to commerce, governmental services, and financial inclusion. It delves into the multifaceted impact that digital technologies have exerted on India's economic

DOI: 10.4018/979-8-3373-1534-8.ch007

landscape, corporate strategies, social dynamics, and cultural paradigms (United Nations, 2020). Technological advances have reshaped the labour force, displacing others and driving societal shifts. This present wave of change will certainly be substantial. The International Labour Organization estimates that 24 million new jobs will be created through renewable energy, electric vehicles, and energy efficiency improvements in existing buildings by 2030 (ILO Report, 2018).

According to surveys conducted by McKinsey and other firms, the majority of employees fear they lack the requisite training or skills to secure a well-paying job. The solution to these trends lies in changing our educational approach, which entails teaching soft skills, flexibility, and re-skilling and up-skilling individuals over time. In today's world, digital technologies like data pooling and artificial intelligence are being to track everything from transportation to bills (McKinsey, 2018). A computerised system can be used to protect and practice human rights as well as misuse them. Governments and corporations may now harvest data for financial and other goals. Personal data would become a valuable asset if there was more legislation around the ownership of personal data. By protecting data-driven technologies properly, individuals can be empowered, human welfare can be improved, and universal rights can be promoted (United Nations, 2020).

According to a Cognizant survey, just 36% of Indian enterprises will see a negative impact on their company's performance by 2020 (ET Bureau, 2018). The study also found that organizations in the Asia-Pacific and the Middle East understood the value of digital technologies and the human-machine labour balance. In India, 52% of organizations believe the pandemic has led to the implementation of new digital work processes. The survey collected responses from 23 countries and 14 industries, comprising over 4,000 senior executives, with 160 participants from India. In the Indian sample, over 13% of the total revenue originates from digital channels, representing the highest figure in the Asia-Pacific region. However, the benefits of AI far exceed the drawbacks. Security is evolving in lockstep with artificial intelligence (CXOtoday, 2021). As a result of the rapid advancement of computer, information, communication, and networking technologies, digital transformation is now influencing every aspect of our lives, and is no longer exclusive to enterprises and the workplace.

The objective is to examine the impacts of digital technologies on Indian businesses and society, with the goal of improving employee attitudes towards digital technology. It is hard to transform using digital because of current resources and capabilities. These include culture, human skills, practices, technology, attitudes, identities, mindsets, and values (Vial, 2019). It believes that the digital divide is about integrating digital technology into meaningful social behaviours and reaping its advantages (Song et al., 2020). To do so, the next generation must learn to design, develop, construct, and build digital technology. Employees must be cautious and proactive when it comes to digital technologies.

The research study is to create a model which describes the main elements for improving employees' attitudes towards digital technology. This study examines the digital transformation and helps to reveal differences between technology challenge, influence, and support. The study focuses on the effectiveness of digital technologies for Indian professionals. Using digital technology efficiently allows users to develop skills that will assist them use digital services. It has been suggested that employees should be encouraged to use digital technology in their regular routines. The study used descriptive and partial least squares analysis to analyse the findings of an Indian business and society survey.

## LITERATURE REVIEW

### Digital Technologies

Digital technology is a fast-expanding scientific area, representing considerable digitization and technological growth. Digital technology is at the core of digital innovation research. It can be implemented by selecting and configuring resources from an offering or by completely rethinking their purpose and use. It transcends industries and businesses (Henfridsson et al., 2018). The only limitation is the creativity of individuals, groups, and organizations when it comes to co-creating services and apps. In the field of technology, generativity refers to the capacity of technology to elicit unanticipated change from large, varied, and uncoordinated audiences (Yoo et al., 2012). Digital innovation has evolved into a network phenomenon as democratised access to digital technology makes it easier for individuals to work with others through informal partnerships without being resisted by established structures (Benkler, 2006). It is necessary for executives to be knowledgeable about new digital technologies in order to recognise the need for change and effectively convey it to their teams.

In the last decade, digital technology has transformed our worldview. Technological advancements are often assumed to be ephemeral and futuristic. Society provides technology and technological progress with meaning by making our lives simpler. The convergence of technology and society has improved the world beyond our expectations. Digital technologies are an integral part of our lives today. Digitisation impacts financial policy, employment, and competition. This paradigm included all technical advances, particularly in information technology. The Indian economy has been transformed by digitization in many ways. Healthcare, law enforcement, art, education, mobility, and religion are all positively impacted by digital technology. During the lockdown, many businesses went digital, and many new enterprises were founded. Digital technologies will continue to be vital in our lives for many years (Patel, 2021).

## Digitalization in Businesses

Digital technologies have had a transformational effect on communication and data processing since their inception on the internet. It enables faster and more complex data processing and dramatically expands reach, access, and empowerment for customers. Small and niche businesses can now benefit from the same communication advantages that large corporations liked in the early days of the internet. So new sectors and jobs grew, and conventional industry borders were blurred. Digital technologies have a big impact on other technologies and industries such as manufacturing, services, and traditional professions. Digital data, communication, and service technologies may provide a company with a competitive advantage (NU Cepal, 2021). Drones and mechanical devices/robots for distribution are predicted to significantly revolutionise the sector in the future. Building a digitally-savvy organization involves knowledge of the technology's impact on the organization's purpose and relevance to stakeholders. Developing a digitally savvy organization requires understanding the technology's impact on the organization's purpose (Nadkarni & Prügl, 2020).

In general, digital technologies have increased communication speed and reduced communication and processing expenses by introducing new digital tools. This technology has accelerated change by enabling small businesses to create new business models and specialised customer offerings. In addition, business executives must manage and harness increasing levels of unpredictability and complexity to create value for customers. The consequences have led to organizational issues of varying magnitude and impact. Their impact on customers, supply chains, and value chains must be understood. They involve understanding changing regulatory fields, new business models, and new skill sets and employment opportunities (Ling Li, 2022). The consequences for managers are that strategic and operational decisions must reflect organizational flexibility and resilience. These consequences include dealing with increased change-pace, complexity, uncertainty, and information overload. Digital technologies impact corporate culture, design, and procedures (Anujayesh Krishna, 2018).

## Theoretical Framework and Hypotheses

Organizations are radically transformed by digital technology, with the rapid rate of change compounding the issues facing Indian business and society. Businesses must have a clear strategy that includes a plan for retraining employees. Whereas past technical revolutions took place over an extended period of time, the pace of digital transformation requires firms to act swiftly. Due to the possibility of inequality and pay deflation, as well as social upheaval, immediate action is required to prepare the workforce for a digital future (WEF, 2021). While the empirical literature on employee attitudes toward the new normal, as discussed above, indicates a range of elements that may influence such attitudes, there is no commonly acknowledged theory or theoretical framework for conducting such studies (Carol, 1996). However, the theoretical model described herein builds on and combines a number of broad theoretical premises that appear to adequately represent the essential justifications for new technology. Figure 1 Proposed Digital Technology Model presents an integrated framework that illustrates how digital transformation impacts employee attitudes within Indian businesses.

## Employees' Attitude towards Digital Technologies

Employee attitudes toward the implementation of digital technology have been examined from a variety of perspectives. Employees' attitudes have been observed with various determinants like technology challenge, support, and influence, to frame the theoretical propositions, informative to employees and good for digitalization. However, there are not many studies to understand the employees' attitude towards digital technologies which would help businesses to plan and create their working content to elicit positive employees' attitude (Beenu, and Rathika, 2019). In the business world, this will demonstrate to an organization that the expected working environment has a significant impact on the attitudes of workers regarding digital workplace transformation.

## Technology Challenge

According to Kitchin (2014), professionals, educators, researchers, policymakers, and the general public are just now realising the severity of the issues posed by digital technology. Professional education tends to approach digital technology as tools: useful to master but obviously subordinate to the skilled professional. Moreover, research on an ageing workforce confronted with constant technological changes must incorporate age as a moderating factor (Steven et al., 2012). Employees must be ready for change and actively participate in it. Rather than technical trends, disruptive inventions, or new consumer behaviour, firms must adjust their culture, attitude, and consequences to the new digital way of working. Culture, mentality, and competencies need attention from employees. If they are not technologically literate, they may struggle to adapt (Khan and Vuopala, 2019). However, if employees are not equipped for digital technology applications, employee experience will not be individualised, engaging, and memorable (Bawden and Robinson, 2009).

H1: Technology challenge negatively effects on employees' attitude toward digital technology.

## Technology Influence

Employees build expectations about possible work design qualities and outcomes based on information supplied by management, and how these expectations impact employees' attitude and support digital technology. This research suggests that assumptions regarding job design elements influence predicted work outcomes. It also contends that projected job outcomes shape an employee's attitude toward digital workplace technology, affecting their readiness to actively support change (Christine and Iris, 2020). A productive work environment is maintained and enhanced via employee connectivity and responsive leadership (Dery et al., 2017). Similarly, social connectivity in digital work contexts has been found to impact performance (Kuegler et al., 2015). Researchers have found a link between self-rule and users' attempts to innovate with information technology, enabling new methods of working (Deng and Joshi, 2016). Performance expectations for a variety of new tools at work are considered to have a substantial impact on users' attitudes towards digital technology. Hence, it is hypothesised:

H2: Technology influence positively effects on employees' attitude towards digital technology.

## Technology Support

Digital technology leads to employees' positive opinions about digital workplace transformation, as well as their intentions to actively support the essential change process (Christian and Iris, 2020). The intention of a person to embrace digital workplace tools is motivated by their self-determination as the sole arbiter of their work. They are gradually complemented, as demonstrated in this research, by an individual's active support and involvement in furthering the transition process. A committed employee is a vital condition for a successful workplace change. It is predicated on employee engagement, and participation in the change process may boost the likelihood of digital technology adoption (Richter et al. 2018). Employees who have invested in some manner in the projected workplace design are more inclined to support transformative initiatives (Barrett and Oborn 2013). Their active support is critical to the success of digital transformation programmes within organizations. On the basis of these findings and in light of the context of digital technologies in the workplace, it is hypothesised that he or she would experience increased levels of happiness while interacting with new technologies in the office that help employees feel connected to supervisors and peers. Hence, it is proposed:

H3: Technology support positively effects on employees' attitude towards digital technology.

*Figure 1. Proposed Digital Technology Model*

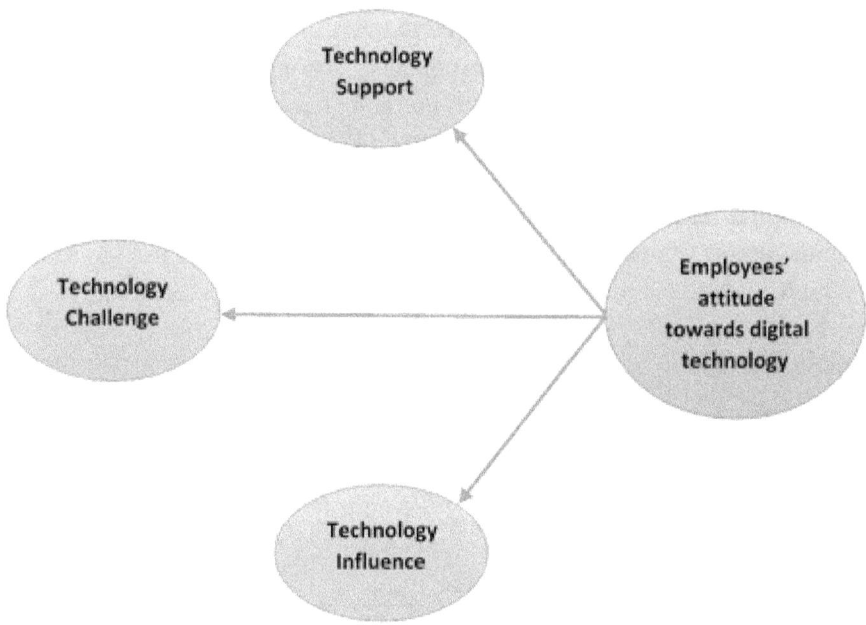


## MATERIALS AND METHODS

### Research Design

The research objective used in this study is to examine employees' attitude towards digital technologies for enlightening digital business and society. This research distributed an online questionnaire using Google Forms. The approach used in this study is descriptive and quantitative. The survey instrument was considered in two parts. The first part of the questionnaire incorporated the demographic profile of the respondents and the second part of the questionnaire includes four constructs applying the research model, i.e., employees' attitude, technology challenge, technology support, and technology influence. All the measurement items were adopted from previous studies and then adapted into the current research perspective (Ho and Lin, 2010; Alaboudi and Alharbi, 2019).

The sample of the study consists of private and public sector employees, business owners, and self-employed in India. A structured and self-administrated online survey was formed and circulated among the respondents to collect information about the employees' attitude towards digital technologies in business and society as well as to understand their acceptance to use it in the future. For data collection, a convenience sampling approach was used. The sample of the respondents taken for the study is 200. To collect information from the respondents, five-point Likert-type scaling that includes 21 items derived from previous literature has been used with the option of (1) strongly disagree, (2) disagree, (3) neutral, (4) agree and (5) strongly agree. The data analysis was done through SmartPLS 3 to balance to analyse employees' attitude towards digital technologies in Indian business and society to accept technology among the users. The partial least square (PLS) model has been used to figure out the complex relationship of one construct to another and its indicators.

## Sample Profile

*Table 1. Results of the Demographic Profile Information*

| *Category* | *Demographic Measures* | *Frequency (f)* | *Percentage (%)* |
|---|---|---|---|
| Gender | Male | 154 | 77.0 |
| | Female | 46 | 23.0 |
| Age | Below 25 years | 46 | 23.0 |
| | 26 to 35 years | 84 | 42.0 |
| | 36 to 45 years | 56 | 28.0 |
| | Above 46 years | 14 | 7.0 |
| Educational Background | Diploma/Professional | 24 | 12.0 |
| | Bachelor's Degree | 66 | 33.0 |
| | Master's Degree | 92 | 46.0 |
| | Ph.D. Degree | 18 | 9.0 |
| Occupation | Public Sector | 42 | 21.0 |
| | Private Sector | 98 | 49.0 |

continued on following page

*Table 1. Continued*

| *Category* | *Demographic Measures* | *Frequency (f)* | *Percentage (%)* |
|---|---|---|---|
| | Business | 32 | 16.0 |
| | Academia | 28 | 14.0 |
| Experience | Less than 5 years | 54 | 27.0 |
| | 6 to 10 years | 80 | 40.0 |
| | 11 to 15 years | 42 | 21.0 |
| | Above 15 years | 24 | 12.0 |

Table 1 shows that 77% of the respondents are male, while 23% are female. Analysing the age group, 42% of the respondents are in the category of 26 to 35 years, 28% of the respondents are in the age group of 36 to 45 years, 23% of the respondents are below 25 years' category, and the remaining 7% are above 46 years old. The level of education indicates that 46% are master's degree holders, 33% are bachelor's degree, 12% are diploma/professional degree holders, and the remaining 9% of the respondents are Ph.D. holders. The occupation displays that the majority of the respondents are working in the private sector and it consists of 49%, 21% of the respondents are working in the public sector, 16% of the respondents are keeping their own business and 14% are academicians. The profile of the experience shows that 40% are in the category of 6 to 10 years, 27% are having less than 5 years, 21% of the respondents fall in the category of 11 to 15 years, and 12% are highly experiences consisting of above 15 years.

## RESULTS AND DISCUSSION

### Reliability and Validity of the Instrument

The reliability extent the quality of an instrument used to collect response from respondents. This instrument consists of two sections as section A deals with demographic profile, and section B includes 21 items designed to measure the impact on the attitudes of employees towards digital workplace transformation. There are many ways to test the reliability and internal consistency of an instrument. The most popular and widely accepted technique to test the reliability of an instrument is Cronbach's alpha. The value of Cronbach's alpha needs to be 0.7 to accept the instrument, greater than 0.8 is measured well and more than 0.9 is measured as extraordinary internal consistency and reliability of the instrument. The results of composite reliability and Cronbach's alpha value are greater than 0.7. It shows that the instrument constructed keeps extraordinary internal consistency and is highly reliable.

The validity of the construct can be studied using two methods of validity measures such as average variance extracted and discriminant validity. If the AVE is above 0.5, it will be accepted that it has adequate convergent validity. The AVE construct score is higher than 0.5 and it shows that the construct used is highly valid. The result of the reliability and validity is given in Table 2.

*Table 2. Results of the Reliability and Validity Measures*

| *Construct* | *Items* | *Factor Loadings* | *Cronbach Alpha* | *CR* | *AVE* |
|---|---|---|---|---|---|
| Employees' Attitude | 5 | 0.844 - 0.977 | 0.944 | 0.957 | 0.818 |
| Technology Challenge | 6 | 0.879 - 0.984 | 0.973 | 0.974 | 0.860 |
| Technology Influence | 5 | 0.732 - 0.870 | 0.870 | 0.905 | 0.656 |
| Technology Support | 5 | 0.741 - 0.861 | 0.863 | 0.902 | 0.648 |

CR = Composite Reliability; AVE: Average Variance Extracted

## Discriminate Validity of the Instrument

The discriminant validity was evaluated to ensure the validity of the instrument. The discriminate validity results should be lesser than the average variance extracted. The discriminate validity has below 0.90 and it confirms that the constructs are valid. The results of the discriminate validity are presented in Table 3.

*Table 3. Discriminate Validity*

| *Factors* | *Employees' Attitude* | *Technology Challenge* | *Technology Influence* | *Technology Support* |
|---|---|---|---|---|
| Employees' Attitude | 0.905 | | | |
| Technology Challenge | -0.227 | 0.927 | | |
| Technology Influence | 0.673 | -0.214 | 0.810 | |
| Technology Support | 0.753 | -0.243 | 0.667 | 0.805 |

## Results of Heterotrait Monotrait Ratio (HTMT) Analysis

The discriminant validity was evaluated to ensure that a reflective construct through the heterotrait and monotrait (HTMT) ratio of correlation (Gefen, 2000). According to the HTMT 0.85 criterion detects the collinearity problems among the multicollinearity (latent constructs). HTMT is the foremost degree of measures for discriminant validity viz. Fornell and Larker criteria, cross-loading, and HTMT criterion is given in Table 4.

*Table 4. Heterotrait and Monotrait Ratio (HTMT)*

| *Factors* | *Employees' attitude* | *Technology Challenge* | *Technology Influence* | *Technology Support* |
|---|---|---|---|---|
| Employees' Attitude | | | | |
| Technology Challenge | 0.176 | | | |
| Technology Influence | 0.710 | 0.180 | | |
| Technology Support | 0.823 | 0.191 | 0.702 | |

## Structural Equation Model

### Measurement Model

A structural equation model is established and the hypothesis set by the researcher is tested using a standardized path. Table 5 determines that the R squared is a statistical measure of the regression line. The coefficient of determination for multiple regression is also called the coefficient of multiple determination. The R2 value for the construct employees' attitude is 0.620 implying 62.0% is due to the latent variable in the model. Likewise, the path coefficient from employees' attitude toward digital technology is resilient and noteworthy. Figure 2 shows the structural model path coefficients. Based on the results of the PLS algorithm, path coefficients are determined. The bootstrapping procedure is applied in this study to evaluate the significance level of partial least square estimation. The study was conducted using bootstrapping using 5000 subsamples. As per the structural model result.

### Hypothesis Testing

A hypothesis test is done by calculating t-values on all the factors based on the proposed digital technology model (Figure 1). The results are displayed in Table 5.

*Table 5. Hypothesis test results*

| *Hypothesis* | *Structural Path* | *Path Coefficient* | *t-values* | *p-values* | *Conclusion* |
|---|---|---|---|---|---|
| H1 | Technology Challenge -> Employees' attitude | -0.030 | 0.366 | 0.714 | Not Supported |
| H2 | Technology Influence -> Employees' attitude | 0.304 | 3.035 | 0.001 | Supported |
| H3 | Technology Support -> Employees' attitude | 0.542 | 4.592 | 0.000 | Supported |

Table 5 shows the values of the path coefficients, t-values, and p-values of the structural model. The featured result identifies that the construct technology support had the positive impact (β= 0.542, t = 4.592) on the employees' attitude towards digital technologies and technology influence (β = 0.304, t = 3.035), correspondingly. Surprisingly, technology challenge (β = -0.030, t = 0.366), has a negative influence on the employees' attitude towards digital technologies. The result depicts that technology influence and technology support are the significant factors that impact the digital technology on business and society. The significant level used in this research is a 5% level of significance. Table 6 depicts the results of the t-value which is acquired by running bootstrapping function on *SmartPLS* 3. The hypothesis test result H2 and H3 has a significant effect on the employees' attitude towards digital technology. H1 do not have a significant influence on the employees' attitude towards digital technologies in the Indian business and society. Hence, digital technology can be established to improve digital businesses that enable employees' accessibility and flexibility.

*Figure 2. The structural model (PLS-SEM)*

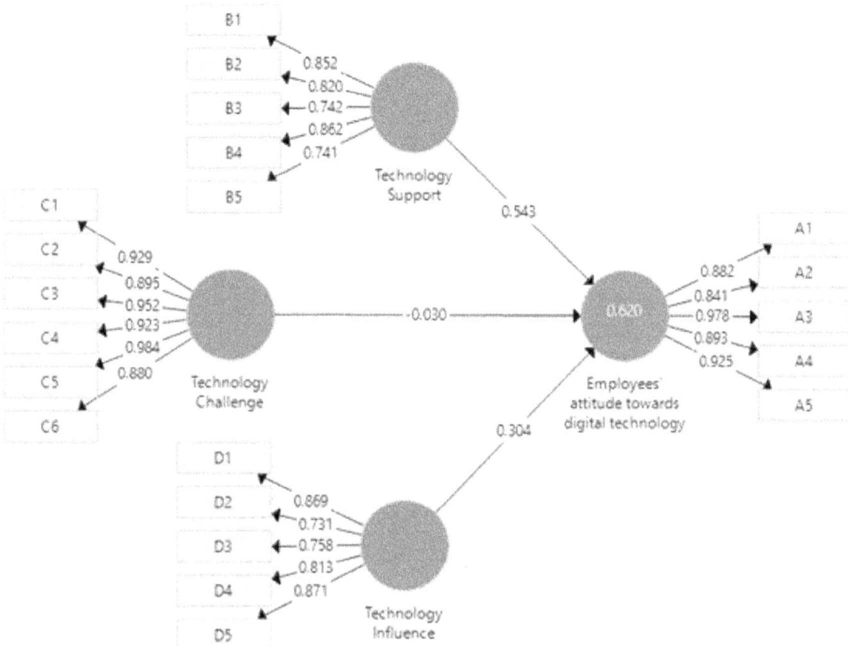


## Model Fit Summary

Table 6 represents the goodness-of-fit measure that shows the standardized root mean square residual which helps to find out whether the measurement model is fit or not. The result shows that SRMR was 0.136 which was fitting below 0.14. It is an indication that the measurement model is a good fit (Henseler, 2013).

*Table 6. SRMR (Standardized root mean square residual)*

| *Fit Summary* | *Saturated Model* | *Estimated Model* |
|---|---|---|
| SRMR | 0.136 | 0.136 |

### Implications

This study provides some valuable insights and suggestions for businesses and society, it emphasises the use of digital technology in India. Overall, the findings of the three factors examined from the employees' perspective to boost digital technology use to be recommended. As a result, the managerial focus should be on delivering the possible technical protection and providing a secure system. Employees' attitudes have been observed with various determinants like technology challenge, influence, and support to frame the theoretical propositions, informative to employees and good for digitalization. The education, healthcare, and banking sectors cannot reach out to citizens because of obstacles such as middlemen, illiteracy, poverty, lack of funds, information, and investments. There are marked differences in the economic and social conditions of the people in rural and urban areas as a result of these challenges (Sharma, 2016).

Organizations undergoing digital transformation face a number of challenges, both from a data integration and implementation perspective and also an end-user experience perspective. Data analytics, data migration, software integration, and cybersecurity skills are lacking in organizations. The use of digital technology has made it possible to collect and analyse advanced data. It is significant to collect data on potential customers, such as their search histories, in order to improve production. Artificial intelligence and blockchain will continue to change business models and companies until 2040 (Musostova et al., 2023). Technology gives society a sense of purpose through technological advancements. Technology and society have led the world to futuristic improvements.

This study will help businesses, professionals, educators, researchers, and policymakers and the society to understand the awareness of digital transformation. The results of digital technologies have a huge impact on business and society. Digitization affects every industry in areas like financial policy, employment, and competition. The Make in India initiative has improved India's digital manufacturing sector. According to economic analysts, the Digital India initiative could boost GDP by around $1 trillion by 2025 (Srivastava, 2020). It can also be detrimental to play an important role in macroeconomic factors such as job creation, labour productivity, business development, and revenue generation.

## CONCLUSIONS

Digital technology and its applications are part of our daily lives. New technology influences consumer behavior, resulting in higher expectations, as well as opportunities for companies to meet those expectations. Businesses can deliver relevant information and messaging to their customers through chatbots powered by AI (Adam et al., 2021). It has a significant impact on customer service and the products that they can offer to their clients. Technology's constant evolution will always impact user behavior. There is an increase in the speed of that change. Companies should prepare by identifying trends now and capitalizing before it is too late with IoT and 5G still in their infancy (Pons et al., 2023). Technological advancements are powerful catalysts for change.

The outcome of this study is accepting fundamental changes to the way a business operates. It helps businesses improve workflow efficiency, secure information, and increase profitability by modernizing legacy processes. According to this study, technological challenges, influence, and support are the most compelling factors affecting employees' attitudes toward digital technology. Technology is influenced by many factors, including job structure and power dynamics at the time of its introduction, as well as work culture and employee relations. Organizational factors dominate employees' perceptions of workplace technology, not technical factors. No matter how complex a technological system may be, it still works.

The use of technology can increase productivity and expand markets for goods and services, which can affect economic growth. This paper has attempted to identify those factors which encourage digital technology for business and society. The back office uses technology to automate back office functions like record keeping, accounting, and payroll. It is also possible for business owners to create secure environments for maintaining sensitive consumer or business information using technology. The finding suggests that the employees' position in the organization and new technologies greatly affect their attitudes. People's attitudes toward new applications and facilities are most strongly influenced by the availability of technological support and influence.

## FUTURE DIRECTIONS

The present study has not collected responses from various categories of people working in different industries. The results of this study help us understand how technology adoption affects the views of different types of employees in a workplace. The study recommends that the use of digital technology and additional influential factors with suitable models will bring new insights on the impact of digitization on businesses and society. In particular, managerial theory and practice would benefit from more research on technology decision-making throughout the entire process. Furthermore, it is demanding that different aspects of digital technology advantages among Indian businesses and people can further be explored.

## REFERENCES


Adam, M., Wessel, M., & Benlian, A. (2021). AI-based chatbots in customer service and their effects on user compliance. *Electronic Markets*, *31*(2), 427–445. DOI: 10.1007/s12525-020-00414-7

Alaboudi, A., & Alharbi, A. S. (2019). *The current state of teachers' ICT use in classrooms in boys' secondary schools in Saudi Arabia* (Doctoral dissertation).

Barrett, M., & Oborn, E. (2013). Envisioning E-HRM and strategic HR: Taking seriously identity, innovative practice, and service. *The Journal of Strategic Information Systems*, *22*(3), 252–256. DOI: 10.1016/j.jsis.2013.07.002

Bawden, D., & Robinson, L. (2009). The dark side of information: Overload, anxiety and other paradoxes and pathologies. *Journal of Information Science*, *35*(2), 180–191. DOI: 10.1177/0165551508095781

Beenu, K. G. H., & Rathika, S. (2019). Attitude of customers towards digital and conventional advertising. *International Journal of Recent Technology and Engineering*, *8*(4S).

Benkler, Y. (2006). *The wealth of networks*. Yale University Press.

Bureau, E. T. (2018). Indian companies weathered the pandemic 'much better' than counterparts: Cognizant study. *The Economic Times.*

Cepal, N. U. (2021). *Digital technologies for a new future.*https://www.cepal.org/sites/default/files/publication/files/46817/S2000960_en.pdf

Christian, M., & Iris, J. (2020). Investigating the elicitation of employees' support towards digital workplace transformation. *Behaviour & Information Technology*. Advance online publication. DOI: 10.1080/0144929X.2020.1818824

CXOtoday. (2021). Indian firms fare better than global counterparts amid COVID-19 pandemic. https://www.cxotoday.com/ai/indian-firms-fare-better-than-global-counterparts-amid-covid-19-pandemic/

Deng, X., & Joshi, K. D. (2016). Why individuals participate in micro-task crowdsourcing work environment: Revealing crowdworkers' perceptions. *Journal of the Association for Information Systems*, *17*(10), 648–673. DOI: 10.17705/1jais.00441

Dery, K., Sebastian, I., & van der Meulen, N. (2017). The digital workplace is key to digital innovation. *MIS Quarterly Executive*, *16*(2), 135–152.

Gefen, D. (2000). E-commerce: The role of familiarity and trust. *Omega, 28*(6), 725–737. DOI: 10.1016/S0305-0483(00)00021-9

Henfridsson, O., Nandhakumar, J., Scarbrough, H., & Panourgias, N. (2018). Recombination in the open-ended value landscape of digital innovation. *Information and Organization*, *28*(2), 89–100. DOI: 10.1016/j.infoandorg.2018.03.001

Henseler, J., & Sarstedt, M. (2013). Goodness-of-fit indices for partial least squares path modeling. *Computational Statistics*, *28*(2), 565–580. DOI: 10.1007/s00180-012-0317-1

Ho, C. T. B., & Lin, W. C. (2010). Measuring the service quality of internet banking: Scale development and validation. *European Business Review*, *22*(1), 5–24. DOI: 10.1108/09555341011008981

International Labour Organization. (2018). 24 million jobs to open up in the green economy. *World Employment and Social Outlook, 2018*, •••. https://www.ilo.org/global/about-the-ilo/newsroom/news/WCMS_628644/lang--en/index.htm

Kuegler, M., Dittes, S., Smolnik, S., & Richter, A. (2015). Connect me! Antecedents and impact of social connectedness in enterprise social software. *Business & Information Systems Engineering*, *57*(3), 181–196. DOI: 10.1007/s12599-015-0379-z

Li, L. (2022). Reskilling and upskilling the future-ready workforce for Industry 4.0 and beyond. *Information Systems Frontiers*. Advance online publication. DOI: 10.1007/s10796-022-10308-y PMID: 35855776

McKinsey Global Institute. (2018). *Notes from the AI frontier: Modeling the impact of AI on the world economy.*

Musostova, D., Berkaeva, A., & Geraskin, Y. (2023). Digitalization as a factor of sustainable development of the economy and economic security. In *SHS Web of Conferences, 172*, 02024. EDP Sciences. DOI: 10.1051/shsconf/202317202024

Nadkarni, S., & Prügl, R. (2020). Digital transformation: A review, synthesis and opportunities for future research. *Management Review Quarterly*, *71*(2), 233–341. DOI: 10.1007/s11301-020-00185-7

Patel, D. (2018). The impact of digital technologies on society. *Digital India.*https://www.gbim.com/the-impact-of-digital-technologies-on-society/

Pons, M., Valenzuela, E., Rodríguez, B., Nolazco-Flores, J. A., & Del-Valle-Soto, C. (2023). Utilization of 5G technologies in IoT applications: Current limitations by interference and network optimization difficulties—A review. *Sensors (Basel), 23*(8), 3876. DOI: 10.3390/s23083876 PMID: 37112216

Richter, A., Heinrich, P., Stocker, A., & Schwabe, G. (2018). Digital work design—The interplay of human and computer in future work practices as an interdisciplinary (grand) challenge. *Business & Information Systems Engineering*, *60*(3), 259–264. DOI: 10.1007/s12599-018-0534-4

Sharma, J. (2016). Digital India and its impact on the society. *International Journal of Research in Humanities & Social Sciences*, *4*(4), 64–70.

Song, F., Shi, N., Shan, F., Zhang, Z., Shen, J., Lu, H., Ling, Y., Jiang, Y., & Shi, Y. (2020). Emerging 2019 novel coronavirus (2019-nCoV) pneumonia. *Radiology*, *295*(1), 210–217. DOI: 10.1148/radiol.2020200274 PMID: 32027573

Srivastava, M. (2020). Digital economy expected to account for 20% of Indian GDP by 2025. https://kr-asia.com/digital-economy-expected-to-account-for-20-of-indian-gdp-by-2025

United Nations. (2020). *The impact of digital technologies.*https://www.un.org/en/un75/impact-digital-technologies

Vial, G. (2019). Understanding digital transformation: A review and a research agenda. *The Journal of Strategic Information Systems*, *28*(2), 118–144. DOI: 10.1016/j.jsis.2019.01.003

World Economic Forum. (2021). *The global risks report 2021* (16th ed.).

Yoo, Y., Boland, R. J.Jr, Lyytinen, K., & Majchrzak, A. (2012). Organizing for innovation in the digitized world. *Organization Science*, *23*(5), 1398–1408. DOI: 10.1287/orsc.1120.0771

# Chapter 8
# The Next Frontier Advancing Towards Industry 5.0

**Praveen Kumar Malik**
https://orcid.org/0000-0003-3433-8248
*Lovely Professional University, India*

**Rajesh Singh**
*Division of Research and Innovation, Uttaranchal Institute of Technology, Uttaranchal University, India*

**Anita Gehlot**
*Division of Research and Innovation, Uttaranchal Institute of Technology, Uttaranchal University, India*

**Aashna Sinha**
*Division of Research and Innovation, Uttaranchal Institute of Technology, Uttaranchal University, India*

**Arshi Naim**
*European Global Institute of Innovation and Technology, Malta*

## ABSTRACT

*Industry 5.0 is projected to bring about a social revolution and the emergence of a human-centric revolution. This new society will emphasize the social and human-centred features of Industry 4.0's tools and technology. Sustainability and human well-being will be essential to Industry 5.0, which enlarges on Industry 4.0's foundations by accenting sustainable, resilient and human-centred design principles. This revolution seeks to create a outline for supporting the cohabitation of industry*

DOI: 10.4018/979-8-3373-1534-8.ch008

*with growing society trends and requirements, therefore simplifying the transition from Industry 4.0 to 5.0. This may offer academics, practitioners, and policymakers with significant insights into Industry 5.0 supply chains, emphasizing their ability to boost social well-being, sustainability and financial growth. It also provides a conceptual basis for upcoming research and applied applications.*

## INTRODUCTION

The industrial sector has gone through interesting changes through the course of time, which have been grouped as industrial revolutions. At the end of the 18th century, the First Industrial Revolution started when people relied on water and steam power to build new machines for industry (Ravasoo, 2014). As electricity and mass production launched in the late 1800s and early 1900s, the Second Industrial Revolution improved how much industries could produce. The introduction of automation thanks to electronics, computers, and information technology in the Third Industrial Revolution formed the basis of the digital age. In the early 21st century, the concept of Industry 4.0 appeared when smart factories started using cyber-physical systems, Artificial Intelligence (AI), the Internet of Things (IoT), cloud computing, and advanced robotics (Folgado et al., 2024).

Although Industry 4.0 has made automation, productivity, and connectivity better than ever before, it has also created some fresh issues. Such issues are workforce displacement caused by automation, possible loss of personal data and cybersecurity, missing personal attention, and increasing damage to the planet. Since climate change, the lack of resources, and unequal wealth among people are getting worse, industry must transform to continue making progress and also to protect human rights, the environment, and social interests (Nguyen et al., 2023).

Industry 5.0 follows Industry 4.0 by putting more focus on people, caring for the environment, and making the process more adaptable. It looks forward to a setup where humans and smart machines cooperate and brains join forces to make a difference, not only in terms of productivity. It is clear from this new way of looking at industry that ethical design, SDGs, and inclusive innovation must be included in core industrial strategies (Khan et al., 2025).

The purpose of this chapter is to examine how Industry 5.0 started and what its key concepts involve. It explains the relevant technologies, significant applications, and effects on people of Industry 5.0. Besides other information, the chapter explains how various organizations and policy-makers can achieve a more flexible, inclusive, and responsible industry. Industry 5.0 is ready to shake up the world's manufacturing by uniting development with human needs and caring for the environment.

## LITERATURE REVIEW

### Evolution from Industry 1.0 to 4.0

Over the course of the industrial revolutions, new technology has helped change both manufacturing and society step by step. It was during the late 18th century First Industrial Revolution that steam and water power began to be used in machines. Workers in textile and agricultural sectors were able to work faster and factories began to appear for the first time (Mokyr, 1990).

Between late nineteenth and early twentieth centuries, electricity, engines powered by internal combustion, and the assembly line started being employed in industry. Thanks to these innovations, it was possible to make more products, especially in steel, chemicals, and cars, which in turn led to more cities and increased well-being (Gordon, 2016). In the 1970s, electronics, programmable logic controllers, and information technology were included in the manufacturing process by the Third Industrial Revolution. The emphasis of this period was on robots and introducing automation and digital control in factories (Phuyal et al., 2020).

The fourth phase, Indsutry 4.0, which is called the Fourth Industrial Revolution, was presented as a key initiative by the German government in 2011. It adds to the use of cyber-physical systems and more digital and intelligent technology in the industry (Xu et al., 2021). Unlike previous times, Industry 4.0 brings a new emphasis on how power, data, and automation affect the whole chain of manufacturing operations. There are four major stages of revolution, according to what figure 1 reveals.

Due to Industry 4.0, smart factories started using real-time monitoring, can make their own decisions, and interact with machines, causing major improvements in both efficiency and responsiveness (Khan et al., 2025). Still, this period has brought new issues and pointed out weak spots in the system, which has led to the development of Industry 5.0.

*Figure 1. Industrial Revolution*

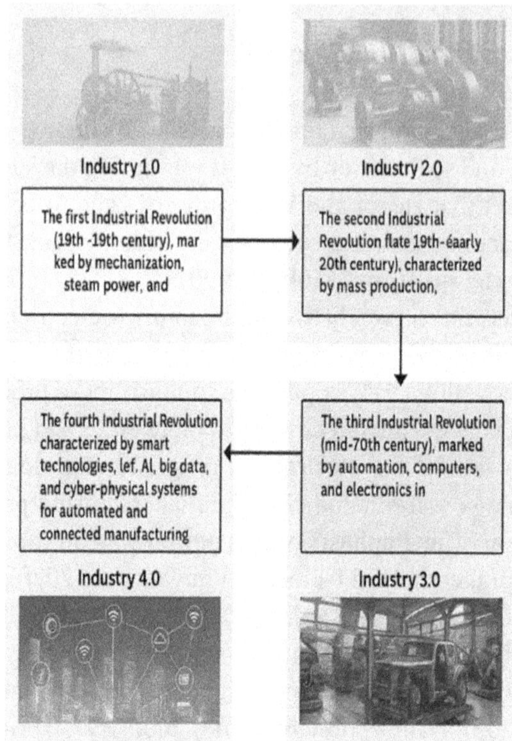


## Key Technologies Driving Industry 4.0

Through the use of industry 4.0 technology, traditional manufacturing is being changed into smart and highly efficient operations. Among all these technologies, AI is especially important. Using AI, quality can be controlled as it happens, maintenance can be predicted, and key decisions can be made in industry thanks to its ability to learn from a lot of data (Soori et al., 2023). IoT allows connected gadgets and sensors to connect and transfer data between devices, giving control and visibility to supply chains that are not centralized. These Cyber-Physical Systems (CPS) make it possible for digital machines and physical operations to communicate and exchange information all the time, which improves accuracy and ability to adjust to new input (Javaid et al., 2023).

Also, cloud computing makes it possible to scale both computing and storage needs, while big data analytics helps identify patterns and trends in huge amounts of data that play a role in making processes run smoothly and meeting demand forecasts (AL-Jumaili et al., 2023). Besides, 3D printing is transforming the way we create and test models and ship made-to-order products, while blockchain helps keep an eye on supply chains.

Still, Grand Challenges have to be connected, as they often exist in separate systems demanding much integration. Besides, concentrating on automation and improving efficiency usually means those important points about workforce, morality, and sustainability linked to the idea of Industry 5.0 are ignored. As Industry 4.0 technology advances, it's more obvious that using only technology isn't enough.

## Limitations of Industry 4.0 and the Shift to Industry 5.0

Even though Industry 4.0 has made industrial processes more advanced with automation and digitalization, it has brought up many key limitations and social issues. It is important to highlight how people's roles are being decreased. With machines assuming more independent tasks, humans may either need to oversee them or lose their jobs, which causes employees to worry about their skillful employment (Beier et al., 2020). As a result of not prioritizing human-centric design, there are offices that work well, but are not friendly. Environmental sustainability is another challenge as well. As a result of using robotics, data centers, and energy-intensive procedures in Industry 4.0, a large carbon impact has been created. Even though improvements have been made by using less energy, it is still unclear how sustainable Industry 4.0 technology is (Javaid et al., 2022). We should not overlook that cybersecurity and data privacy problems are on the rise because of the popularity of connected devices and cloud networks. This is why the concept of Industry 5.0 has emerged as a better and greener approach to industry. The European Commission (2021) points out that Industry 5.0 concentrates on including human workers in the production process by making them work alongside intelligent machines (Rame et al., 2024). The global economy also supports robustness, so production systems can survive big disturbances like pandemics or conflicts overseas, and it agrees with the United Nations Sustainable Development Goals (SDGs) on sustainability. Industry 5.0 supports using technology to promote people's creativity, enhance their morality, and secure a balance with the environment. New tools such as quantum computing, digital twins, materials inspired by nature, and edge AI are being used to help make things more personalized, transparent, and resource saving. It has been argued that this model proves to be a significant shift in both technology and industrial ideas, moving from automated focus to valuing new methods of innovation (Rashid & Kausik, 2024).

## TRANSITIONING INDUSTRY 4.0 TO INDUSTRY 5.0

Industry 4.0, has resulted in a trend toward intelligent mechanization, dropping the need for human interaction. For years, researchers have focused on the IoT. As the world enters the Fifth Industrial Revolution, there is a need for more critical examination of its potential benefits, problems, opportunities, trends and social impacts (Nagy et al., 2018). Industry 5.0 is a trend designed to revolutionizing production processes and productivity by integrating advanced technologies including the IoT, robotics, and AI, with the primary goal of promoting economic growth and competitiveness; however, incorporating environmental and social concerns remains a challenge (Ghobakhloo et al., 2022). Edge computing and other IoT technologies have the ability to generate a circular economy by digitalizing processes such as water distribution and smart manufacturing. However, because of their substantial carbon footprint, these technologies do not completely contribute to the IoT sector's long-term development. The Green IoT (G-IoT) paradigm tries to decrease this footprint, but it is inextricably related to Edge Artificial Intelligence (Edge AI), which uses more energy (Fraga-Lamas et al., 2021). Despite these obstacles, the Green IoT paradigm has promise for lowering the carbon footprint of the IoT sector. Table 1 explains industry 5.0 and sustainable industrial revolution.

*Table 1. Industry 5.0 and Sustainable Industrial Revolution*

| Topic | Description | Reference |
|---|---|---|
| Green IoT and Edge AI | Green IoT aims to reduce IoT's carbon footprint, but Edge AI increases energy consumption | (Alzoubi & Mishra, 2024) |
| Human-Centric AI in Industry 5.0 | Industry 5.0 integrates AI and human collaboration for user-friendly human-machine interfaces | (Tóth et al., 2023) |
| Human-Machine Interface in Industry 5.0 | Industry 5.0 emphasizes sustainability and human-centricity through HMI and BCI technologies. | (Yang et al., 2024) |
| Cognitive Cobots and Smart Manufacturing | Industry 5.0 blends AI, cobots, and communication infrastructure for efficient manufacturing. | (Zafar et al., 2024) |
| Industrial Evolution from 4.0 to 5.0 | Industry 5.0 builds on Industry 4.0, focusing on sustainability and human intelligence | (Rame et al., 2024) |

Industry 5.0, is projected to transform human-centric AI by combining human and AI progress. The move is in difference to Industry 4.0, which emphasizes automation and output through cyber-physical systems and IoT. AI skills such as natural language processing and computer vision will enable the development of user-friendly human-machine interfaces (Liu et al., 2024). To encounter the changing desires of Industry 5.0, the workforce will be re-skilled using AI-enabled training and simulation technologies as per shown in table 1. The progression from

Industry 4.0 to Industry 5.0, with an emphasis on the intersection of humanistic perspectives and technology improvements in Continuing Engineering Education (CEE). It encourages biomanufacturing and human-centricity while incorporating sustainability and resilience principles. The interplay of Human-Machine Interface (HMI) and Brain-Computer Interface (BCI) spans the 5th Industrial Revolution (Zizic et al., 2022). This vision necessitates the strategic orchestration of technical enablers such as cognitive cobots, AI, and communication infrastructure to link the cyber and physical worlds. Conferring to the Industry 5.0 framework, clever factories should include digitization, forecast, sustainability, and human centrality into their operations. This entails improving production processes for energy efficiency, safety, and new technology (Singh et al., 2025). Technological advancements can improve Human-Robot Interaction (HRI) in the industrial sector. Traditional industrial robots, such as cobots, are still commonly utilized due to safety restrictions and a lack of particular equipment, squandering the potential for true Human-Robot Collaboration (HRC). The goal is to provide integrated procedures, tools, standards, and systems for executing projects.

Digital twins represent a technical innovation that will pave the way for the incorporation of digital and physical systems. They give real-time information and insights, making it easier for humans to collaborate with autonomous systems. Industry 5.0 stresses the importance of flexible manufacture to encounter individual customer demands, and digital twins can help with rapid and effective changes to manufacturing processes (Javaid et al., 2022). Digital twins, or virtual replications of physical systems, employ AI and real-time data to optimize operations, foresee faults, and develop new products. They improve industrial processes by using smart automation, predictive maintenance, real-time monitoring, and IoT integration, resulting in greater throughput, decreased downtime, and cost savings. As automation progresses to Industry 5.0, digital twins enable human-machine cooperation, tailored manufacturing, and worker safety (Zhong et al., 2023). They can learn and interact with humans and machines in an adaptable manner, allowing for significant innovation and customization. Industry 5.0 to provide a robust, networked industrial environment. It improves human-machine interactions by merging robots and intelligent devices. IoT services include digital domains for corporate administration, real-time production monitoring, and machine status tracking. The objective is to strike a balance between productivity and efficiency in industries such as heavy manufacturing, oil and gas, and warehouse management, which rely largely on heavy machinery. The deployment of robots, cobots, and automation in manufacturing has considerably increased the sector's development and economics. However, successful implementation necessitates a healthy association between persons and technologies (Martini et al., 2024). As Industry 5.0 stresses sustainability, efficiency, and human-cantered enterprise, human-robot/machine interaction (HRMI) is flattering more important in the industrial industry as shown in table 2.

*Table 2. Showing Key Technologies and Innovations in Industry 5.0*

| Topic | Description | Reference |
|---|---|---|
| Digital Twins | AI-driven virtual replicas optimize industrial processes. | (Zhong et al., 2023) |
| Industry 5.0 and IIoT | IIoT enhances human-machine interactions in industries. | (Yang et al., 2024) |
| Human-Robot/Machine Interaction (HRMI) | Ensures efficient human-machine collaboration in Industry 5.0. | (Martini et al., 2024) |
| Blockchain-integrated Federated Learning (Block chained FL) | A privacy-focused data-sharing model with security challenges. | (Wang et al., 2024) |
| Sustainability in IR 5.0 | Promotes eco-friendly practices to reduce industrial impact. | (Rame et al., 2024) |

Industry 5.0 focuses on customized data sharing to drive innovation and improve customer satisfaction in everyday products. Blockchain-integrated federated learning (Blockchained FL) is a data-sharing paradigm that respects user privacy (Wang et al., 2024). However, most blockchain-enabled FL sharing models face complexity and unpredictability, such as insufficient data quality trade-offs and incentives that favour immediate profit over long-term device dynamics, resulting in poor-quality model training and security deterioration. Although fast technological advancement has enhanced wireless communication, it has also had environmental consequences like e-waste, greenhouse gas emissions, and resource depletion (Wang et al., 2024). The 'IR5.0 Eco-Future' program focuses on incorporating eco-friendly practices into technological development. Enablers, individuals, management, technology, and innovation processes play important roles in both traditional and digital technology businesses. It discovers context-specific treatments prompted by these processes, resulting in long-term industry 4.0 and industry 5.0 results. Generative AI, including ChatGPT, is accelerating Industry 4.0, Industry 5.0 by automating production and monitoring processes as per shown in figure 2. It supports user experience optimization, intelligent decision support systems, and human-robot collaboration (Islam et al., 2025).

*Figure 2. Development from Industry 4.0 to Industry 5.0*

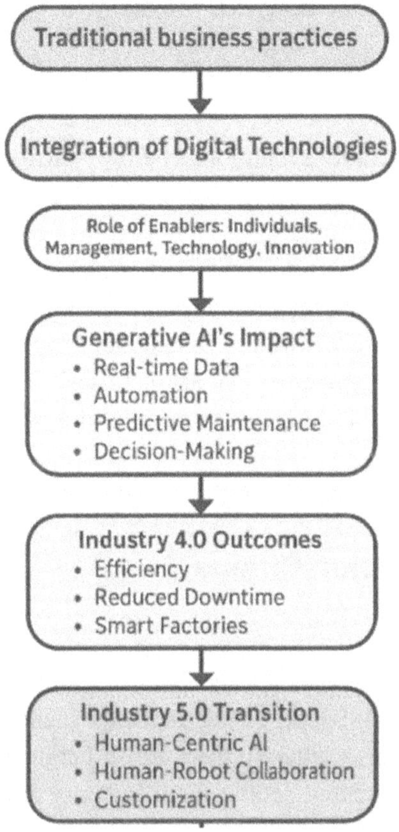


## RESEARCH METHODOLOGY

To put it together, the chapter relies on a narrative review style to gather information on the industrial revolutions, the essential technologies of Industry 4.0, and the rise of Industry 5.0. Using narrative was preferred to provide a complete and coherent account of economics, instead of a full meta-analysis, so that a mix of diverse inputs and ideas could easily be included (Roos & Reccius, 2024).

### Literature Search Strategy

I looked at research by searching in both scientific and technical databases, as well as digital repositories such as Scopus, Web of Science, IEEE Xplore, SpringerLink, and Google Scholar. The focus was placed on sources published between

2015 and 2025 so that the review would look at recent studies and ones relevant to Industry 4.0 and up-to-date Industry 5.0 theories.

## Keywords Used

The keywords and operations mentioned below were used to carry out the search process:

- "Industry 4.0", "technologies", and "cyber-physical systems"
- Include the word "Industrial Revolution" with either "history" (or) "evolution"
- "Artificial Intelligence" as well as "manufacturing"
- "Internet of Things" as well as "smart factories"
- Industry 5.0 should be centered on people and sustainability as well as be resilient.
- "Industry 4.0 has weaknesses" OR "the move to Industry 5.0 is necessary"

## Inclusion and Exclusion Criteria

Inclusion criteria:

- Researchers should look at peer-reviewed articles, papers from conferences, and official papers established by agencies or organizations.
- Research studies written in the English language.
- Papers that presented details about Industry 4.0 or 5.0 technologies and frameworks.
- Publications that were created starting from 2015.

Exclusion criteria:

- Materials that only presented facts without a wider look at the industry.
- Avoid using articles or sources that lack a peer-review for analysis; however, articles of established organizations (such as the European Commission) can be used.
- Information in papers that came out before 2015 and has very little historical significance.

## Data Extraction and Synthesis

After choosing the papers based on the screening, the chosen ones were carefully reviewed, and important themes were taken from them in the context of the phases of industrial change, key Industry 4.0 technologies, social and technical barriers, and the formation of Industry 5.0.

## RESULTS

Recent literature makes it clear that industry is moving from an automatic, industrial-focused Industry 4.0 to an industry that concerns sustainability, well-being, and toughness, which they call Industry 5.0. Experts are growing in agreement that Industry 5.0 focuses on three major principles: putting people first, environmental protection, and being strong. Leading groups, such as the European Commission (2021) and Nahavandi (2022) (van Erp et al., 2024), keep highlighting these themes by mentioning how intelligent systems should be integrated with human capabilities, along with the significance of ethical ideas in the future manufacturing environment. Looking at how the two industries differ, it becomes obvious how they are distinct from each other. Although Industry 4.0 focuses on improving efficiency by using AI, IoT, big data, and cyber-physical technology, Industry 5.0 brings people back into the process by teaming machines and humans. Creating a balance between clever manufacturing, promoting people's well-being, helping them learn, and caring for the environment is important. In this manner, Industry 4.0 works towards efficient processes, while Industry 5.0 strives for better outcomes in society and for the environment as well as for the economy. Lately, a number of frameworks and strategic models have appeared to explain the structure of Industry 5.0. According to the European Commission (2021), future factories should feature technology meant to help rather than replace people, and all areas should focus on working conditions that benefit and protect workers. Moreover, research is underway on digital twins, edge computing, quantum computing, and green AI to see their benefits for improving manufacturing and being more eco-conscious. Bio-inspired robotics, adaptive supply chains, and models of circular economy are now part of the wider concept of Industry 5.0. All things considered, the industry is moving from faster and more efficient operations to ones centered on meaningful innovation.

## DISCUSSION

- The integration of digital twins, IoT, AI, and big data analytics offers a promising approach to social growth, combining economic gain with personal and communal well-being while paving the way for a more intelligent, intuitive, and sustainable society.
- Chatbots and virtual assistants provide sympathetic customer support, increasing efficiency and creativity. Human contact and cooperation are prioritized in Industry 5.0, guaranteeing that technical improvements are in line with human values and social well-being, hence increasing productivity and creativity.

- Future studies should employ a complete analysis and data mining techniques to acquire a deeper knowledge of Industry 5.0 and its perspective among the research community.
- Digital twins and human contact are critical for the Industry 5.0 modification, which improves human-machine interactions and cohabitation in production environments.
- The 5G/6G communication networks strive for low latency and huge capacity, necessitating increased edge computing power and the convergence of G-IoT and Edge-AI as core technology enablers for the forthcoming 6G mobile network.

## CONCLUSION

Smart workplace technology can help to build skills that are important for achieving sustainability goals. Stakeholders should work together to identify and fund skills necessary for long-term transformation in industry. The study seeks to create a preliminary approach to Industry 5.0, incorporating digital technology and adopting a holistic approach to production. This requires sophisticated, adaptable and interconnected digital twin systems that will enhance efficiency, innovation and competitiveness. The aim is to promote safe collaboration between people, robots and machines by transferring control from the real world to the virtual world, resulting in a fully integrated and efficient virtual controller for the industrial workplace. Generative AI is transforming Industry 4.0 by improving precision manufacturing, streamlining supply chains, and allowing product design innovation. By analyzing vast datasets, AI improves processes, reduces downtime, and increases efficiency, driving production to new heights. As society regains focus, AI collaborates with humans to make industrial processes more sustainable and durable. It improves human creativity, decision-making, and personalization in production, bridging the integration gap between human and machine intelligence. AI-powered solutions in Industry 5.0 increase quality of life in fields like healthcare, education, and urban planning, while also encouraging inclusion and equitable technology sharing. This article employs text mining algorithms to comprehend and categorize Industry 5.0, a phrase developed between 2016 and 2022. The report emphasizes the significance of these technologies in the present corporate landscape.

## REFERENCES

AL-Jumaili, A. H. A., Muniyandi, R. C., Hasan, M. K., Paw, J. K. S., & Singh, M. J. (2023). Big Data Analytics Using Cloud Computing Based Frameworks for Power Management Systems: Status, Constraints, and Future Recommendations. *Sensors 2023, Vol. 23, Page 2952, 23*(6), 2952. DOI: 10.3390/s23062952

Alzoubi, Y. I., & Mishra, A. (2024). Green artificial intelligence initiatives: Potentials and challenges. *Journal of Cleaner Production, 468*, 143090. DOI: 10.1016/j.jclepro.2024.143090

Beier, G., Ullrich, A., Niehoff, S., Reißig, M., & Habich, M. (2020). Industry 4.0: How it is defined from a sociotechnical perspective and how much sustainability it includes – A literature review. *Journal of Cleaner Production, 259*, 120856. DOI: 10.1016/j.jclepro.2020.120856

Folgado, F. J., Calderón, D., González, I., & Calderón, A. J. (2024). Review of Industry 4.0 from the Perspective of Automation and Supervision Systems: Definitions, Architectures and Recent Trends. *Electronics 2024, Vol. 13, Page 782, 13*(4), 782. DOI: 10.3390/electronics13040782

Fraga-Lamas, P., Lopes, S. I., & Fernández-Caramés, T. M. (2021). Green IoT and Edge AI as Key Technological Enablers for a Sustainable Digital Transition towards a Smart Circular Economy: An Industry 5.0 Use Case. *Sensors 2021, Vol. 21, Page 5745, 21*(17), 5745. DOI: 10.3390/s21175745

Ghobakhloo, M., Iranmanesh, M., Mubarak, M. F., Mubarik, M., Rejeb, A., & Nilashi, M. (2022). Identifying industry 5.0 contributions to sustainable development: A strategy roadmap for delivering sustainability values. *Sustainable Production and Consumption, 33*, 716–737. DOI: 10.1016/j.spc.2022.08.003

Gordon, R. J (2016). *The Rise and Fall of American Growth: The U.S. Standard of Living since the Civil War*. https://econpapers.repec.org/RePEc:pup:pbooks:10544

Islam, M. T., Sepanloo, K., Woo, S., Woo, S. H., & Son, Y. J. (2025). A Review of the Industry 4.0 to 5.0 Transition: Exploring the Intersection, Challenges, and Opportunities of Technology and Human–Machine Collaboration. *Machines 2025, Vol. 13, Page 267, 13*(4), 267. DOI: 10.3390/machines13040267

Javaid, M., Haleem, A., Singh, R. P., & Suman, R. (2023). An integrated outlook of Cyber–Physical Systems for Industry 4.0: Topical practices, architecture, and applications. *Green Technologies and Sustainability, 1*(1), 100001. DOI: 10.1016/j.grets.2022.100001

Javaid, M., Haleem, A., Singh, R. P., Suman, R., & Gonzalez, E. S. (2022). Understanding the adoption of Industry 4.0 technologies in improving environmental sustainability. *Sustainable Operations and Computers*, *3*, 203–217. DOI: 10.1016/j.susoc.2022.01.008

Khan, M. I., Yasmeen, T., Khan, M., Hadi, N. U., Asif, M., Farooq, M., & Al-Ghamdi, S. G. (2025). Integrating industry 4.0 for enhanced sustainability: Pathways and prospects. *Sustainable Production and Consumption*, *54*, 149–189. DOI: 10.1016/j.spc.2024.12.012

Liu, S., Li, P., Wang, J., & Liu, P. (2024). Toward industry 5.0: Challenges and enablers of intelligent manufacturing technology implementation under the perspective of sustainability. *Heliyon*, *10*(15), e35162. DOI: 10.1016/j.heliyon.2024.e35162 PMID: 39157342

Martini, B., Bellisario, D., & Coletti, P. (2024). Human-Centered and Sustainable Artificial Intelligence in Industry 5.0: Challenges and Perspectives. *Sustainability 2024, Vol. 16, Page 5448*, *16*(13), 5448. DOI: 10.3390/su16135448

Mokyr, J. (1990). *The lever of riches: technological creativity and economic progress*. Oxford University Press.

Nagy, J., Oláh, J., Erdei, E., Máté, D., & Popp, J. (2018). The Role and Impact of Industry 4.0 and the Internet of Things on the Business Strategy of the Value Chain—The Case of Hungary. *Sustainability 2018, Vol. 10, Page 3491*, *10*(10), 3491. DOI: 10.3390/su10103491

Nguyen, T. T., Grote, U., Neubacher, F., Rahut, D. B., Do, M. H., & Paudel, G. P. (2023). Security risks from climate change and environmental degradation: Implications for sustainable land use transformation in the Global South. *Current Opinion in Environmental Sustainability*, *63*, 101322. DOI: 10.1016/j.cosust.2023.101322

Phuyal, S., Bista, D., & Bista, R. (2020). Challenges, Opportunities and Future Directions of Smart Manufacturing: A State of Art Review. *Sustainable Futures : An Applied Journal of Technology, Environment and Society*, *2*, 100023. DOI: 10.1016/j.sftr.2020.100023

Rame, R., Purwanto, P., & Sudarno, S. (2024). Industry 5.0 and sustainability: An overview of emerging trends and challenges for a green future. *Innovation and Green Development*, *3*(4), 100173. DOI: 10.1016/j.igd.2024.100173

Rashid, A., & Kausik, M. D. A. K. (2024). AI revolutionizing industries worldwide: A comprehensive overview of its diverse applications. *Hybrid Advances*, *7*, 100277. DOI: 10.1016/j.hybadv.2024.100277

Ravasoo, A. (2014). Interaction of bursts in exponentially graded materials characterized by parametric plots. *Wave Motion*, *51*(5), 758–767. DOI: 10.1016/j.wavemoti.2014.01.006

Roos, M., & Reccius, M. (2024). Narratives in economics. *Journal of Economic Surveys*, *38*(2), 303–341. DOI: 10.1111/joes.12576

Singh, K. A., Patra, F., Ghosh, T., Mahnot, N. K., Dutta, H., & Duary, R. K. (2025). Advancing food systems with industry 5.0: A systematic review of smart technologies, sustainability, and resource optimization. *Sustainable Futures : An Applied Journal of Technology, Environment and Society*, *9*, 100694. DOI: 10.1016/j.sftr.2025.100694

Soori, M., Arezoo, B., & Dastres, R. (2023). Internet of things for smart factories in industry 4.0, a review. *Internet of Things and Cyber-Physical Systems*, *3*, 192–204. DOI: 10.1016/j.iotcps.2023.04.006

Tóth, A., Nagy, L., Kennedy, R., Bohuš, B., Abonyi, J., & Ruppert, T. (2023). The human-centric Industry 5.0 collaboration architecture. *MethodsX*, *11*, 102260. DOI: 10.1016/j.mex.2023.102260 PMID: 37388166

van Erp, T., Carvalho, N. G. P., Gerolamo, M. C., Gonçalves, R., Rytter, N. G. M., & Gladysz, B. (2024). Industry 5.0: A new strategy framework for sustainability management and beyond. *Journal of Cleaner Production*, *461*, 142271. DOI: 10.1016/j.jclepro.2024.142271

Wang, N., Yang, W., Wang, X., Wu, L., Guan, Z., Du, X., & Guizani, M. (2024). A blockchain based privacy-preserving federated learning scheme for Internet of Vehicles. *Digital Communications and Networks*, *10*(1), 126–134. DOI: 10.1016/j.dcan.2022.05.020

Xu, X., Lu, Y., Vogel-Heuser, B., & Wang, L. (2021). Industry 4.0 and Industry 5.0—Inception, conception and perception. *Journal of Manufacturing Systems*, *61*, 530–535. DOI: 10.1016/j.jmsy.2021.10.006

Yang, J., Liu, Y., & Morgan, P. L. (2024). Human–machine interaction towards Industry 5.0: Human-centric smart manufacturing. *Digital Engineering*, *2*, 100013. DOI: 10.1016/j.dte.2024.100013

Zafar, M. H., Langås, E. F., & Sanfilippo, F. (2024). Exploring the synergies between collaborative robotics, digital twins, augmentation, and industry 5.0 for smart manufacturing: A state-of-the-art review. *Robotics and Computer-integrated Manufacturing*, *89*, 102769. DOI: 10.1016/j.rcim.2024.102769

Zhong, D., Xia, Z., Zhu, Y., & Duan, J. (2023). Overview of predictive maintenance based on digital twin technology. *Heliyon*, *9*(4), e14534. DOI: 10.1016/j.heliyon.2023.e14534 PMID: 37025897

Zizic, M. C., Mladineo, M., Gjeldum, N., & Celent, L. (2022). From Industry 4.0 towards Industry 5.0: A Review and Analysis of Paradigm Shift for the People, Organization and Technology. *Energies 2022, Vol. 15, Page 5221, 15*(14), 5221. DOI: 10.3390/en15145221

# Chapter 9
# The Virtual Economy:
## Opportunities and Challenges in the Digital Age

**Ravi Kant Sharma**
*Galgotias University, India*

**Pankaj Singh**
https://orcid.org/0000-0002-7559-9762
*Galgotias University, India*

## ABSTRACT

*In the digital age, the virtual economy has become a disruptive force that is changing consumer behavior, economic structures, and industries. The virtual economy, which is made possible by digital platforms, blockchain technology, and artificial intelligence, consists of digital experiences, goods, and services that are not limited by physical boundaries. The growth of digital transactions has given consumers, businesses, and entrepreneurs access to previously unheard-of options, ranging from e-commerce and digital banking to virtual assets like cryptocurrencies. The accessibility of the virtual economy, which permits worldwide involvement with little barriers to entry, is one of its main benefits. Digital marketplaces open up new sources of income, and the gig and create economies enable people to make money off of their talents. The blockchain-based technologies and decentralized finance provide financial inclusion to marginalized communities. This chapter offers perspectives on how virtual economy may affect the societal advancement and economic expansion in the digital era.*

DOI: 10.4018/979-8-3373-1534-8.ch009

## INTRODUCTION

A financial system that operates within an online environment that allows users to create, buy, sell, and trade virtual labor and goods is known as a virtual economy (Acemoglu & Restrepo, 2019). A few essential components characterize these economies. The concept of economy has transcended traditional boundaries in the modern era, giving rise to what we now refer to as virtual economies (Anand & Singh, 2025). These economies are found within virtual worlds, games, and web-based platforms where users engage in financial transactions that mimic real-world transactions (Arvidsson, 2020). The way we collaborate, create, and exchange virtual environments has been impacted by virtual economics, which have become an essential aspect of modern computerized life (Barlow, 2018). This chapter explores the concept of virtual economies, including their components, recommendations, and challenges.

With stages ranging from online gaming biological systems to decentralized finance (DeFi) and advanced commerce, virtual trade has grown to be a significant part of the current computer scene (Barzilai-Nahon, 2021). This section delves deeply into the architecture of the virtual economy, focusing on the production, exchange, and adaptation of virtual labor and goods. It also examines the motivations for these interactions, member behavior, and the impact on businesses and clients (Benkler, 2022). The feed is typically found within a biological system or digital stage. These companies occasionally mirror traditional companies, such as market interest productivity, finance (both online and offline), and private companies (Brynjolfsson & McAfee, 2021). Regardless, they also possess unique characteristics, such as creating resources with essentially no cooperation, establishing new groups, and gaining possession (Castells, 2020).

It discusses the difference between a non-commercial setting where blockchain innovation enables shared exchanges and democratization, and a virtual business basis that is typically governed by a stage (such as a gaming environment). The exchange of virtual labor and goods, ranging from in-game items and skins to sophisticated craftsmanship like non-fungible tokens (NFTs) and online services like distributed storage or virtual administrations, is one of the characteristics of virtual trade (Choudary, 2023). Solid products (reusable items) and consumables (expendable goods) are two categories of virtual products. Unlike real devices, virtual devices are typically designed with cheap production costs, giving designers flexible estimation (Cohen, 2021). They contribute to the advancement of the computerized biological system's customer experience. Social collaboration is required, for example, because virtual items in games often result in situations, the development of skills, or the personality of the symbol (Cowen, 2020). For instance, NFTs handle accountability for resources that can be traded across commercial centers and often provide compensation in relation to decentralized stages and virtual items. Make money using virtual labor and goods (Daugherty & Wilson, 2021).

To generate revenue from computerized goods, stages and engineers employ various strategies, keeping in mind application purchases, membership models, micro transactions, and sovereignties from optional transactions (David, 2020). For example, the freemium business model in computer games, where users can experiment for free but purchase premium content, has become a vital source of income (De Filippi & Wright, 2022). Engineers can create computerized resources (NFTs) and sell them on the open market using widely used platforms like Ethereum and Binance Shrewd Chain. Craftspeople, designers, and gamers can now purchase from auxiliary deals through blockchain-based sovereignties, creating new revenue sources (Dolata, 2021). Due to mental, social, and mechanical factors, customer behavior in the virtual economy may differ from that in the traditional economy (Lessig, 2022). Customers may purchase virtual goods for a variety of purposes (e.g., recreation, adventure, or just as a guide to the surrounding area). This section examines how gamification, scarcity, and social incentives are used by programmed stages to influence consumer interest (Manyika, & Sneader, 2023). For example, limited edition or limited release products are commonly utilized to create excitement and boost sales. Whether purchasing a computerized system, a dispersed storage membership, or a menial assistant product, it is essential to provide innovation that is useful, flexible, and meets the needs of the modern consumer (Parker et al. 2023). Virtual enterprises have many incredible opportunities for growth and advantage, but they also present challenges, particularly in the areas of administration, protected invention privileges, and specialized issues (Einav et al. 2023).

This looks a possible problems in a decentralized environment, such as lack of control, deception, and command over business (Goldfarb & Tucker, 2022). By examining the consequences of possession in virtual settings, such as stages that have the ability to alter or remove virtual goods, it also raises questions regarding advanced proprietorship (Greenstein, 2021). Due to the rapid advancement of innovation and shifting consumer behaviour (Jain, & Singh, 2023). Since virtual labor and goods continue to play a significant role in the global computerized economy, comprehending the complexity of these frameworks is fundamental for businesses, buyers, and legislators (Hagiu & Wright, 2023). This section provides a general framework for examining the virtual economy, looking into the opportunities and challenges it offers, and figuring out how labor will ultimately fare in the computer age (Graham, 2020).

There are also a lot of difficulties with the virtual economy. Businesses and customers are at danger from cybersecurity threats, data privacy issues, and regulatory uncertainties (Jain et al. 2025). Because differences in internet access and technology proficiency prevent many people from participating, the digital divide is still a major problem (Kenney & Zysman, 2023). Furthermore, in order to guarantee equity, openness, and moral behavior, the quick development of digital business models calls for flexible regulatory frameworks. Maintaining a sustainable and inclusive digital ecosystem will require striking a balance between innovation and regulation as the virtual economy grows (Kushwaha et al. 2023b). Businesses, consumers, and policymakers must work together to solve issues and optimize the benefits of digital transformation (Kumar et al. 2025).

By facilitating decentralized and digital-first marketplaces, the virtual economy—a byproduct of digital transformation—has altered conventional economic systems (Mazzucato, 2021). The potential and constraints of this changing economic paradigm have been thoroughly examined by academics and industry professionals, with particular attention paid to how it will affect labor markets, financial systems, business models, and regulatory frameworks (Kushwaha et al. 2024). This survey of the literature summarizes important viewpoints from academia and business regarding the advantages and disadvantages of the virtual economy (Kushwaha et al. 2023a).

## Kinds of Virtual Economies

Given its structure and rationale, virtual economies can be divided into several categories (Prüfer & Schottmüller, 2021) mentioned below:

**Gaming Economies**: A lot of online games, especially massively multiplayer online games (MMOs), contain robust economies that allow users to trade goods, money, and services. Games like "Fortnite" and "Universe of Warcraft" have created intricate financial systems that allow users to engage in trade, often with verifiable financial consequences (Ragnedda, 2022).

**Social Media Economies**: Due of powerful advertising, platforms such as Instagram and TikTok have created economies where customers can modify their content through partnerships and organizations. Devotee value and commitment metrics create a remarkable financial landscape propelled by cordial exchanges (Saraswat & Singh, 2025).

**Virtual worlds** such as "Second Life" and "Decentral and" allow users to buy, trade, and cultivate virtual land. These transactions may include significant sums of money; for example, some virtual properties may sell for millions of dollars, reflecting the growing profits from speculating on virtual land.

**Cryptocurrency Economies**: Decentralized virtual economies have emerged as a result of blockchain innovation. Digital currencies like Bitcoin and Ethereum open up new avenues for trade and venture capital, enabling pooled trades without the need for middlemen.

## Effects of Online Marketplaces

The emergence of virtual economies has important implications for several stakeholders, such as controllers, designers, and clients.

**Financial Open Doors**: Clients are given new avenues for the money age through virtual economies. Many have made gaming or video production their full-time careers, using their skills to earn real money through online transactions.
**Innovation and Innovativeness**: Virtual economies' client-produced nature fosters growth and creativity. Customers are encouraged to create unique content, which leads to a variety of goods and services that enhance the online environment.
**Regulatory Challenges**: Controllers have challenges due to the rapid growth of virtual economies. Careful consideration must be given to issues like tax evasion, extortion, and the tax assessment from virtual trades. Administrative institutions should adapt as these economies continue to grow in order to ensure equitable practices and protect customers.
**Social Communication**: Social cooperation among customers is improved by virtual economies. Within these economies, collaboration and exchange can spur the growth of networks, fostering ties that transcend geological boundaries.

## Types of Virtual Goods

Virtual goods can be divided into a number of categories according to their features and purposes:

**Cosmetic Items**: These are optional extras that improve a character's or avatar's appearance without changing gameplay. Video game skins, clothes, and accessories are a few examples. In games like "Fortnite" and "League of Legends," where users frequently spend real money to modify their avatars, cosmetic items are especially popular.
**Functional products**: In contrast to cosmetic products, functional goods offer real advantages inside a platform or game. Weapons, equipment, and power-ups that improve a player's skills or advancement can fall under this category. In-game economies are frequently driven by the need for useful objects as players try to get better at the game.
**Virtual Real Estate**: Users can purchase, sell, and build virtual land in "Second Life" and "Decentraland," among other virtual worlds. The market for virtual real estate has grown to be very profitable, with some homes fetching

high prices. Possessing virtual land enables users to organize events, produce experiences, and make money from rentals.

**Digital Art and Collectibles**: The idea of digital art and collectibles has been completely transformed by the emergence of non-fungible tokens (NFTs). NFTs are distinct digital assets that signify the ownership of a particular work of art, song, or video. As a result, artists and collectors now have new avenues for making money off of their works that were previously unattainable.

**Subscriptions and Services**: A lot of platforms give users access to premium features, content, or experiences through subscription-based services. Premium subscriptions in online gaming, streaming, or instructional platforms can fall under this category. Subscriptions improve user engagement and give developers a consistent source of income.

## Opportunities in the Virtual Economy

**Expansion of Digital Marketplaces:** By removing regional restrictions and facilitating international trade, the emergence of e-commerce platforms like Amazon, Alibaba, and Shopify has stimulated economic growth (Brynjolfsson & McAfee, 2017). The growth of microbusinesses has also been aided by digital marketplaces, which give small businesses and individuals direct access to customers. Global e-commerce revenues are expected to reach $7.4 trillion by 2025, according to Statista (2022), demonstrating the exponential rise of digital trade.

**Cryptocurrency and Blockchain-Based Transactions:** Decentralized financial systems made possible by blockchain technology have improved digital transactions' efficiency, security, and transparency (Nakamoto, 2008). Peer-to-peer transactions have been made easier by cryptocurrencies like Ethereum and Bitcoin, which have decreased dependency on conventional banking systems (Böhme et al., 2015). By providing financial inclusion to underbanked communities, Decentralized Finance (DeFi) has further transformed lending, borrowing, and asset management (Zhang et al., 2021).

**Digital Labor and the Gig Economy:** With the help of online marketplaces like Upwork, Fiverr, and Uber, the gig economy has revolutionized work arrangements by providing remote, flexible, and freelancing options (Kalleberg, 2018). People can make money from specific abilities like software development and content creation thanks to the digital labor market. According to research by Sundararajan (2022), platform-based work encourages entrepreneurship while lowering operating expenses for companies.

**The Rise of the Metaverse and Virtual Assets:** Through digital real estate, virtual goods, and immersive experiences, virtual environments—including the metaverse—offer new economic potential (Ball, 2022). The importance of digital ownership has been illustrated by the sale of NFTs (Non-Fungible Tokens), and sectors including gaming, entertainment, and the arts are using blockchain-based authentication to make money off of their intellectual assets (Dowling, 2022).

## Challenges in the Virtual Economy

**Regulatory Uncertainty and Policy Gaps:** Regulations have not kept up with the quick development of digital business models, which has led to problems with consumer protection, intellectual property rights, and taxes (Zetzsche et al., 2017). There are worries about money laundering, fraud, and financial instability because cryptocurrencies and DeFi platforms function in mainly uncontrolled environments (FATF, 2021). Adaptive regulations that strike a balance between innovation and control are difficult for policymakers to create (Gans, 2019).

**Cybersecurity and Data Privacy Concerns:** Vulnerabilities to cyberthreats, such as identity theft, hacking, and data breaches, have increased due to the rise in digital transactions (Choo, 2011). Concerns regarding data privacy must be addressed by businesses that gather user data, especially in light of strict laws like the General Data Protection Regulation (GDPR) (Kuner et al., 2019). Security and trust are still essential to the survival of virtual economies.

**Issues with Accessibility and the Digital Divide:** The virtual economy is not widely available, despite its potential. For underprivileged populations, disparities in financial resources, digital literacy, and internet connectivity erect obstacles (Hilbert, 2016). In developing nations, where inadequate infrastructure prevents access to online marketplaces and digital finance, the digital gap is more noticeable (World Bank, 2021).

**Job Displacement and Labor Market Disruptions:** Disruptions to the labor market and job displacement. Automation and artificial intelligence pose a threat to established job sectors, even as digital labor platforms generate opportunities. Nearly 47% of American employment could be automated, according to research by Frey and Osborne (2017), which raises worries about job displacement and economic inequality. In order to support impacted workers, the shift to a digital economy necessitates legislative interventions and workforce reskilling (Bessen, 2019).

## LITERATURE REVIEW

The virtual economy offers transformative potential for economic participation, innovation, and global trade. However, it also presents significant challenges related to regulation, equity, and sustainability. A balanced approach that promotes innovation while ensuring fair labor practices, environmental responsibility, and digital inclusion is essential for harnessing the full benefits of the digital age.

### Opportunities in the Virtual Economy

The concept of the *virtual economy* refers to economic activities that occur within digital environments, often mediated by the internet, digital platforms, and virtual goods. Castronova (2008) was one of the early scholars to define virtual economies as those arising within online games and virtual worlds where virtual goods are created, exchanged, and monetized. Over time, this definition has expanded to encompass e-commerce, the gig economy, crypto assets, and digital platforms (Lehdonvirta & Ernkvist, 2011). Digital transformation has blurred the boundaries between the physical and virtual realms. It highlights that digital platforms such as Amazon, YouTube, and Uber have reshaped entire industries, accounting for trillions in global economic activity. Furthermore, global e-commerce revenues surpassed $5 trillion, illustrating the immense scale of the digital economy.

Digital platforms facilitate innovative business models, such as freemium services, direct-to-consumer models, and creator economies. According to Brynjolfsson and McAfee (2014), digital technologies reduce marginal costs, enabling rapid scaling and personalized offerings. The internet allows businesses and individuals to reach global audiences. Entrepreneurs in developing countries can now access customers, investors, and collaborators across borders. The virtual economy offers flexible work arrangements, providing opportunities for marginalized groups, including women, persons with disabilities, and those in rural areas. Blockchain technology underpins new forms of ownership and value creation, such as Non-Fungible Tokens (NFTs) and Decentralized Finance (DeFi), enabling peer-to-peer transactions without intermediaries (Tapscott & Tapscott, 2016).

### Challenges in the Virtual Economy

Virtual economies often outpace regulation. Issues include tax evasion, digital monopolies, and consumer protection. Zuboff (2019) discusses "surveillance capitalism," where dominant tech firms exploit personal data with minimal oversight. Access to the virtual economy requires digital infrastructure and skills. The digital divide exacerbates inequalities, leaving behind those without access to technology or literacy (UNCTAD, 2022). Gig workers and digital freelancers face unstable income, lack of benefits, and limited bargaining power. Scholz (2017) warns of "platform capitalism" where platforms extract value while externalizing labor costs. Virtual economies, especially blockchain and data centers, have substantial carbon footprints. For example, Bitcoin mining consumes more energy annually than some countries (Cambridge Bitcoin Electricity Consumption Index, 2023).

## RESEARCH METHODOLOGY

This chapter uses a secondary data-based research approach to examine the virtual economy's prospects and limitations. In order to provide a thorough grasp of the digital economic landscape, the process include an organized assessment of the body of available literature, reports, and case studies. This methodology provides a structured, reliable approach to analyzing the virtual economy solely through secondary data. Let me know if you need any refinement.

### Research Approach and data collection

Based on qualitative and descriptive research, the report examines trends, possibilities, and difficulties in the virtual economy using secondary data sources. By combining knowledge from several reliable sources, the method guarantees a wide viewpoint. These sources are used to get secondary data like Academic Literature, Research studies, conference proceedings, and peer-reviewed journal articles about virtual transactions, online monetization techniques, and digital economies. Industry Reports and White Papers, Reports on digital markets, blockchain, and e-commerce from groups including the World Economic Forum (WEF), McKinsey, Deloitte, and PwC. Government and Regulatory Reports, Legislative frameworks, policy papers, and taxation rules pertaining to online trading and virtual assets.

### Analytical Framework

To interpret the collected data, the following analytical frameworks are applied:

**Thematic Analysis:** Identifying recurring themes related to opportunities (e.g., digital entrepreneurship, decentralized finance) and challenges (e.g., regulatory risks, market volatility).
**Comparative Analysis:** Examining different virtual economy models (e.g., gaming-based economies vs. blockchain-driven economies).

**Trend Analysis:** Studying emerging trends in digital commerce, virtual assets, and the gig economy to forecast potential developments.

## The Effects of Online Products and Services

Economic models, social interactions in digital settings, and user engagement have all been significantly impacted by the growth of virtual goods and services.

**Increased User Engagement**: By offering players incentives to engage in online activities, virtual products and services greatly increase user engagement. Users may spend time and money on a game or platform because they want to obtain uncommon or distinctive items. As users display their virtual belongings to their friends, social interactions frequently serve to promote this involvement.

**Economic Models**: A variety of economic models have emerged inside digital ecosystems as a result of the sale of virtual products and services. For instance, microtransactions—where users can buy virtual goods to improve their experience—are frequently used in free-to-play games. This paradigm keeps games accessible to a wider audience while enabling developers to make money off of them.

**Social Status and Identity**: In online communities, virtual items are frequently used as status markers. To build their reputation and personality among their peers, players could try to obtain uncommon objects or display their collections. As users bond over common interests and experiences, the social component of virtual products promotes a feeling of community and belonging.

**Creativity and ingenuity**: Users' creativity and ingenuity have been sparked by the desire for virtual products. Numerous platforms promote user-generated content, enabling people to produce and market their own virtual goods. In addition to improving the digital environment, this gives people more freedom to express their originality and spirit of entrepreneurship.

**Cross-Platform Integration**: As virtual goods have grown in popularity, there is now more cross-platform integration. Now, users may use or move their virtual assets between games and surroundings, making the digital experience more cohesive. Because consumers may utilize their investments in a variety of circumstances, this cross-platform flexibility raises the value of virtual commodities.

## RESULTS

### Risks and Difficulties with virtual services

Notwithstanding their advantages, virtual products and services come with a number of hazards and difficulties that need to be considered:

**Virtual Economy Marketing Techniques, Virtual Economy Marketing Techniques:** The emergence of virtual economies has changed the digital business scene and opened up new doors for consumers, developers, and producers. As these economies grow, a number of monetization techniques have been developed that enable both individuals and companies to profit from the special qualities of virtual goods and services. This paper investigates the various monetization techniques used in virtual economies, their efficacy, and the consequences for all parties concerned.

**Comprehending Virtual Economies:** It is crucial to comprehend what virtual economies are before diving into monetization tactics. An economic system that operates in a digital setting and allows users to produce, purchase, sell, and exchange virtual products and services is known as a virtual economy. These economies, which function according to separate rules and dynamics but frequently resemble real-world economic principles, are distinguished by digital assets, user-generated content, and distinctive currency systems. Virtual economies are present in a number of settings, such as social media sites, online games, virtual worlds, and blockchain-based settings. The capacity to enable transactions involving digital products and services—which can include everything from digital art and cryptocurrencies to in-game items and virtual real estate—is what unites these platforms.

**Small-scale transactions:** Small, incremental purchases made by consumers on a digital platform are known as microtransactions. This tactic is common in free-to-play games where users can purchase in-game currency, virtual goods, or cosmetics to improve their gaming experience. By enabling creators to commercialize their games without requiring an upfront payment, microtransactions enable them to reach a wider audience. When a game has a big user base, microtransactions can bring in a sizable sum of money. One such example is "Fortnite," which has made billions of dollars from the selling of battle passes and skins. However, finding a balance between monetization and user experience is essential to microtransactions' efficacy. Excessively aggressive monetization may cause reaction and player discontent.

**Models of Subscriptions:** Users who use subscription models pay a regular price to gain access to premium features, content, or services. This tactic is frequently used in instructional platforms, streaming services, and online gaming. In addition to giving consumers access to special advantages, subscriptions can give developers a consistent source of income. Since players are more inclined to stick with a platform if they are paying for continuous access, subscription models can encourage sustained user engagement. Games like "World of Warcraft" and websites like "Twitch" have effectively used subscription models to build devoted fan bases. However, the problem lies in consistently giving value to consumers to prevent churn.

**Models of Freemium:** The freemium business model offers both the opportunity to purchase premium features or content and free access to a basic version of a product. This tactic is frequently employed in online games and mobile apps, where players can play for free but are urged to subscribe for more features. Because the free product reduces the barrier to entry, freemium models have the potential to draw a sizable user base. "Candy Crush Saga" and "Clash of Clans," where a sizable percentage of players turn into paying clients, are successful examples. Making ensuring the free version is interesting enough to keep customers interested while also offering strong incentives to upgrade is a difficult task.

**Advertising within games:** By incorporating commercials into the gameplay, in-game advertising enables developers to make money from outside companies. Product placements within the game, sponsored content, and banner adverts are some examples of this. Particularly for free-to-play games, in-game advertising can be a very profitable monetization tactic. It enables developers to make money without charging players directly. However, how well this tactic is implemented will determine how effective it is; invasive advertisements can ruin gaming and result in bad user experiences. Effective implementations can increase engagement and benefit advertisers, as seen by "Fortnite"'s branded events.

**Sales of Virtual Goods:** One simple monetization technique that takes advantage of the demand for digital assets is selling virtual items directly to users. This can include virtual real estate, useful things, and cosmetic items. To make these transactions easier, developers might build a marketplace on their platform. Selling virtual goods may be very lucrative, particularly if the products are made to be distinctive and appealing. Skins and weapon cases have been a successful way for games like "Counter-Strike: Global Offensive" to make money. Developers must, however, make sure that users find the products valuable and that the pricing is viewed as reasonable.

**Community assistance and crowdsourcing:** Creators are able to raise money directly from their communities through crowdfunding websites like Kickstarter and Patreon. For independent developers and content producers who might not have access to conventional funding sources, this approach works very well. Strong support from supporters can result from crowdfunding's ability to foster a sense of community ownership and investment in a project. A successful campaign can bring in a sizable sum of money, as seen by games like "Shovel Knight." The difficulty, though, is in keeping backers informed and fulfilling commitments.

**Consumer Conduct in Online Marketplaces:** The way customers engage with digital goods and services has changed since the emergence of virtual economies. Understanding consumer behavior in these economies is becoming more and more crucial for researchers, developers, and marketers as they continue to expand. This essay examines the several aspects that affect consumer behavior in virtual economies, the social and psychological dynamics at work, and the consequences for companies that operate in these online marketplaces.

**Comprehending Virtual Economies:** An economic system that operates in a digital setting and allows users to produce, purchase, sell, and exchange virtual products and services is known as a virtual economy. These economies, which function according to separate rules and dynamics but frequently resemble real-world economic principles, are distinguished by digital assets, user-generated content, and distinctive currency systems. Virtual economies are present in a number of settings, such as social media sites, online games, virtual worlds, and blockchain-based settings.

## Elements That Affect Consumer Behavior

**The perceived worth of digital goods:** Consumer behaviour is significantly influenced by how virtual items are perceived to be worth. In contrast to tangible things, virtual items' worth is frequently arbitrary and impacted by social standing, usefulness, and rarity. For example, even if two in-game items have similar purposes, a rare one could be worth more than a common one. Customers are frequently prepared to pay more for goods that provide them a competitive edge or raise their standing in the community. For instance: Players are prepared to shell out a substantial sum of money for uncommon skins in video games such as "Counter-Strike: Global Offensive," not only because of their visual attractiveness but also because of the prestige they bestow within the gaming community.

**Community dynamics and social influence:** In virtual economies, social connections and community dynamics have a big influence on consumer behavior. Users frequently seek advice from their friends about what to purchase, which causes fads and trends in online communities. Purchase decisions may be influenced by the desire to blend in or stand out in a community, as buyers aim to obtain goods that are well-liked or respected by their peers. For instance, when limited-time skins are introduced in "Fortnite," players are frequently prompted to buy them in order to follow community trends because they create a sense of urgency and exclusivity.

**Engagement and Gamification:** The use of game design components outside of games to increase user engagement is known as gamification. Gamification can affect consumer behavior in virtual economies by enhancing the pleasure and rewards of the purchasing experience. Features that encourage users to get more involved with the economy include leaderboards, achievements, and incentives for doing activities. For instance, gamification strategies are used in many mobile games to promote in-game purchases. Completing challenges or levels may earn player's incentives, which may increase their spending on virtual items.

**Affective Elements:** Consumer behavior is heavily influenced by emotions, especially in virtual markets. Purchasing decisions can be influenced by a variety of factors, such as the thrill of acquiring a new item, the fulfilment of finishing a collection, or the disappointment of losing out on a limited-time bargain. These emotional cues are frequently used by marketers to craft gripping stories about virtual goods. As an illustration, the marketing techniques used by firms such as Riot Games for "League of Legends" frequently arouse feelings by emphasizing the excitement of competition and the delight of personalization, so motivating users to spend money on virtual goods.

**Exclusiveness and Scarcity:** Exclusiveness and scarcity are strong inducers of consumer behavior. Virtual commodities are frequently more desirable when they are thought to be scarce. The psychological idea of scarcity, which holds that individuals appreciate things that are harder to find, is the foundation of this theory.

**Content Generated by Users and Personalization:** The capacity to design and personalize virtual products improves customer happiness and engagement. Customers can express their uniqueness and creativity through user-generated material, which strengthens their emotional bond with the products they buy. Because consumers are more inclined to spend money on products that express their individuality, personalization can have a big impact on their purchasing decisions.

**Security and Trust:** Consumer behavior is heavily influenced by trust and security, especially in virtual markets where real money is frequently exchanged. When customers have faith in the platform's security and the authenticity of the products being sold, they are more inclined to make a purchase. Consumer participation in virtual economies may be discouraged by problems like fraud, scams, and data breaches.

**Virtual Economies' Psychological Dynamics:** A psychological phenomenon known as the endowment effect occurs when people place a higher value on identical objects they own than they do on those they do not.

## ANALYSIS

In virtual economies, especially those involving virtual currencies and digital assets, market volatility is a major problem. Depending on supply and demand, user mood, and outside variables like platform policy changes or game upgrades, the value of virtual items can vary significantly (Singh & Agrawal, 2020). Both developers and consumers may experience financial instability as a result of this unpredictability. Consumers may suffer unanticipated financial losses as a result of market volatility (Singh & Agrawal, 2022). Because of changes in the game environment or community preferences, players who spend real money on virtual products may discover that their worth drastically declines (Singh, 2014). This uncertainty can cause annoyance and discontent, especially for people who consider their virtual assets to be a type of financial asset. Managing the economy of their virtual environments presents difficulties for developers (Singh, 2024b). A game's economy may discourage new players and cause user engagement to drop if it becomes overly unstable. In order to preserve consumer investments, developers must find a balance between fostering a vibrant economy that promotes trade and maintaining stability. This frequently necessitates constant observation and resource-intensive in-game economy modifications (Singh & Agrawal, 2019).

Market volatility has important ethical ramifications. It is the duty of developers to design transparent, equitable economic structures free from consumer exploitation. This entails giving players precise information regarding the dangers of virtual products and making sure they are aware of the possibility of value swings. Developers also need to think about how their choices will affect the general state of the virtual economy and the welfare of their user base. (Singh & Kushwaha, 2024)

## Overview of Consumer Protection

In virtual economies, where consumers frequently spend real money on digital goods and services, consumer protection is a crucial factor. Strong safeguards must be put in place since customers may become vulnerable due to the anonymity of online transactions and the absence of real items (Singh & Kushwaha, 2025). The frequency of fraud and scams is one of the main issues facing consumer protection. Users may become victims of fake markets that promise virtual goods but provide nothing, counterfeit goods, or phishing assaults. These frauds have the potential to cause large financial losses as well as damage the reputation of the virtual economy (Singh & Agrawal, 2021). Efforts to protect consumers are complicated by the lack of comprehensive regulatory frameworks for virtual economies. Virtual economies frequently function in a legal murky area, in contrast to traditional markets where consumer rights are clearly established (Singh et al. 2021a). Customers may be left without options in the event of fraud or complaints due to this lack of control. It is the ethical duty of platform operators and developers to safeguard users from fraud and maintain a secure trading environment. This entails putting strong security measures in place, outlining precise rules for transactions, and giving assistance to users who run into problems (Singh et al. 2021b). Developers should also keep a close eye out for fraudulent activity on their platforms and act quickly to stop it.

In virtual economies, where users frequently divulge personal information in order to register, make transactions, and interact with other users, data privacy is a major concern (Singh et al. 2023a).. Important ethical concerns regarding user consent, data security, and misuse potential are brought up by the acquisition and use of this data. It's possible that many users are unaware of the entire extent to which their data is being gathered, utilized, and disseminated. In order for users to engage in the virtual economy, developers frequently incorporate long terms of service agreements, which might be challenging to understand. A gap between user expectations and real data practices may result from this lack of transparency (Singh et al. 2023b).

Data leaks are a serious problem as well. Payment information and personal identifiers are among the sensitive user data frequently stored in virtual economies. Users may be at serious risk of financial fraud and identity theft as a result of a breach (Singh et al. 2024a). Developers are ethically obligated to put user privacy and data security first. Users' informed consent regarding data collecting procedures must be obtained, robust security measures must be put in place to safeguard user data, and data usage must be open and transparent. The ability to remove accounts or choose not to share data are two other ways that developers should give users control over their data (Singh et al. 2024b).

Real-world economies are impacted by the growth of virtual economies, especially when virtual currencies and goods become more prevalent in daily life. The effects of virtual economies on conventional economic systems and consumer behavior are called into question by this integration. Spending habits may change as customers spend more money on virtual goods, which could have an impact on the traditional retail and service sectors.

## Possibilities in the Digital Economy

**Job Creation and Economic Growth:** Significant economic growth and employment creation might be fueled by the virtual economy. Businesses are moving more and more of their activities online, which opens up new markets and opportunities for creators and entrepreneurs. The virtual economy gives people a variety of ways to make money off of their abilities, from influencers and content producers to game developers and digital artists. People from all origins can participate and prosper in a more inclusive economy thanks to the democratization of economic involvement.

**Creativity and Innovation:** An atmosphere that encourages creativity and innovation is fostered by the virtual economy. Without the restrictions frequently present in traditional marketplaces, people can experiment with novel concepts, goods, and services when entry barriers are minimal. For instance, user-generated content, which enables users to produce and market their own digital items, has emerged as a key component of many virtual economies. In addition to enhancing the virtual environment, this promotes cooperation and idea exchange, creating a thriving creative ecosystem.

**Accessibility and Worldwide Reach:** The worldwide reach of the virtual economy is one of its biggest benefits. A genuinely global marketplace is created by digital platforms that allow companies and customers to interact across geographic borders. Small companies and independent artists can now reach audiences they might not have been able to reach through more conventional channels thanks to this accessibility. As a result, the virtual economy can promote a sense of global connection by bridging gaps between various groups and cultures.

**Novel Business Structures:** New business concepts that go against conventional wisdom have emerged as a result of the growth of the virtual economy. Microtransactions, freemium business models, and subscription services are a few ways that companies are adjusting to the digital environment. These models open up possibilities for recurring revenue streams and permit variable pricing methods. Further increasing the opportunities for ownership and monetization in the digital sphere are decentralized finance (DeFi) and non-fungible tokens (NFTs), which have been established by the integration of blockchain technology.

**Improved Interaction with Customers:** Virtual economies offer special chances to improve customer interaction. Users may find a more engaging and joyful environment thanks to gamification, social interactions, and customized experiences. Companies can use these components to establish trusting bonds with their clients, which will promote repeat business and loyalty. Businesses can modify their offers to satisfy the changing demands and preferences of their audiences by comprehending consumer behavior in virtual economies.

## Difficulties in the Digital Economy

**Speculation and Market Volatility:** The virtual economy presents chances for financial gain, but it is also marked by speculation and market volatility. Virtual currencies and goods can have significant price swings, putting both consumers and companies at risk financially. Potential investors may be turned off by this volatility, which also breeds market ambiguity. Developers need to put policies in place to keep their economies stable and shield users from the negative consequences of abrupt pricing adjustments.

**Fraud and Consumer Protection:** Because virtual economies are digital, there are worries regarding fraud and consumer protection. Users may experience financial losses and a decline in platform trust as a result of falling for scams, phishing attempts, or fake goods. It is the duty of developers and platform operators to put strong security measures in place, give consumers clear transaction instructions, and provide assistance when they run into problems. Building consumer confidence and promoting involvement in the virtual economy need creating a secure and reliable environment.

**Security and Privacy of Data:** Data security and privacy become critical issues as virtual economies rely more and more on user data for engagement and customisation. In order to create accounts and make purchases, users frequently divulge personal information, which raises ethical concerns regarding data collecting and use. By getting informed consent, putting robust security measures in place, and being open and honest about data practices, developers can put user privacy first. Ignoring these issues can result in identity theft, data breaches, and a decline in customer confidence.

**Difficulties with Regulation:** Virtual economies have grown faster than regulatory frameworks, which has made things more complicated for both consumers and corporations. Inconsistencies in intellectual property rights, taxation, and consumer protection may result from a lack of thorough rules. Legislators must endeavour to create precise regulations that balance innovation and consumer protection while addressing the particularities of virtual economies. To provide a just and sustainable environment for all participants, cooperation between regulators and industry stakeholders is crucial.

**Effect on Economies in the Real World:** As consumer spending moves toward digital goods and services, the growth of virtual economies may have an impact on economies in the real world. Traditional retail and service sectors may be impacted by this transition, which could alter employment trends and economic systems. Companies need to embrace these developments in order to adapt.

**Moral Issues in Virtual Economies:** Stakeholders in virtual economies must negotiate a distinct set of difficulties and moral dilemmas as they develop and grow. These issues, which can have a big impact on developers, customers, and regulators, come from the nexus of technology, consumer behavior, and economic principles. Market instability, consumer protection, data privacy, the effect of virtual goods on real-world economies, and the moral implications of user-generated content are the five main issues and ethical concerns in virtual economies that are examined in this essay.

## DISCUSSION

Virtual economies have potential, but they also come with risks and obstacles. A number of factors, including supply and demand, game upgrades, and user mood, can have a significant impact on the value of virtual products and currencies (Singh et al. 2025a). Users that make large investments in virtual assets may suffer large financial losses as a result of this instability. Since internet transactions are anonymous, fraudulent activity may be made easier. The economy may become distrustful as a result of users falling for fraud, phishing, or fake goods. A lot of virtual economies are linked to particular games or platforms. Policy changes

Virtual goods and services have become essential elements of online economies in the digital world, impacting how people communicate, collaborate, and conduct business in virtual spaces (Singh et al. 2025b). Our understanding of worth and ownership in the digital world has changed as a result of these digital assets, which might include everything from in-game goods and skins to virtual real estate and digital art (Singh & Agrawal, 2022). The importance of virtual goods and services, their different forms, their effects on user involvement and economic models, and the difficulties they pose are all examined in this essay (S.

Digital products that can be bought, sold, or traded through online marketplaces, video games, or virtual worlds are known as virtual goods. Virtual goods, in contrast to physical goods, are valuable in the context of the digital world but lack a physical form. On the other side, virtual services include things like gaming, virtual events, and digital content production that can be accessed or used online. The usefulness, uniqueness, and social standing that virtual goods and services bestow upon their owners are frequently the factors that determine their worth. For example, a unique digital artwork can represent ownership of a one-of-a-kind production, while a rare in-game item may improve a player's performance or visual appeal. Therefore, in digital ecosystems, virtual goods and services are essential in determining user experiences and interactions.

## CONCLUSION

In conclusion, the digital age presents both opportunities and challenges for the virtual economy. In the digital era, the virtual economy has become a disruptive force that is changing how people and companies communicate, collaborate, and conduct business. The virtual economy offers a multitude of chances for creativity, business, and community development due to the growth of online platforms, gaming environments, and blockchain technology. But it also presents serious difficulties that call for thoughtful analysis and decisive handling. It is critical to acknowledge both the possible advantages and the inherent risks of virtual economies as we traverse this changing terrain.

Advanced objects, which can range from in-game items, skins, and symbols to virtual land and digital currencies, are the center of virtual economies. In contrast to real products, these items usually lack any significant structure other than their ability to be valued in the virtual environment. Client-produced content, in which users or clients create their own work and goods, is a thriving component of many virtual economies. This can recall entire virtual worlds, virtual artwork, or even unique skins for video games. The ability of customers to contribute to the economy strengthens loyalty and fosters a sense of accomplishment. In-game monetary standards or digital currencies are two examples of the common money frameworks seen in virtual economies. These monetary standards blur the boundaries between virtual and real economies because they function with transactions and can occasionally be exchanged for actual cash. Commercial hubs where customers can trade goods and labor are typically a part of virtual economies. These commercial hubs may be real, like those found in video games, or they may be unofficial, like third-party websites where customers exchange virtual goods. Although virtual economies are free to operate, they are frequently governed by the rules established by game or stage designers. This keeps in mind rules for trading, business ownership, and product manufacturing, all of which might affect the overall soundness of the economy.

Industries, cultures, and financial systems are changing as a result of the virtual economy, which presents both new opportunities and difficulties. A linked world where companies, consumers, and artists may participate in smooth transactions, remote work, and decentralized financing has been made possible by digital transformation, which is fueled by e-commerce, blockchain, artificial intelligence, and the metaverse. This change has opened up new revenue sources for sectors like gaming, retail, education, and entertainment by facilitating economic inclusion, innovation, and personalized experiences.

But even with its quick growth, the virtual economy has serious problems that need to be solved. Because governments find it difficult to enact clear regulations pertaining to digital assets, cryptocurrencies, and decentralized platforms, regulatory ambiguity is still a significant worry. Both individuals and corporations are at danger from cybersecurity threats such as fraud, data breaches, and digital identity theft.

Furthermore, because impoverished areas lack the infrastructure necessary to fully engage in the virtual economy, the digital divide hinders equitable access to it. A balanced strategy is necessary to optimize the virtual economy's advantages while reducing its hazards. To increase confidence and sustainability in digital transactions, businesses need to invest in safe technologies, moral AI, and compliance frameworks. Global regulatory frameworks that support innovation while guaranteeing consumer safety and financial transparency must be established by policymakers. Investments in digital infrastructure and literacy can also reduce economic inequality and promote more egalitarian participation.

In the future, the distinction between the physical and digital economies will become even more hazy due to the integration of Web3, AI, and immersive technology. While those that oppose may find it difficult to stay relevant, those who adjust to this changing environment will have a competitive edge. In the end, the virtual economy is the next big thing in economic development, and it will take cooperation to fully and fairly utilize its potential.

Reliance on existing data, which may not reflect the latest real-time changes in digital markets. Possible bias in industry reports, as some may be influenced by corporate interests. Limited access to proprietary data, restricting insights into confidential financial or operational metrics. Ensuring credibility by selecting data from reputed sources. Avoiding misrepresentation or selective bias in interpreting findings. Properly citing all references to maintain academic integrity.

## REFERENCES


Acemoglu, D., & Restrepo, P. (2019). Automation and new tasks: How technology displaces and reinstates labor. *The Journal of Economic Perspectives*, *33*(2), 3–30. DOI: 10.1257/jep.33.2.3

Anand, A., & Singh, P. (2025). Interdisciplinary Studies: Issues, Perspectives, and Challenges From National Education Program 2020. *Integrating Personalized Learning Methods Into STEAM Education*, 233-260.

Arvidsson, A. (2020). Changemakers: The industrious future of the digital economy. *Polity*, *52*(4), 555–573. DOI: 10.1177/0032329220932426

Barlow, J. P. (2018). The economy of ideas: Selling wine without bottles on the global net. *Harvard Business Review*, *96*(3), 84–96.

Barzilai-Nahon, K. (2021). Toward a theory of network gatekeeping: A framework for exploring information control. *Journal of the American Society for Information Science and Technology*, *59*(9), 1493–1512. DOI: 10.1002/asi.20857

Benkler, Y. (2022). Peer production and cooperation. *The Journal of Economic Perspectives*, *36*(2), 69–92. DOI: 10.1257/jep.36.2.69

Brynjolfsson, E., & McAfee, A. (2014). *The second machine age: Work, progress, and prosperity in a time of brilliant technologies*. WW Norton & company.

Brynjolfsson, E., & McAfee, A. (2021). The business of artificial intelligence. *Harvard Business Review*, *99*(4), 60–73.

Castells, M. (2020). The rise of the network society: The information age - Economy, society, and culture. *Information Communication and Society*, *23*(5), 629–645. DOI: 10.1080/1369118X.2020.1718177

Castronova, E. (2008). Synthetic worlds: The business and culture of online games. In *Synthetic Worlds*. University of Chicago press.

Choudary, S. P. (2023). Platform scale: How an emerging business model helps startups build large empires with minimum investment. *Journal of Business Models*, *11*(1), 34–52.

Cohen, J. E. (2021). The regulatory state in the information age. *Theoretical Inquiries in Law*, *22*(1), 1–32. DOI: 10.1515/til-2019-0002

Cowen, T. (2020). The age of average: Why the digital economy rewards similarity. *Journal of Economic Behavior & Organization*, *175*, 357–372. DOI: 10.1016/j.jebo.2020.04.015

Daugherty, P. R., & Wilson, H. J. (2021). Human + machine: Reimagining work in the age of AI. *MIT Sloan Management Review*, *62*(3), 1–9.

David, P. A. (2020). The digital technology boomerang: New intellectual property rights threaten global "open science". *World Development*, *128*, 104829. DOI: 10.1016/j.worlddev.2019.104829

De Filippi, P., & Wright, A. (2022). Blockchain and the law: The rule of code. *Harvard Journal of Law & Technology*, *35*(1), 1–45.

Dolata, U. (2021). The digital economy: Promises, perils, and policy challenges. *The Review of Policy Research*, *38*(4), 403–420. DOI: 10.1111/ropr.12433

Einav, L., Levin, J., & Sundaresan, N. (2023). Economics in the age of big data. *Science*, *379*(6634), 658–660. DOI: 10.1126/science.ade7362 PMID: 25378629

Goldfarb, A., & Tucker, C. (2022). Digital economics. *Journal of Economic Literature*, *57*(1), 3–43. DOI: 10.1257/jel.20171452

Graham, M. (2020). Regulate, replicate, and resist - The conjunctural geographies of platform urbanism. *Urban Studies (Edinburgh, Scotland)*, *57*(7), 1397–1415. DOI: 10.1177/0042098019879693

Greenstein, S. (2021). How the internet became commercial: Innovation, privatization, and the birth of a new network. *Information Economics and Policy*, *54*, 100881. DOI: 10.1016/j.infoecopol.2020.100881

Hagiu, A., & Wright, J. (2023). The status of middlemen in the digital economy. *Journal of Economics & Management Strategy*, *32*(1), 5–24. DOI: 10.1111/jems.12515

Jain, P., & Singh, P. (2023). Lethargic approach in expediting financial reporting system: A case of construction firm. *International Journal of Teaching and Case Studies*, *14*(1), 1–10. DOI: 10.1504/IJTCS.2023.131663

Jain, P., Singh, P., & Kushwaha, R. (2025). Investigating change management as a novel digital wallet service innovation in banks: A case of Indian bank. *International Journal of Teaching and Case Studies*, *15*(1), 67–81. DOI: 10.1504/IJTCS.2025.143552

Kenney, M., & Zysman, J. (2023). The platform economy: Restructuring the space of capitalist accumulation. *Cambridge Journal of Regions, Economy and Society*, *16*(1), 55–76. DOI: 10.1093/cjres/rsaa001

Kumar, A., Singh, P., Parihar, A.S. & Saraswat, S. (2025), "Analyzing the impact of psychological capital on the entrepreneurial intentions of students through mediating role of individual expectancy using PLS-SEM", *International Journal of Social Economics*, Vol. ahead-of-print No. ahead-of-print. .DOI: 10.1108/IJSE-03-2024-0252

Kushwaha, J., Sharma, A., & Singh, P. (2023). b. Exploration and prioritization of enablers to organization work–family balance planning for working sole Indian mothers integrating fuzzy Delphi and AHP. *International Journal of Social Economics*, *50*(3), 398–418. DOI: 10.1108/IJSE-05-2022-0348

Kushwaha, J., Singh, P., & Kushwaha, R. (2024). Predicting working sole Indian mothers' satisfaction towards work–family balance integrating the Kano model and weighted average method. *International Journal of Social Economics*, *51*(6), 725–740. DOI: 10.1108/IJSE-02-2023-0074

Kushwaha, J., Singh, P., & Sharma, A. (2023). a, "Modelling the enablers of work-family balance among working single mothers integrating expert-mining and TISM-MICMAC". *Journal of Organizational Effectiveness: People and Performance*, *10*(4), 501–526. DOI: 10.1108/JOEPP-05-2022-0106

Lehdonvirta, V., & Ernkvist, M. (2011). Knowledge map of the virtual economy: Converting the virtual economy into development potential.

Lessig, L. (2022). The future of ideas: The fate of the commons in a connected world. *The Yale Law Journal*, *131*(5), 1289–1332.

Manyika, J., & Sneader, K. (2023). AI, automation, and the future of work: Ten things to solve for. *The McKinsey Quarterly*, *2*(1), 1–15.

Mazzucato, M. (2021). The value of everything: Making and taking in the global economy. *The Journal of Economic Perspectives*, *35*(4), 223–246. DOI: 10.1257/jep.35.4.223

Parker, G. G., Van Alstyne, M. W., & Choudary, S. P. (2023). Platform revolution: How networked markets are transforming the economy. *Strategic Management Journal*, *44*(3), 567–589. DOI: 10.1002/smj.3456

Prüfer, J., & Schottmüller, C. (2021). Competing with big data. *The Journal of Industrial Economics*, *69*(4), 967–1003. DOI: 10.1111/joie.12259

Ragnedda, M. (2022). The third digital divide: A Weberian approach to digital inequalities. *New Media & Society*, *24*(1), 28–47. DOI: 10.1177/1461444820986813

Saraswat, S. & Singh, P. (2025), "The moderating role of perceived trust in predicting the adoption intention of consumers toward online pharmacies using the application of UTAUT model", *International Journal of Pharmaceutical and Healthcare Marketing*, Vol. ahead-of-print No. ahead-of-print. DOI: 10.1108/IJPHM-05-2024-0042

Singh, A. K., Meshram, S. A., Khandelwal, V., Tiwari, P., & Singh, P. (2024c). Implementation of Industry-Based Perspective in Technical Education: A Qualitative and Quantitative Analysis. *Industrial and Manufacturing Designs: Quantitative and Qualitative Analysis*, 321-346. https://doi.org/.DOI: 10.1002/9781394212668.ch11

Singh, P. (2014). Unorganized sector and India's informal economy: challenges and prospects. *New Man International Journal of Multidisciplinary Studies, 1*(2), 209-218. https://www.newmanpublication.com/dash/issueworkfiles/71.pdf?1721465584

Singh, P. (2024). a. Weather index insurance viability in mitigation of climate change impact risk: A systematic review and future agenda. *Journal of Science and Technology Policy Management, 15*(1), 142–163. DOI: 10.1108/JSTPM-07-2021-0102

Singh, P. (2024b). Analyzing Responsible Leadership Effectiveness for Attainment of Sustainable Development Goals in Developing Countries Using Environmental Scanning Techniques. In: Bhattacharya, S., Venkatesh, V.G., Chatterjee, S.R. (eds) Responsible Corporate Leadership Towards Attainment of Sustainable Development Goals. Responsible Leadership and Sustainable Management. Springer, Singapore. https://doi.org/DOI: 10.1007/978-981-97-1386-8_13

Singh, P., & Agrawal, G. (2019). Efficacy of weather index insurance for mitigation of weather risks in agriculture: An integrative review. *International Journal of Ethics and Systems, 35*(4), 584–616. DOI: 10.1108/IJOES-09-2018-0132

Singh, P., & Agrawal, G. (2020). Development, present status and performance analysis of agriculture insurance schemes in India: Review of evidence. *International Journal of Social Economics, 47*(4), 461–481. DOI: 10.1108/IJSE-02-2019-0119

Singh, P., & Agrawal, G. (2021). Modelling the barriers of weather index insurance service adoption integrating expert mining and ISM Fuzzy-MICMAC. *Benchmarking: An International Journal, 29*(8), 2527-2554. . https://doi.org/.DOI: 10.1108/BIJ-04-2021-0183

Singh, P., & Agrawal, G. (2022). Mapping the customer centric weather index insurance service design using quality function deployment. *The TQM Journal, 34*(6), 1800–1822. DOI: 10.1108/TQM-08-2021-0236

Singh, P., Agrawal, G., & Kushwaha, R. (2021a, November) Evaluating Customer Satisfaction of Weather Index Insurance Service Quality Using Kano Model. In *2021 IEEE Bombay Section Signature Conference (IBSSC)* (pp. 1-6). IEEE. **DOI:** DOI: 10.1109/IBSSC53889.2021.9673370

Singh, P. & Kushwaha, R. (2024), "Envisaging Indian farmers' desires from agricultural index insurance integrating rank sum weighting method and MoSCoW technique: an approach to requirements prioritization", International Journal of Social Economics, Vol. ahead-of-print No. ahead-of-print. .DOI: 10.1108/IJSE-03-2024-0233

Singh, P., & Kushwaha, R. (2025), "Conceptualizing the corporate-social accountability and sustainability assessment of weather-risk management policy towards stakeholders using qualitative insights: a case of weather index insurance", *Qualitative Research in Financial Markets*, Vol. ahead-of-print No. ahead-of-print. DOI: 10.1108/QRFM-05-2024-0113

Singh, P., Kushwaha, R., & Kushwaha, J. (2021b). Assessing visitors' satisfaction about museum service quality using Kano model. In *2021 IEEE 9th Region 10 Humanitarian Technology Conference (R10-HTC)* (pp. 1-6). IEEE. DOI: DOI: 10.1109/R10-HTC53172.2021.9641707

Singh, P., Kushwaha, R., & Kushwaha, J. (2023a). Analyzing Accountability of Weather Index Insurance Service in Attainment of Sustainable Development Goals: A Sustainable Accounting Perspective Review. In Pratik Arte et al. (Eds): *Sustainable International Business, 2023*. Springer Nature, 2023. doi: , 531340. DOI: 978-3-031-43784-7

Singh, P., Kushwaha, R., & Kushwaha, J. (2023b). Evaluating Farmer Participation in Weather Index Insurance in a Socio-Demographic Perspective: Evidence From the Crop Insurance Data of India. In A. Popoola, H. Magidimisha-Chipungu, & L. Chipungu (Eds.), *Handbook of Research on Managing the Urban-Rural Divide Through an Inclusive Framework* (pp. 69-88). IGI Global. DOI: 10.4018/978-1-6684-6258-4.ch005

Singh, P., Kushwaha, R., & Kushwaha, J. (2024a). Foresight the future of Industry 5.0 in perspective of sustainable services in developing countries. *Modern Technologies and Tools Supporting the Development of Industry 5.0*, 381. DOI: .DOI: 10.1201/9781003489269-11

Singh, P., Kushwaha, R., & Kushwaha, J. (2025b). Retail Innovation's Impact on Competitiveness in the Retail Sector: Evidence From SAP-LAP Analysis. In *Retail Innovations in Business Models* (pp. 255-278). IGI Global Scientific Publishing.

Singh, P., Kushwaha, R., & Kushwaha, J. (2025a), “A Qualitative Study on Analysing the Role of Mimicking Technology in Electronic Customer Relationship Management: A Case of Service Sector”, Kumar, A., Ciddikie, M.D., Kashyap, A.K. and Akram, H.W. (Ed.) *Marketing 5.0*, Emerald Publishing Limited, Leeds, pp. 103-119. DOI: 10.1108/978-1-83797-815-120251008

Singh, P., Kushwaha, R., Kushwaha, J., & Singh, A. K. (2024b). A Qualitative Study to Rank Non-Conventional Energy Sources for Industrial Sustainability and Energy Management Decisions Using MoSCoW Prioritization Method. *Industrial and Manufacturing Designs: Quantitative and Qualitative Analysis*, 105-137. https://doi.org/DOI: 10.1002/9781394212668.ch4

Sundararajan, A. (2022). The sharing economy: The end of employment and the rise of crowd-based capitalism. *The Journal of Economic Perspectives*, *36*(2), 93–112. DOI: 10.1257/jep.36.2.93

Tapscott, D., & Tapscott, A. (2016). *Die Blockchain-Revolution: Wie die Technologie hinter Bitcoin nicht nur das Finanzsystem, sondern die ganze Welt verändert.* Plassen Verlag.

Zuboff, S. (2023). The age of surveillance capitalism: The fight for a human future at the new frontier of power. *Business Ethics Quarterly*, *33*(1), 156–182. DOI: 10.1017/beq.2022.41

# Chapter 10
# Integrating Technology Across Sectors:
## Innovations in Public Administration, Engineering, and Business

**Navin Kumar**
https://orcid.org/0000-0002-3856-0607
*Bharat College of Law, India*

**Smriti Tandon Gupta**
https://orcid.org/0000-0003-4999-5881
*Graphic Era University, India*

## ABSTRACT

*This chapter explores the transformative impact of technology integration across public administration, engineering, and business sectors. It highlights how advancements such as e-Government platforms, smart cities, and blockchain are reshaping governance and service delivery, enhancing transparency and efficiency. In engineering, digital tools, AI, and automation are revolutionizing infrastructure and manufacturing, driving sustainability and operational optimization. The business sector benefits from innovations like big data analytics, cloud computing, and FinTech, improving decision-making and customer engagement. The convergence of these technologies presents opportunities for cross-sector collaboration, addressing global challenges such as climate change and urbanization. However, this integration also brings challenges related to cybersecurity, data privacy, and the need for a skilled workforce. The chapter emphasizes the importance of responsible and inclusive technology management to harness its full potential for societal advancement.*

DOI: 10.4018/979-8-3373-1534-8.ch010

## INTRODUCTION

The pace at which technology is evolving in contemporary life has emerged as a key driver of innovation and increased efficiency in major sectors. The power of transformation in technology is seen in public administration, engineering, and commerce, with each enjoying benefits from improved processes and outcomes. Yet the innovations also come with unique challenges that necessitate solutions. This section delves into innovative technological applications in these industries, identifying opportunities and challenges they offer. Technology in the public administration sector, in the form of e-governance initiatives, has reformed the delivery of services with digital portals that streamline transactions and enhance transparency. But issues of privacy, security, and digital disparity need to be solved to guarantee equitable access and safeguard sensitive information. In engineering, advances such as computer-aided design (CAD), building information modeling (BIM), and automation have transformed project management and construction, improving accuracy and efficiency of operations. Emerging technologies like AI, IoT, and robotics have the potential to fuel sustainable development in infrastructure but demand ongoing skill enhancement and adaption to evolving practices.

In business, the implementation of digital marketing, e-commerce, and data analytics has revolutionized customer interactions and operational models. AI and machine learning assist companies in processing big data to forecast consumer actions and enhance market positioning. Nevertheless, companies need to navigate issues such as cybersecurity threats, digital skills requirements, and ethical issues around the use of data. The convergence of public administration, engineering, and business carries great promise for developing end-to-end solutions, especially in sectors such as green urban development. Yet, technological development needs to be adopted cautiously, with regard to the social and moral implications, so that equitable and ethical progress is made.

### Objective of the study

The objective of this study is to examine the change-making influence of technological innovations in the public administration, engineering, and business domains. Through an analysis of how these sectors incorporate cutting-edge technologies, the research hopes to identify their contribution to systemic efficiency, sustainability, and adaptive governance. In a fast-changing digital world, the research hopes to find out how convergence of technologies across sectors can facilitate creative innovations, enhance resource utilization, and enhance service delivery, leading ultimately to more responsive and robust systems that respond to the needs of modern society and the challenges of the international community.

## Research questions

1) How does the combination of e-governance and AI revolutionize transparency and citizen participation in public administration?
2) What are the effects of IoT and BIM implementation on engineering sustainability and infrastructure construction?
3) How do new digital technologies such as blockchain and big data analytics reshape business models and operating strategies?
4) What are ethical and cybersecurity issues in integrating different sectors with technology?
5) How can public-private collaborations promote innovation and mitigate differences in technological access and deployment?

## Innovations in Public Administration

### The Role of Technology in Public Administration

The core governance framework includes public administration as it manages policy implementation and provision of resources to citizens. Public service delivery needs increased significance due to the fact that society changes while public need changes during times. Emerging technological advancements make significant changes in public institution functioning through engendering transparency and accountability as well as boosting participatory effect from citizens (Scholl, H.et al 2014). Public administration can be more effective along with increased responsiveness by using e-governance with data analysis and artificial intelligence (AI). Further, technology does not just make operations more efficient; it fosters settings prioritizing transparency and public accountability. Current citizens are able to view government budget information and decision-making processes online more readily than ever. People's trust in governmental relations and civic participation increases when citizens retain access to governmental process. Public officials are held to greater accountability when citizens are well-informed, hence improving governance results.

### E-Governance: Concepts and Evolution

In e-governance, computer technology links sites to improve administration of government services. The web-based system of delivering services is a revolutionary way through which individuals are able to access various forms of information and services without physically visiting various government offices (Siau, K., & Long, Y. 2018). Implementation of e-governance has resulted in great changes across other countries, where they have established web-based resources for filling taxation, portal-based license renewals, and applications for public services. Incorporating technology into government functions results in more streamlined operations, high efficiency, and improved citizen service, which generate a more responsive, accountable public administration.

## The Emergence of E-Governance

Public authorities came up with the e-governance idea towards the close of the 20th century by learning about the potential of the Internet in terms of improving the delivery of services. Under its initial development stage, e-governance initiatives provided government policies, laws, and departmental contact information via electronic means (Choudhury, P., & Kauffman, R. J. 2019). Governments are still evolving serious online services whereby citizens are able to interact with public agencies via real-time interfaces. Nation-states have modern websites and mobile applications that allow citizens to do transactions, file forms, and get responses on their applications avoiding lines. Such platforms have additional features like online paying applications, appointment scheduling or in-box, and chat to access all required concerning e-governance: Examples of e-governance advantages and benefits obtained from digital government operations (Ghosh, R., & Scott, J. 2018). With effective simplification, e-governance designs streamlined bureaucratic procedures that reduce the delivery time for individuals within and outside the government. In the classic bureaucracy, citizens have to go through a lengthy process involving several physical office appearances to fill in paperwork. With e-governance, citizens can make online transactions that do not require them to physically attend document-based procedures. E-governance provides equal access to citizens of various geographical locations and underprivileged areas. Online services by governments create avenues of access that enable citizens to receive vital services irrespective of location (Kauffman, R. J., & Walden, E. A. 2018). The presence of large rural areas necessitates such actions because transport infrastructure is negligible in these areas. By means of e-governance, citizens are provided direct access to government processes from their residences. An electronic infrastructure becomes financially efficient for governments at different levels by necessitating low physical trips to institutions by citizens. This process would ease the operational load of buildings and public institution employees. Process automation also lowers administrative expenses, which can be redirected towards other significant sectors such as education, healthcare, and public safety (Barlow, J., & Maull, R. 2008).

## Challenges in Implementation of E-Governance

Implementation of e-governance encounters different implementation challenges. E-governance faces three major obstacles: digital capability standards, global internet connectivity, and deprivation of positive digital opportunities. Accurate online service utilization would only occur with a limited percentage of citizens, so governments should strive to guarantee basic skills and resources across the population for effective service consumption. The lack of ability to gain necessary skills or gain necessary resources hinders seniors, low-income individuals, and rural residents from being able to access any e-governance service due to low internet access (Damanpour, F., & Schneider, M. 2009). The government must offer education through training workshops, workshops, and other support network abilities to enable citizens to gain expertise in operating digital platforms. Event agencies need to develop blended online platforms to extend services to citizens who do not have access to the Internet or prefer face-to-face services. Citizens require robust cybersecurity systems that safeguard digital information and restore trust in electronic platforms. Government authorities collecting vast amounts of citizen data are vulnerable to constant cyber-attacks. The key necessity for the success of e-governance is based on secure online services that secure citizen data and create trust in digital platforms. Investment in advanced cybersecurity technology and regular audits supported by particular data protection norms are imperative actions governments have to undertake (Kauffman, R. J., & Riggins, F. J. 2012).

## LITERATURE REVIEW

Technology integration across industries has been universally recognized as a power driver for changing efficiency, transparency, and innovation. Researchers such as (Janssen and Estevez, 2013) point out that e-governance, aided by artificial intelligence, encourages data-based decision-making and transparency in government institutions. Their study identifies how digital governance enables greater citizen participation and administrative responsiveness. In engineering, utilization of the Internet of Things (IoT) and Building Information Modeling (BIM) has demonstrated improved sustainability and operational performance. As per (Eastman et al. 2011), BIM improves the visualization, simulation, and coordination of the construction process, greatly minimizing errors and expenses. In the same way, (Gubbi et al. 2013) discuss how IoT enables real-time monitoring and predictive maintenance of infrastructure and generates long-term resilience and efficiency. Business digital transformation has also witnessed a tremendous change with the emergence of block-chain and big data analytics. (Tapscott and Williams, 2010) argue that blockchain introduces unprecedented levels of trust and transparency in business transactions, while (Chen et al. 2019) demonstrate how big data analytics supports personalized marketing, strategic planning, and customer satisfaction.

Despite the benefits, concerns around data privacy, algorithmic bias, and digital inequality persist. (Eubanks 2018) highlights the risks of automated decision-making systems in public welfare, calling for ethical standards and regulatory oversight. Consequently, researchers support inclusive digital policy and enhanced cybersecurity policies. Lastly, public-private partnerships (PPPs) are becoming fundamental facilitators of technological innovation. PPPs can fill gaps in resources and expertise, enhancing scalable and innovative responses to emergent issues, according to (Hodge and Greve 2007). Generally, the existing literature emphasizes that effective integration of technology is a function not only of technological instruments but also of governance, ethics, and collaboration.

## Data Analytics in Public Decision-Making

Data analytics has become a strong, evidence-based operational tool for decision-making by public administrators. Large data analyses allow governments to make informed decisions about citizen needs while maximizing resources, resulting in improved public service delivery systems. With proper use of data, public organizations are able to learn patterns, measure the impacts of policy and project future challenges (Sarker, S., & Sahay, S. 2004). One of them has been the presence of predictive analytics in public administration: With analysis of historical data and processing of statistical algorithms, government agencies come up with forecasts that inform trends in key areas such as public health, crime, and traffic flow. Predictive analytics software reflects the impending threat of certain diseases by making predictions that enable health officials to utilize needed resources at the brink of the outbreak when it is about to reach a critical point. Through processing of data from the police force, law enforcers forecast dangerous regions, enabling them to utilize their resources to maximum public safety effect (Van der Meer, F. 2017). Data analytics provides the ability to select evidence-based decisions through factual data rather than constructing decisions from assumptions. When practices adopt this process, they deliver higher quality governance outcomes with evident accountability because decision-makers are held accountable for patient outcomes. Data analytics implementation requires organizations to commit effort to quality data upkeep, privacy protection, and ethics conduct management. To safeguard people's rights and public confidence, governments have to construct accurate regulations for collecting data and establishing how it will be warehoused and used (Wirtz, B. W., & Göttel, V. 2016).

## Artificial Intelligence and Automation in Governance

Using artificial intelligence and automation increases public administration's effectiveness and erases human errors in operations. Current technologies are serving multiple functions in process optimization, adding to the quality of services and efficiency of resources. Most government services currently employ chat bots as virtual helpers that answer citizen questions through instant response systems to handle typical questions while supporting users at procedure steps. Consequently, the experience for the user improves as public servants handle tasks limited to human-level judgment and knowledge. Utilizing machine learning algorithms is essential in executing public administrative activities, fraud detection, and risk assessments (Zeng, M., & Glaister, K. W. 2018). The algorithms analyze patterns of data to find fraudulent activities so government agencies can act against such risks in anticipation. The automation systems assist organizations in making decisions through data-driven recommendations. End. Implementation of artificial intelligence with automated processes generates high rewards but has particular implementation barriers. AI implementation is confronted by unemployment replacement effects and issues of ethical limitations and methodological opacity of AI decision systems. Operational personnel within public agencies require adequate training and guidance when applying AI technologies in order to adjust smoothly to automated systems and acquire new duties (Albury, D. 2005).

## Key Challenges in Public Administration

For public administration to realize its technological advantages to the fullest, organizations need to navigate a myriad of integration challenges. The biggest issue that arises today is privacy of data. State agencies hold massive individual data, which increases the likelihood of security breaches and misuse of data considerably. In order to ensure citizens' data protection, public institutions have to enhance security measures by implementing robust cybersecurity and strict privacy regulations. Digital access inequities pose a major hurdle since they divide societies into those who have digital capabilities and those who have nothing more than print media and low-technology devices. Neither marginalized groups nor communities receive access to internet services since their resource shortage renders it infeasible for them to learn the relevant skills. Thus, this strategy worsens their conditions. Government response requires three things to fix this issue: internet access infrastructure, digital literacy initiatives for marginalized groups, and funds to support both initiatives (Mancero Arias, G. 2022). Bureaucratic organization offers resistance to innovative technological methods, which prevents its acceptance. The official public administration procedures establish hurdles to technological implementation due to the fact that they oppose employee mind set changes and behavioural changes. Public agencies must develop organizational cultures that facilitate innovation in addition to the adoption of emerging technology and new techniques by employees.

## The Future of E-Governance

Various innovative solutions signal that e-governance will grow in promising new directions to develop public administration systems. Citizen interactions by the public sector are destined for a major overhaul through nascent technology platforms like Artificial Intelligence, blockchain and IoT. AI data processing accurately forecasts the needs of citizens, which helps governments to provide customized services more effectively. Block chain technology enhances security in transactions by bringing real-time control measures to public service systems (Parycek, P., & Scholl, H. J. 2013). Tour government's digitalization of systems provides scope for individuals to become engaged constructively in democratic processes. E-governance is a platform for users facilitating citizens to communicate their feedback and active participation in decision-making. Through channels of citizen feedback, governments construct frameworks of inclusive governance that support public engagement.

## Innovations in Engineering

Engineering has long been leading the pace of technological change, constantly being refined to become more efficient in its operations and come up with innovative solutions to complicated issues. Contemporary technological advancements such as Building Information Modeling (BIM), the Internet of Things (IoT), and 3D printing have greatly changed the manner in which engineers design, build, and coordinate projects. These technologies are transforming the field of engineering by making processes more streamlined, collaboration better, and more sustainable and efficient practices possible across the industry (Salter & Gann, 2003).

## Technology's Role in Modern Engineering

Technology has revolutionized the work of engineers, altering the ways engineers undertake problem-solving and project implementation. The use of digital tools and approaches has made it possible for them to have increased precision, effectiveness, and cooperation among stakeholders. With increasingly complex projects, engineers depend more on technology to accomplish tight deadlines and budget requirements and maintain high-quality deliverables. Digital platforms stimulate real-time communication and data sharing, reducing errors and improving project implementation. Engineers currently use artificial intelligence (AI) and data analytics to make decisions, recognize patterns, and anticipate project outcomes, ultimately streamlining designs and increasing sustainability (Cropley, 2015). These technologies allow engineers to produce enhanced solutions while enabling more sustainable engineering practices.

## Building Information Modeling (BIM)

BIM is revolutionary technology in construction and engineering. By creating detailed digital models of buildings, BIM allows engineers and architects to collaborate more effectively, visualizing the entire project in 3D and detecting potential issues before construction begins. This technology enhances project management by improving coordination and reducing errors. BIM provides valuable insights into environmental impact, energy efficiency, and structural integrity, allowing teams to optimize designs early in the planning process. Its capacity to monitor progress and materials in real-time makes sure that projects remain on track and on budget while reducing waste and enhancing sustainability (Whitford & Meier, 2013).

## The Internet of Things (IoT) in Engineering

The IoT has revolutionized engineering by linking different devices and sensors that observe and control infrastructure. Engineers apply the use of IoT technology to constantly evaluate structures' health levels and detect problems like moisture, temperature variations, and vibrations. These sensors produce real-time data that assists in the early detection of possible issues, enabling proactive maintenance and prevention of expensive repairs. IoT-based systems also improve safety by tracking construction sites and the conditions of workers, minimizing accidents. Additionally, IoT helps in the creation of smart cities by streamlining resource management, energy consumption, and public services, leading to sustainable cities (Massey, 2019).

## The Impact of 3D Printing on Engineering Solutions

3D printing, also known as additive manufacturing, is transforming the design and manufacturing of engineering products. 3D printing allows quick prototyping and the production of intricate shapes that other manufacturing technologies cannot produce. 3D printing minimizes material wastage by creating products layer by layer and speeds up the process of developing products. Engineers are able to test and improve their designs quickly, compressing delivery times for products. This technology is also revolutionizing the construction sector, with big 3D printing utilized to build complete structures made of concrete and other materials. With the capability to print personalized components as needed, inventory and shipping are no longer necessary, saving costs and enhancing project efficiency (Gopinathan & Noh, 2018).

## Challenges in Engineering Innovation

Despite the advantages, the implementation of new technologies in engineering has multiple challenges. High expense, the absence of qualified human capital, and intellectual property concerns are the main obstacles to innovation. The accelerated rate of technological progress has exceeded the ability of training institutions to equip the country with the required expertise, resulting in a skill shortage in emerging areas like AI, data science, and nanotechnology (Desmarchelier et al., 2019). In addition, the reluctance of the engineering industry to adapt can decelerate technological implementation, as companies stand to incur high initial investments and alterations to current processes. Moreover, intellectual property issues, including ownership of data and patent rights, can trigger conflict among interested parties and hinder cooperation. Last but not least, environmental sustainability is still an issue of concern, where engineers must embrace more sustainable strategies to lower ecological footprints while still addressing increasing infrastructure needs (Matei & Savulescu, 2014).

## Innovations in Business

### The Role of Technology in Business Transformation

The application of technology in contemporary business has resulted in a paradigm shift in the way organizations function, communicate, and interact with customers. Technological progress has ensured remarkable boosts in operational effectiveness, productivity, and customer satisfaction. Advanced technologies help firms respond to market changes, changing customer needs, and streamline in-house processes (Ongaro, Gong, & Jing, 2021). Efficiency is one of the most important effects that technology has on business operations. Automated tools and software applications simplify processes and cut down on human mistakes, enabling employees to concentrate on more valuable undertakings. Enterprise Resource Planning (ERP) software combines different functions like finance, HR, and supply chain management to form an integrated platform that enhances decision-making and coordination of operations. Moreover, technology has also redefined the customer experience by delivering more personalized interactions. Through data analytics, organizations are able to comprehend consumer behavior, predict demands, and devise customized products and services. Organizations now use customer data to create customized experiences, enhance product offerings, and interact with customers across multiple channels, increasing customer satisfaction (Mitra, 2023). With the era of the digital age, organizations can provide global services without having a physical store, which supports new business models such as subscription services, sharing economy platforms, and digital marketplaces.

## E-Commerce and Digital Transformation

E-commerce has transformed the way that businesses connect with customers, changing product promotion, sales, and distribution processes. With online payment services, businesses are able to connect to the entire world, selling their products and services across countries (Park et al., 2022). Online stores enable businesses to showcase products, track orders, and process safe payments, making it easier for consumers to shop. The convenience of internet shopping has stimulated explosive expansion of e-commerce, and analysts foresee this trend to persist. Efficient logistics management and supply chain monitoring are crucial aspects of e-commerce achievement, providing timely delivery and customer satisfaction (Kondratenko et al., 2020). In addition, digital transformation includes customer relationship management (CRM), whereby companies use information to understand customer tastes, improve marketing efforts, and foster long-term relationships with customers.

## Data-Driven Business Decision-Making

Data analytics forms the core of contemporary business strategies since data allows organizations to make data-driven decisions founded on market trends and consumer patterns. Businesses today gather enormous volumes of data from different channels, including sales history, customer feedback, and social media trends, to determine areas of improvement and expansion. Predictive analytics enables companies to forecast trends in the future and thereby lead the market (Olcay, 2021). Data-driven decision-making is also made in marketing strategies, where companies utilize customer segmentation to send out customized messages, enhancing interaction and conversion rates. By constantly tracking key performance indicators (KPIs) and refining plans according to real-time information, businesses are able to stay nimble and responsive to changing market conditions.

## Artificial Intelligence in Business Strategy

Artificial Intelligence (AI) is facilitating change throughout business functions, from customer service automation to supply chain optimization. AI technologies such as machine learning, natural language processing, and computer vision allow businesses to perform tasks that were previously done by humans. Chatbots powered by artificial intelligence are increasingly employed for customer service, enhancing the response time and lowering operational expenses (Damanpour, 2017). Moreover, artificial intelligence systems assist organizations in streamlining inventory management through past sales and future demand prediction by analyzing historical sales data. This avoids stockouts and overstocking instances, enhancing supply chain effectiveness. AI also supports personalization of the customer, whereby machine learning algorithms recommend products based on individual choice, enhancing the sales and loyalty of the customer (Heeks, 2006).

### Key Challenges in Business Innovation

Despite the many advantages brought by technology, businesses encounter challenges in adopting and implementing it. One of the major concerns is cybersecurity, with businesses depending more on digital infrastructure and off-site data storage, and hence being targets of cyberattacks (Sinni, 2017). Safeguarding customer information and corporate data is essential, and businesses need to spend money on effective cybersecurity solutions to offset risks. In addition, the accelerated rate of technological innovation necessitates companies to innovate continuously, necessitating a sizeable investment in research and development. This failure may result in loss of competitive edge. Automation and AI also necessitate adaptation regarding job replacement, necessitating companies to focus on reskilling and upskilling workers (Gupta et al., 2019). Ethical issues of data privacy, algorithmic prejudice, and transparency also raise challenges for organizations embracing AI and data analytics that need to be carefully managed to achieve responsible and transparent use of technology.

## Intersections and Cross-Sector Collaborations

### The Importance of Cross-Sector Collaboration

Ecological interdependence in society generates complicated and multidimensional issues which cross sectoral boundaries. The convergent response to climate change, public health emergencies, and urban expansion and economic disparities requires all-of-government collaboration between public administration, engineering, and business assets. Solutions development in response to pressing societal needs requires purposive collaboration between the public administration and engineering and business disciplines. Technology provides the underpinning support in bringing disparate industries together for collaboration. Public administration is able to draw engineering skills to develop efficient and sustainable infrastructure by utilizing design and implementation processes (Rocha & Zavale, 2021). The collaboration between the private sector and public institutions generates technological innovation alongside market sensing, resulting in better universal services. New innovative technology occurs through collaborative industry efforts to achieve higher economic growth along with higher quality of life solutions for the population. Pluralism in problem-solving strategies propels the primary benefit of integrating various sectors. Various organizational attributes and their essential principles and competencies facilitate more efficient and practical solutions. Public administrators solving urban transportation issues are capable of interpreting current regulations to society, engineers develop effective transport systems, and companies devise mobile solutions for immediate transit information. By this collaborative strategy, various sectors obtain innovative outcomes and preserve solutions that adapt to actual conditions (Feller, J. et al 2011). The collaborative efforts between various public sectors help develop more efficient organization in resource distribution. Sharing resources, such as expertise and knowledge, allows sectors to avoid repeating efforts as they collaborate to attain greatest impact from their efforts. Obesity prevention campaigns in public health are enhanced by active collaboration among food stakeholders, educational institutions, and health departments. They are able to develop comprehensive programs that simultaneously address the issue from several fronts, like healthy school meals, community facilities for exercise, and policy-driven encouragement of healthier foods in the market.

## Smart Cities: A Case Study in Integration

Smart cities offer a compelling research that illustrates how public administration is bounded by engineering and business to solve urban issues using new technology. Cities employing advanced technologies and data analytics systems strive to create smarter urban configurations for citizens and provide improved public services while creating sustainable urban communities. Technology facilitates integration into urban land development frameworks and physical infrastructures with service delivery platforms to enhance residents' quality of life by making major improvements. IoT is a key element since it enables sensor devices to send time-critical information during real-time deployments. Urban city planners can optimize traffic movement and reduce congestion levels through IoT sensors observing traffic movements (Attour, A., & Chaupain-Guillot, S. 2020). The data-driven approach propels better decision-making and optimizes transportation efficiency while reducing the environmental impact.

Transportation infrastructure in cities becomes more efficient at work when the green lights of traffic lights adjust their timing order according to real-time traffic data. Modern city developments rely significantly on analysis through data science approaches. Information from various sources provides city authorities with the ability to make more informed decisions through better comprehension of citizens' needs and tastes, hence more opportunities for efficient and responsive public service. The data analysis facilitates public transport planning operations. The city authorities obtain insights into peak traffic hours and busy routes, which allow them to allocate resources more effectively and provide enhanced service reliability (Mergel, I. 2016). Predictive analytics makes projections to determine upcoming urban challenges, i.e., population increase along with infrastructure decay, so cities can implement readiness strategies and careful investments before they become problems. The defining quality of smart cities is necessary engagement and participation by citizens. A technological platform connecting citizens with government officials raises resident satisfaction and local policymaking participation. Under mobile applications, citizens are able to report malfunctioning streetlights and bad road conditions to authorities in person.

Mobile app features will allow residents to create individual reports regarding streetlight failure and potholed roads. With these innovations, Better city responsiveness and improved delivery of services are possible. Urban dwellers will be able to conduct self-initiated change in their urban surroundings through technological innovations (Pakhnenko, O., & Kuan, Z. 2023). The creation of smart cities necessitates public-private partnerships because these encourage technological innovation as they propel investment in technology systems. Public institutions acquire expertise knowledge, funding possibilities, and precious resources to implement innovative solutions through public-private collaborations. Collaborations between technology companies and public authorities facilitate the development of innovative waste management systems integrated with sensor-based technology for improved composition control and route optimization, which lowers operation costs. Together, these actors increase public service operation efficiency and stimulate economic growth through technology investment creating jobs. Smart cities illustrate how multi stakeholder collaboration can deliver strong solutions to urban challenges. Through technological integration for urban planning and public service delivery, cities are able to enhance the quality of life for residents in an economically prosperous and sustainable context. Effective execution of innovative city programs hinges on collaboration among public administrators, engineering professionals, and business executives. Through this strategy, the holistic nature of development strategies becomes evident (Attour, A., & Chaupain-Guillot, S. 2020).

## Public-Private Partnerships for Innovation

Public-private partnerships (PPPs) are fundamental mechanisms that drive technological investment in all sectors. The mechanisms allow partners to combine shared standard services exp, expertise, and risks into resource-sharing platforms that facilitate the establishment of innovative processes for societal issues. The PPPs are instrumental in technology integration because they facilitate improved public service delivery and drive regional socioeconomic development. Public-private partnerships work by bridging public agencies with private sector skill and capital. Public institutions in most cases function below the level of finance they need and lack the expertise needed to apply sophisticated complex solutions. Public agencies are able to realize the application of these technology solutions by engaging in joint venture with private sector companies to leverage private capabilities. For realization objectives, a city wishing to implement a cutting-edge grid energy management system can enter into an agreement with an energy solution technology firm (Frikha, T. et al 2023). As an organization partner, the city gains entry to its expertise at a mutual cost. Through PPP arrangements, governments are able to obtain.

Equipment to build required infrastructure when state-budget funds are inadequate for such projects. By working in partnership with private investors, public entities tap into various funding streams to create large-scale projects that enhance public infrastructure.

The public transportation agency can make transit lines a reality by partnering with private sector financing partners who put up money and accept revenue-sharing agreements. By this model, public services enjoy quicker delivery with ongoing improvement. The economic benefits of PPPs are a key motive, but the potential for triggering innovation through collaborative knowledge methods cannot be underestimated. Collective interfaces between public and private agencies lead to sharing of ideas and best practices and create novel solutions to manage complex societal problems. A local authority collaborates with a technology startup to build solutions for city mobility. The joint work will lead to the development of a real-time mobile system that displays public transport options required to facilitate citizens' mobility across the city. The synergistic working style energizes the creative process, which yields more robust problem-solving results. Public-private collaborations enhance the delivery of public services through the development of more levels of accountability along with increased transparency (Davis, P. W., & McKenzie-Rundle, P. 1984). Involving private partners in the provision of public services enables governments to set performance standards that bind public and private parties to account for their performance. Better outcomes in service delivery come through higher transparency by government institutions due to greater responsibility between parties. Public and private waste management alliances enable authorities to make performance-based contracts that encourage their contractors to meet specified service quality specifications, thus improving resident well-being.

Effective public-private partnership requires both strategic planning and sustainability management strategies. Open communication, well-delineated responsibilities, and mutual clarity of goals are instrumental in fostering collaboration between public and private partners so that they are aligned on goals (Scholl, H. J., & Scholl, M. C. 2014). Public trust should also be sustained through openness in the procurement process and constant monitoring of project performance to ensure that partnerships yield the desired benefits.

## Future Directions and Policy Implications

### Embracing Emerging Technologies

In this age of unprecedented technological progress, various stakeholders, including public administration, engineering, and commerce, need to be adaptive and adopt innovation in the system. Blockchain, AR, quantum computing, and various other technologies are not buzzwords but changing phenomena that can redefine operational models, bring efficiency, and enhance service delivery.

### Blockchain Technology

Bitcoin and other cryptocurrencies run on the main platform offered by blockchain technology. The advantages of this technology extend far beyond the financial sector. This decentralized system has complete transparency, secure storage, and total immutability of record books (Bablani, A., & Tripathi, D. 2018). This technology offers unparalleled characteristics appropriate for government governance. Such procedures, such as the registration of land titles, may become quicker while voting procedures and requirements of managing public records also become available using block chain technology. Blockchain technology provides unverifiable ledgers that enhance public trust in all government processes with limited corruption. Block chain technology allows engineering firms to implement supply chain management by tracking products in real time. Apart from increased accountability and effectiveness, transparency in block chain makes it possible for stakeholders to verify authenticity for products. Smart contracts and block chain technology make it possible for companies to do away with middlemen since automatic execution of transactions takes place once conditions set in advance are met using this mechanism (Siau, K., & Long, Y. 2018).

### Augmented Reality (AR)

Revolutionary Augmented Reality technology integrates digital components with physical surroundings to generate enhanced human engagement in our surroundings. AR facilitates a natural transition from digital and real worlds by using computer-generated aspects, which superimpose images, audio, and data onto living spaces.

The visual and interactive material added by Augmented Reality overlays with the real world rather than providing an isolated virtual reality simulated environment. Technology has rapidly emerged as industry-shaping power in the areas of entertainment, healthcare, education, and retail as it transforms user experiences (Ghosh, R., & Scott, J. 2018). Entertainment and gaming is the major domain where AR has its most powerful applications. In Pokémon and games like that, the AR technology shows how it creates engaging user interaction through its interactive gameplay mechanics. The games are able to bring together digital characters and pieces of scenery with real life surroundings. Virtual reality technology offers consumers tailored entertainment by connecting digital content from one physical space to another. AR technology is advancing exponentially in the film and television sector. Filmmakers are able to preview special effects and virtual sets during production using this technology. The procedure enhances storytelling presentation and cuts production timelines and costs by providing simultaneous visualization and rapid editing functionality (Choudhury, P., & Kauffman, R. J. 2019).

The healthcare industry gains greatly from AR technology for consumers, such as physicians employed in healthcare services. Operating room personnel gain from AR since it allows them to visualize complex anatomical structures as they see critical procedural information overlaid onto real patient bodies. Surgeries find more accuracy and lower risk factors due to this method. Students in medicine can utilize augmented reality in order to receive training tools that generate simulated clinical conditions, offering secure procedure practice environments. AR applications are designed by medical developers that give interactive visualizations to patients to facilitate better comprehension of treatment and diagnosis. The system brings more effective communication results between medical professionals and patients. Business and academic practices both see significant transformative impacts with the implementation of AR. Education will be enhanced with AR since it brings dynamic methods of providing more engaging learning experiences., Students are exposed to historical events via AR applications since interactive 3D modeling and animation technology provide engaging presentations of ancient civilizations. Science education is more effective with AR technology since it brings challenging scientific topics such as molecule structures and human body functions in comprehensible visual forms (Kauffman, R. J., & Walden, E. A. 2018). Adopting educational uses with real-time interactive components aids in learning for students as they become more inquisitive and hone creative thinking abilities.

Application of AR technology transforms the shopping experience by providing improved options to customers who shop. Consumers can utilize AR apps for virtual try-ons of clothes, demonstrations of accessories, and virtual makeup trials at home without the need to go out. Budget furniture store IKEA uses AR technology, allowing customers to visualize how items look and work inside their homes prior to purchasing. The ambiguity that usually accompanies online purchasing becomes less with this method, which results in better customer satisfaction. With innovative in-store experiences facilitated by AR technology, customers are advantaged by easy navigation features and interactive visual content. AR technology continues to play a critical role in operational processes in the manufacturing and industrial industries (Mazzucato, M. 2013). Augmented reality allows employees to perform complex operations via virtual interface attachments that walk them through equipment or machinery operations to improve accuracy and accelerate their execution. Technicians learn real-time data coupled with step-by-step maintenance guidelines through AR tools during repair activities, making their work easier.

Automotive: Design work at the frontline is made feasible through real-time prototype view modification. Accelerating development cycles allows more individuals to innovate while teams deliver improved outcomes when collaborating. The full potential of AR is hindered by barriers that require solutions before maximal advantages can be attained (Parycek, P., & Scholl, H. J. 2013). The development and implementation of AR solutions are confronted by mounting expenses as a primary challenge. The worldwide deployment of cutting-edge AR technology is confronted with limitations since key hardware and software infrastructure components are not accessible in some regions across the globe. AR apps face major security and privacy challenges when accessing the personal data and real-time location history of users.

Technologists, politicians, and their business allies need to establish standardized regulations to create safe and moral practices for this AR arrangement. The future is promising for AR in the years to come. Hardware enhancements, such as the creation of more sophisticated AR glass and wearable devices, will enhance the availability and use of AR applications. Combination with other technologies such as artificial intelligence and 5G connectivity also will enhance what is possible and widen the use cases available. For example, AR applications based on AI would offer personalized and context-aware experiences and 5G networks would enable real-time, seamless interactions. As AR develops further, it has the potential to transform the way we live, work, and engage with the world around us, fueling innovation and improving life quality in different areas.

## Quantum Computing

Quantum computing makes this technological vision possible by using basic quantum mathematics that allows devices to solve issues that classic computers are not able to.

Quantum computer processing units execute information using qubits instead of conventional binary bits of 0s or 1s but at times even retain the capacity to be in states of superposition. Quantum computers are able to perform complex computations beyond and levels achieved by standard computers since entanglement qua, quantum interference, and quantum superposition combine within the device (Hassija, V., et al. 2020). The two fields that illustrate the superiority of quantum computing are fast big-number factorization methods for protection against security encryptions and simulation of chemical and physical quantum systems, which result in research findings regarding materials and molecular configurations. The fields of application for quantum computers include optimization solution machine learning algorithms, and cryptographic procedures along with applications in logistics drug discovery and artificial intelligence research areas (Bardin, J. C. et al 2021). Quantum computing is an unsettled technology that has to overcome challenges in becoming commercially viable. Any interaction with the environment by quantum states results in the decay of quantum states in decoherence systems. Storing stable qubits becomes problematic due to this aspect. Among the most prominent technologies under investigation currently for quantum computers, we have superconducting circuits as well as trapped ions, photonic systems, and topological qubits. Every one of these technologies has unique benefits and technological constraints.

Quantum error correction works as a quantum technology since quantum computation inherently generates errors and needs a fault-tolerant system to be constructed. The competition to create working quantum computing platforms is being propelled by corporations like IBM Goo, glue, and Microsoft, together with start-up companies Rigetti and IonQ, while a defining achievement referred to as quantum supremacy was achieved when Google was able to show that its machine would beat current classical supercomputers in 2019 tests (Lubinski, T. et al. 2023). The business viability of quantum computers is far in the distance since predictions vary widely regarding when they can become available for use in industry. Everyone from all industries and governments pours big money into quantum research since this technology promises to replace major sectors, such as national security and global finance. Because of quantum key distribution, quantum communication offers similar encryption techniques that are unbreakable for quantum security protocols. The global quantum supremacy race has initiated technical ethics issues and geopolitical rivalry among global powers. Education and development of specialist manpower in quantum science have become imperative as this science involves profound specialized expertise. The elementary design of quantum computing is innovative technology that transcends conventional computing processes to address intricate issues that transform future technologies (Rietsche, R. et al 2022). Organizations in all industries must create an innovative business culture. Increased world problems require an organizational culture facilitating experimental work, collaborative work, and learning. Creation of novel solutions becomes feasible when such methods are followed.

## Encourage Experimentation and Experimentation

Organizations need to establish safe spaces where employees can experiment and try things out without fear of failing. Teams engage in innovation labs and hackathons augmented outside of their core projects to support experimental ventures. Organizations empower employees to speak their minds authentically and creatively when they cultivate a thought process that transmutes failure into learning experiences instead of lost opportunities. Both Google and Amazon had employee passion project time allocation initiatives, which created novel product and service ideas.

## Integration across disciplines

Cross-functional teams, bringing together disparate minds and skills, should be encouraged by organizations. Silo-busting and getting interdisciplinary collaborators on board now yield more holistic solutions to complex problems. For instance, in engineering, the product (meeting a particular technical need) must engage emotionally with customers. Engineers, designers, and marketers have to collaborate to develop products that satisfy both needs. In public administration, improved policy implementation and service delivery are advantages that accrue from interaction between government elements (Klein, J. T. 2010). Firms should spend in training schemes that provide their employees with the skills and aptitudes they require in order to adapt their professional capacities and knowledge based on emerging technologies or industry needs.

They achieve this through seminars and courses on the internet and by motivating their staff members to participate in other study courses that will enhance their repertoire of talents. Aside from this, firms ought to promote a growth mentality under which workers are more likely to go in pursuit of new ideas or reflect on the things learned from their previous encounters (Repko, A. F. 2008). With this kind of approach, you will discover that it develops individuals' abilities and injects an enthusiastic innovation spirit into the entire organization. Rewards: Rewarding and Referring Innovation to create an environment of innovation, there should be rewarding of innovative work in various forms, e.g., reward programs, performance management systems, and career paths rewarding performance and risk. Methods of rewarding and acknowledging innovation: A firm can express gratitude to its staff members for their innovative contributions via official reward programs simultaneously with urging them on to seek creative solutions. This serves not only to make workers realize how much the firm values their ideas but also increases the level of its culture (Rhoten, D., & Parker, A. 2004).

## Strengthening Policy and Regulatory Frameworks

As new technologies increasingly reshape other sectors, governments need to set policies and regulatory frameworks to steer this technology integration without compromising due process on ethical issues and data security. Enlisting trust from the public would boost innovation and people's rights.

## Facilitating Sectorial Integration

Government initiatives for this purpose are very important. This may be in the form of creating supportive policies through research and technology infrastructure investment or collaborating with various sectors of public-private enterprises. When organizations make investments in innovative technologies, governments must provide grants or tax relief so that these get promoted on a broader scale with and without both creating jobs. Moreover, through encouraging knowledge and resource sharing in public-private collaborations, agencies have the potential to facilitate companies to bring new technology to the market more quickly. There are ethical issues to be considered. This could even include asking yourself if you must create policies that integrate wider social values into current ones. Governments need to set an example and oversee the development of emerging technologies that will impact everyone in some way.

A good example could be artificial intelligence, which might have numerous other uses in surveillance and data protection, for instance, if they were given too much human intuition. Cases in Point: The regulations governing AI systems need to be transparent, responsive, and impartial. This is particularly important in public administration, where artificial intelligence systems make decisions that can influence people significantly with far-reaching consequences. Through the consideration of ethical issues, governments can establish trust among the people and help technology be utilized responsibly.

## Ensuring Data Privacy and Security

While organizations gather a lot of data and interpret it for insights, the government must enforce regulations that safeguard individuals' data in the best possible way. These encompass data protection legislation that demands that organizations seek permission prior to collecting data, such as forms for acquiring data (Iftekar, S. F. et al 2023). Governments further need to ensure transparency in data practice such that people understand how they are utilising data and are given the facility to be able to control information. Minding data privacy by way of appropriate regulation and law increases people's confidence to have more individuals embracing technological progress without apprehension of being exploited. Balance between Innovation and Regulation Although policies and regulations that promote technology integration are necessary to establish, balancing innovation with regulation is similarly crucial. Excessive regulatory rules can close the avenues of innovation and slow the emergence of new technologies.

Consequently, governments ought to engage a light-touch regulatory style, enabling experimentation and adjusting as the technologies keep changing (Quan, H. et al 2020). Innovation provision against consumer protection fosters an environment with plenty of opportunities for technological growth by creating regulatory sandboxes that enable organizations to manage conditions for testing new business models and technologies and keeping not too onerous regulations. Managerial and social implications of the study: The managerial and social implications of this research underscore the urgent necessity of leaders embracing adaptive approaches and cultivating innovation-oriented cultures that capitalize on technological synergies between public administration, engineering, and business domains.

In this manner, organizations can more effectively overcome the imperatives of a fast-changing digital world while advancing effective governance, sustainable infrastructure, and inclusive economic development.

From the managerial point of view, the research emphasizes the need to incorporate new technologies like AI, IoT, and blockchain to facilitate decision-making, maximize resource allocation, and provide transparency and accountability. At the same time, it underlines the necessity of resolving ethical challenges, including algorithmic bias and data privacy, and establishing robust cybersecurity practices to protect sensitive data. Socially, the implications cover the provision of equal access to the technologies, especially in poor or underserved communities. The leaders need to address digital disparities and make technology adoption serve all citizens and not widen existing social gaps.

Moreover, development of public-private partnerships is crucial to propel inclusive solutions that can be scaled and treat systemic issues, thus empowering the communities. Finally, the study recommends aligning organizational practices with sustainable global goals to make a long-term socio-technical ecosystem that not just improves citizen-oriented services but also socially and economically resilient. Prioritizing these factors, leaders can contribute towards building a more transparent, sustainable, and equitable future for all.

## Research answers and findings of the study

1) The convergence of AI and e-governance increases transparency, increases citizen participation, and facilitates decision-making, leading to increased accountability and facilitating data-based policy interventions.

2) The use of IoT and BIM in engineering expedites sustainability through maximization of resource management, enhancement of structural performance, and facilitation of real-time monitoring of the condition and performance of infrastructure.
3) Business models are revolutionized by digital innovations, for example, blockchain and big data analytics, that improve operational effectiveness, maintain data integrity, and facilitate customer-centric strategies that lead to growth.
4) The fast growth of digital technologies creates ethical and cybersecurity issues like algorithmic biases, data breaches, and fair access, and thus necessitates the creation of solid frameworks to protect sensitive information.
5) Public-private partnerships drive innovation through the convergence of expertise, pooled resources, and minimized implementation gaps, allowing scalable solutions to work on sophisticated systemic problems.

The research finds that an integration of technology in public administration, engineering, and business greatly improves efficiency, transparency, and innovation. Public administration is promoted through the adoption of e-governance and AI, which induces accountability, enhances citizen participation, and allows policymaking with data. In engineering, technologies like IoT and BIM improve sustainability by minimizing resource consumption and maximizing real-time monitoring of infrastructure. The business world gains from digital technologies like blockchain and big data analytics, which promote operational efficiency and customer-driven growth. With digital growth comes concerns about ethics and cybersecurity, including algorithmic bias, data privacy violations, and unequal access to technology. They are met with vigorous regulatory frameworks and inclusive digital policies. The study also emphasizes the significant role of public-private partnerships in innovation and responding to complex societal issues. Cross-sector collaboration, facilitated by ongoing learning and ethical governance, is key to using technology as an instrument for inclusive and sustainable development.

## CONCLUSION

In summary, the interfusion of technology in public administration, engineering, and business has huge potential to power societal change and innovation. By leveraging strengths in each area, this merge can result in enhanced operational effectiveness, higher-quality service delivery, and more effective solutions to urgent global issues like climate change, public health emergencies, and socioeconomic inequalities. New technologies—artificial intelligence, big data analytics, and the Internet of Things (IoT)—are not only making operations more efficient but also powering more responsive government, smart infrastructure, and customer-oriented business models. To be successful, they need more than adoption; they need a robust innovation culture, learning, and organizational adaptability. In addition, ethical and responsible use of technology needs to be supported by clear policies and collaborative governance approaches. Public-private partnerships have proven effective, especially in crises such as the COVID-19 pandemic, where concerted action hastened life-saving innovation. Education and skills development are also important in helping society keep up with technological change. Providing people with the required digital competencies can bridge existing gaps and create inclusive economic growth.

Concurrently, efforts must be directed towards eliminating the digital divide so that technology is made available and affordable for all groups in the population. Sustainability also presents itself as a major advantage of integrating technology, with cutting-edge tools for environmental conservation and efficiency of resources. Yet with rising digital dependence comes the pressing need to better manage cybersecurity threats, data protection issues, and ethical considerations—particularly the application of AI and automated decision-making. Finally, the success of integrating technology hinges on visionary stewardship, cross-industry cooperation, and proactive assurance that innovation is aligned with society's values. In this way, we can make technology a genuine facilitator of sustainable, ethical, and inclusive development in the coming years.

## REFERENCES


Albury, D. (2005). Fostering Innovation in Public Services. *Public Money & Management, 25*(1), 51–56.

Attour, A., & Chaupain-Guillot, S. (2020). Digital Innovations in Public Administrations: Technological or Policy Innovation Diffusion? *Journal of Innovation Economics & Management, 31*(1), 61–75. DOI: 10.3917/jie.pr1.0061

Bablani, A., & Tripathi, D. (2018). "A review on methods applied on P300-based lie detectors." In *Advances in Intelligent Systems and Computing* (pp. 302-315). Springer. https://doi.org/DOI: 10.1007/978-981-10-8569-7_26

Bardin, J. C., Slichter, D. H., & Reilly, D. J. (2021). Microwaves in Quantum Computing. *IEEE Journal of Microwaves, 1*(1), 403–427. Advance online publication. DOI: 10.1109/JMW.2020.3034071 PMID: 34355217

Barlow, J., & Maull, R. (2008). The Role of Technology in Public Service Innovation. *Public Management Review, 10*(3), 345–367.

Chen, Z., Li, Z., Li, J., Liu, C., Lao, C., Fu, Y., Liu, C., Li, Y., Wang, P., & He, Y. (2019). 3D printing of ceramics: A review. *Journal of the European Ceramic Society, 39*(4), 661–687. Advance online publication. DOI: 10.1016/j.jeurceramsoc.2018.11.013

Choudhury, P., & Kauffman, R. J. (2019). The Role of Digital Technologies in Business Model Innovation. *Journal of Business Research, 100*, 1–10.

Cropley, D. H. (2015). Promoting creativity and innovation in engineering education. *Psychology of Aesthetics, Creativity, and the Arts, 9*(2), 137–144. DOI: 10.1037/aca0000008

Damanpour, F., & Schneider, M. (2009). Phases of the Adoption of Innovation in Organizations: Effects of Environment, Organization, and Top Managers. *British Journal of Management, 20*(1), 1–21.

Davis, P. W., & McKenzie-Rundle, P. (1984). The Social Organization of Lie-Detector Tests. *Urban Life, 13*(2), 122–134. DOI: 10.1177/0098303984013002003

Desmarchelier, B., Djellal, F., & Gallouj, F. (2019). Innovation in public services in the light of public administration paradigms and service innovation perspectives. *Revue Européenne d'Économie et Management des Services, 8*, 91–110. DOI: 10.15122/isbn.978-2-406-09862-1.p.0091

Feller, J., Finnegan, P., & Nilsson, O. (2011). Open innovation and public administration: Transformational typologies and business model impacts. *European Journal of Information Systems*, *20*(3), 320–335. DOI: 10.1057/ejis.2010.65

Frikha, T., Ktari, J., Zalila, B., Ghorbel, O., & Amor, N. B. (2023). Integrating blockchain and deep learning for intelligent greenhouse control and traceability. *Alexandria Engineering Journal*, *79*, 1457–1465. DOI: 10.1016/j.aej.2023.08.027

Ghosh, R., & Scott, J. (2018). The Impact of Digital Technologies on Business Models. *Journal of Business Research*, *100*, 1–10.

Gopinathan, J., & Noh, I. (2018). Recent trends in bioinks for 3D printing. *Biomaterials Research*, *22*(1), 11. Advance online publication. DOI: 10.1186/s40824-018-0122-1 PMID: 29636985

Gupta, D., Pandey, A. K., & Garg, A. (2019). Innovations in public administration in India. *Economic and Political Weekly*, *54*(29), 13–25.

Hassija, V., Chamola, V., Saxena, V., Chanana, V., Parashari, P., Mumtaz, S., & Guizani, M. (2020). Present landscape of quantum computing. *IET Quantum Communication*, *1*(2), 42–48. Advance online publication. DOI: 10.1049/iet-qtc.2020.0027

Heeks, R. (2006). Implementing and Managing eGovernment: A Global Perspective. *International Journal of Public Sector Management*, *19*(4), 301–310.

Iftekar, S. F., Aabid, A., Amir, A., & Baig, M. (2023). Advancements and limitations in 3D printing materials and technologies: A critical review. *Polymers*, *15*(11), 2519. Advance online publication. DOI: 10.3390/polym15112519 PMID: 37299318

Kauffman, R. J., & Walden, E. A. (2018). The Role of Digital Platforms in Business Model Innovation. *Journal of Business Research*, *100*, 1–10.

Klein, J. T. (2010). *Creating interdisciplinary campus cultures: A model for strengthening interdisciplinary education*. Jossey-Bass.

Kondratenko, V., Okopnyk, O., Ziganto, L., & Kwilinski, A. (2020). Innovation Development of Public Administration: Management and Legislation Features. *Marketing and Management of Innovations*, *1*(1), 56–64. DOI: 10.21272/mmi.2020.1-06

Lubinski, T., Johri, S., Varosy, P., Coleman, J., Zhao, L., Necaise, J., Baldwin, C. H., Mayer, K., & Proctor, T. (2023). Application-oriented performance benchmarks for quantum computing. *IEEE Transactions on Quantum Engineering*, *4*, 1–32. Advance online publication. DOI: 10.1109/TQE.2023.3253761

Mancero Arias, G. (2022). Innovation in engineering. *Athenea*, *3*(7), 33. DOI: 10.47460/athenea.v3i7.33

Massey, A. (2019). Editorial: Innovation in public administration to leave no one behind. *Public Money & Management*, *39*(7), 495–497. DOI: 10.1080/09540962.2019.1665377

Matei, A., & Savulescu, C. (2014). Enhancing the capacity for innovation of public administration. An exploratory study on e-Governance, ICT, knowledge management in Romania. *Theoretical and Applied Economics*, *21*(11), 63–78.

Mazzucato, M. (2013). *The Entrepreneurial State: Debunking Public vs. Private Sector Myths*. Anthem Press.

Mergel, I. (2016). Digital Service Teams in Government. *Public Administration Review*, *76*(4), 565–577.

Mitra, S. (2023). Innovations in Indian public administration. *Public Money & Management*, *43*(8), 523–526. DOI: 10.1080/09540962.2022.2087946

Olcay, G. A. (2021). International Journal of Innovation and Technology Management-Special Issue in Honor of Professor M. Atilla Öner (1955-2018) Managing Emergent Technologies. *International Journal on Management of Innovation & Technology*, *18*(1), 1–25. DOI: 10.1142/S0219877021020016

Ongaro, E., Gong, T., & Jing, Y. (2021). Public administration, context and innovation: A framework of analysis. *Public Administration and Development*, *41*(1), 47–56. DOI: 10.1002/pad.1902

Pakhnenko, O., & Kuan, Z. (2023). Ethics of Digital Innovation in Public Administration. *Business Ethics and Leadership*, *7*(1), 113–121. DOI: 10.21272/bel.7(1).113-121.2023

Park, S., Shou, W., Makatura, L., Matusik, W., & Fu, K. (2022). 3D printing of polymer composites: Materials, processes, and applications. *Matter*, *5*(1), 43–76. Advance online publication. DOI: 10.1016/j.matt.2021.10.018

Parycek, P., & Scholl, H. J. (2013). The Role of Open Government Data in Public Administration. *International Journal of Public Administration*, *36*(5), 345–356.

Quan, H., Zhang, T., Xu, H., Luo, S., Nie, J., & Zhu, X. (2020). Photo-curing 3D printing technique and its challenges. *Bioactive Materials*, *5*(1), 110–115. Advance online publication. DOI: 10.1016/j.bioactmat.2019.12.003 PMID: 32021945

Repko, A. F. (2008). *Interdisciplinary Research: Process and Theory*. SAGE Publications.

Rhoten, D., & Parker, A. (2004). Risks and rewards of an interdisciplinary research environment. *Science*, *306*(5704), 2046–2047. DOI: 10.1126/science.1103628 PMID: 15604393

Rietsche, R., Dremel, C., Bosch, S., Steinacker, L., Meckel, M., & Leimeister, J. M. (2022). Quantum computing. *Electronic Markets*, *32*(4), 2525–2536. Advance online publication. DOI: 10.1007/s12525-022-00570-y

Rocha, J. A. O., & Zavale, G. J. B. (2021). Innovation and Change in Public Administration. *Open Journal of Social Sciences*, *9*(6), 123–136. DOI: 10.4236/jss.2021.96021

Salter, A., & Gann, D. (2003). Sources of ideas for innovation in engineering design. *Research Policy*, *32*(8), 1161–1181. DOI: 10.1016/S0048-7333(02)00119-1

Sarker, S., & Sahay, S. (2004). The Role of Information Technology in Public Administration: A Review of the Literature. *Public Administration Review*, *64*(3), 1–10.

Scholl, H. J., & Scholl, M. C. (2014). The Role of Information Technology in Public Administration. *Public Administration Review*, *74*(4), 487–498.

Siau, K., & Long, Y. (2018). Artificial Intelligence in Business: A Review of the Literature. *Journal of Business Research*, *100*, 1–10.

Sinni, G. (2017). "Participatory Design for Public Services. Innovation in Public Administration." *Design Journal*, *20*(sup1), 120-138. DOI: 10.1080/14606925.2017.1352841

Tapscott, D., & Williams, A. D. (2010). *Macrowikinomics: Rebooting Business and the World*. Portfolio.

Van der Meer, F. (2017). The Role of Technology in Public Administration: A Review of the Literature. *Public Administration Review*, *77*(4), 1–10.

Whitford, A. B., & Meier, K. J. (2013). Methodological Innovations in Public Administration: A Symposium. *Journal of Public Administration: Research and Theory*, *23*(2), 231–246. DOI: 10.1093/jopart/mut012

Wirtz, B. W., & Göttel, V. (2016). Business Model Innovation: A Review of the Literature. *International Journal of Management Reviews*, *18*(3), 1–20.

Zeng, M., & Glaister, K. W. (2018). The Role of Digital Technologies in Business Model Innovation. *Journal of Business Research*, *100*, 1–10.

## ADDITIONAL READING


Aversano, L., Canfora, G., de Lucia, A., & Gallucci, P. (2002). Business process reengineering and workflow automation: A technology transfer experience. *Journal of Systems and Software*, *63*(1), 5–15. DOI: 10.1016/S0164-1212(01)00128-5

Bason, C. (2010). *Leading Public Sector Innovation: Co-creating for a Better Society*. Policy Press. DOI: 10.2307/j.ctt9qgnsd

Brynjolfsson, E., & McAfee, A. (2014). *The Second Machine Age: Work, Progress, and Prosperity in a Time of Brilliant Technologies*. W. W. Norton & Company.

Cummings, T. G., & Worley, C. G. (2014). *Organization Development and Change*. Cengage Learning.

Kettunen, P., & Kallio, J. (2019). Digital Transformation in Public Administration: A Systematic Literature Review. *Government Information Quarterly*, *36*(2), 101–112.

Lopez-Vega, H., & Moodysson, J. (2023). Digital Transformation of the Automotive Industry: An Integrating Framework to Analyse Technological Novelty and Breadth. *Industry and Innovation*, *30*(1), 58–72. DOI: 10.1080/13662716.2022.2151873

Reddick, C. G. (2010). Citizen Interaction with e-Government: A Comparison of the United States and Canada. *International Journal of Public Administration*, *33*(1), 1–15.

# Chapter 11
# Smart Systems and IoT in Engineering in Business Optimization

**Bekir Yurdugüzel**
https://orcid.org/0000-0001-6588-6074
*Kırşehir Ahi Evran University, Turkey*

**Beste Desticioğlu Taşdemir**
https://orcid.org/0000-0001-8321-4554
*Alparslan Institute of Defense Sciences and National Security, National Defense University, Turkey*

**Merve Asiloğullari Ayan**
https://orcid.org/0000-0001-7522-0911
*Alparslan Institute of Defense Sciences and National Security, National Defense University, Turkey*

## ABSTRACT

*This book chapter explores how intelligent systems and the Internet of Things (IoT) technologies contribute to business optimization in engineering. It highlights the transformative impact of Industry 4.0 through concepts such as smart factories, digital twins, and cyber-physical systems. While addressing the operational benefits of AI, real-time data analytics, and automation, it also discusses major challenges like cybersecurity risks and lack of standardization. The chapter offers strategic recommendations for effective technology adoption, emphasizing sustainability, organizational adaptation, and future-oriented innovation.*

DOI: 10.4018/979-8-3373-1534-8.ch011

## INTRODUCTION

The convergence between traditional industrial practices and digital technologies is radically reorganizing the business world. The smart systems and Internet of Things (IoT) are identified as key factors in enhancing business processes, making them faster, more adaptive, and environmentally friendly, among other benefits. The influence of this change extends beyond just a few impacts on manufacturing processes and how businesses develop their decision-making frameworks. IoT-driven systems provide a wide range of potential benefits, from reducing manufacturing defects to enhancing traffic management. In addition, the integration of IoT systems, big data analytics, artificial intelligence (AI), machine learning (ML), and automation increases the effectiveness of processes and provides significant advantages over competing businesses.

The IoT and AI enhance the effectiveness of automation, allowing production processes to swiftly adapt to ever-changing market conditions. By incorporating ML and AI technologies into these processes, automation has been achieved, leading to a reduction in human involvement. The increased flexibility of production lines and the widespread adoption of automation within the manufacturing sector enable businesses to respond more effectively to shifting market conditions. Furthermore, integrating IoT-based systems into the manufacturing process allows companies to optimize energy and resource management (Belli et al., 2019).

Industrial IoT (IIoT) technology and smart manufacturing systems are increasingly popular as they enable more flexible and scalable manufacturing operations. As production lines become digitalized, it becomes possible to identify manufacturing defects much earlier, before they occur. If we were to describe the product's situation, we would outline those steps. Additionally, we can perform quality control processes more effectively (Menon et al., 2018). The integration of smart systems in production areas also enhances the efficient use of energy and resources. Reducing energy consumption, minimizing machine operating hours, lowering maintenance requirements, and optimizing supply chain operations through data-driven decision-making (Shukurova, 2025) are significant factors in achieving sustainability goals.

The advantages of smart manufacturing technologies extend beyond production processes; they also provide significant benefits for supply chain management (SCM). The IoT-based SCM represents a modern approach that enables real-time inventory tracking, enhances logistics planning accuracy, and reduces issues related to excess or insufficient stock (Zahid et al., 2025). Furthermore, these technologies boost efficiency when applied to production quality control and maintenance processes.

In recent years, it has been observed that the use of digital transformation technologies in business processes has become widespread. With the use of these technologies in production processes, it is possible to carry out operations more efficiently and quickly. This chapter concentrates on the efficient use of digital technologies to enhance business processes and offers a thorough understanding of the essential elements of industrial digitalization.

In the first section, the structural dynamics of smart factories, real-time data analytics, the sectoral adaptation of smart technologies, and AI-based decision support systems are discussed. In the second section, we examined the concepts of IoT and Industry 4.0 in the context of digital twins, autonomous systems, and cyber-physical systems, using specific examples from sectors such as agriculture, logistics, healthcare, and energy. Additionally, smart traffic management systems and other urban-scale solutions are also discussed. In the third section, in addition to summarizing the competitive advantages these technologies provide to businesses, important issues such as technological standardization, integration problems, security vulnerabilities, and data privacy risks are also addressed.

This multidisciplinary study has been created with contributions from researchers with various expertise, aiming to provide a theoretical contribution to the existing literature as well as to offer a different and creative perspective for real-world industry applications.

## USE OF INTELLIGENT SYSTEMS IN BUSINESS OPTIMIZATION

### 1. Technologies Used in Smart Systems

Industry 4.0 can be defined as "the management and organization level of the new value chain in the life cycle of products" or "the integration of networked software and sensors to accurately predict, plan, and control the results of complex systems and machines." The technologies associated with Industry 4.0 aim to optimize the value chain (Mrugalska and Wyrwicka, 2017). Key objectives of Industry 4.0 applications include ensuring flexibility within the production chain, establishing communication between products and machines, monitoring materials and products, applying human-machine interaction approaches, optimizing smart factories using IoT technologies, and introducing new types of business models (Lu, 2017). Concepts such as IoT, Enterprise Resource Planning (ERP), Radio Frequency Identification (RFID), cloud computing, Cyber-Physical Systems (CPSs), smart factories, and smart products have emerged alongside the technologies introduced by Industry 4.0. The implementation of smart technologies facilitates quicker decision-making in business processes and enhances profitability by improving process efficiency (Sarker, 2022).

Intelligent computer systems are widely used in health care, energy production and distribution, manufacturing, industrial systems, transportation systems, supply chains, security, the environment, and social activities. For example, robotic systems, warehouse inventory automation, consumer personalization, use of AI in e-commerce, image-based advertising, and route optimization are areas where smart technologies are applied (Sarker et al., 2020).

The unified management of machine programs and standardization of processes with cloud computing systems that emerged with Industry 4.0 are effective in reducing human errors and ensuring more efficient production. In this section, smart systems that have entered our lives with Industry 4.0 and the areas of use of these smart systems are discussed.

## 1.1. Internet of things (IoT)

IoT can be defined as a network structure in which machines or systems collect information through data communication among themselves and make decisions using the collected information, without the need for any manual entry of data or human intervention. IoT is a communication network that includes devices, software, and systems that enable monitoring, controlling, and analyzing the data obtained from physical events occurring in different areas, such as patient monitoring systems, production processes, transportation, recycling processes, energy networks, shopping malls, smart buildings, and smart factories. The purpose of IoT is to enable communication between objects using the internet. These objects can communicate with the internal server and external environment by using embedded systems.

The emergence of new-generation hardware equipped with various low-cost and energy-efficient sensors has led to the widespread adoption of IoT technology in areas such as monitoring, detection, intelligent management, and control systems. It is evident that IoT and cloud computing technologies are being integrated across diverse fields, including smart traffic management, smart transportation systems, smart home systems, automated vehicle management, and remote patient care in healthcare services.

The widespread use of sensors and smart devices developed with IoT technology has enabled its application in various fields, including logistics, production, healthcare, urban planning, supply chain management, and agriculture (Kahadam et al., 2024). IoT allows for the collection and sharing of real-time data, which can subsequently be used to train AI and ML systems, enhancing their performance. Furthermore, managing the devices and machinery used in production through a network can minimize the need for human oversight.

## 1.2. Cyber-physical systems (CPS)

Another effective system for establishing smart factories is Cyber-Physical Systems (CPS). CPSs are designed to detect and control all aspects of production processes, as well as to find solutions to potential problems. Technologies that facilitate communication and coordination between the cyber world and real life are collectively referred to as CPS (Timm and Lorig, 2015). CPS employs specialized algorithms and software to enable systems to interact with each other, fulfilling necessary functions within the system. It establishes a connection between the real world and the virtual world through the use of sensors, RFID technology, and actuators. Data regarding objects is collected using RFID readers; this information is initially processed in the interface before being transferred to information systems. Consequently, CPS and IoT systems can collaborate by maintaining communication both within and outside the factory. CPS effectively addresses dynamic production deficiencies, enhancing both efficiency and effectiveness within the industry (Lu, 2017).

Smart machines share data about changes in demand, losses, potential issues, and current stock levels with CPS applications. RFID-compatible sensors and control devices facilitate communication between machines, the factory, the network, and individuals. The central element of the system is the smart production robots. These robots, connected to the production system, oversee the entire production process and ensure the optimization of operations (Guban and Kovacs, 2017).

## 1.3. Cloud computing technology

Another component of smart factories is cloud computing technology. Information obtained from smart machines and objects used in smart factories is instantly transferred to the cloud system. Cloud computing is a set of services where all owned applications, programs, and data are stored on a virtual server and can be easily accessed from any device connected to the Internet. Storing data on servers on the internet is done through cloud computing. Using cloud computing technology, faster data transfer, more storage space, and cost savings are possible. By using cloud computing, it is possible to access the required data at the desired place and time. Cloud computing technology reduces costs, eliminates infrastructure complexity, expands the workspace, and enables working from anywhere at a much cheaper cost without requiring installation (Gündüz and Daş, 2018).

## 1.4. Components of IoT and smart systems

The development of IoT systems in all areas is effective in the widespread use of smart devices recently (Chen et al., 2022). In the competitive business world, it is very difficult to compete with other companies without using smart systems. The supplier must deliver raw materials and products to the final destination. It is crucial to monitor the locations of these products and materials using smart transportation systems. RFID technology, which uses radio waves in the flow of information and data, allows the tracking and analysis of objects with tags integrated into processors located in objects. The RFID tag and the RFID reader can exchange data without physical contact. RFID technology allows for the remote tracking of vehicles carrying materials. The information received from RFID tags by readers is stored, and another process can be started using this data. The ERP system controls the relationships between processes. The ERP system sheds light on information such as device status, stock level, new order status, and delays in the production line, using data from RFID readers (Atzori et al., 2010). RFID technology is not only used in solving problems in the supply chain, but is also frequently used in warehouse and inventory management in production systems (Lee and Ozer, 2007). Thus, it is possible to achieve optimization in supply chain and inventory management with RFID technology (Thames and Schaefer, 2016).

The IoT enables physical entities to communicate, exchange data, and make decisions collaboratively, which enhances their capabilities to see, hear, think, and act. Through this technology, traditional objects become intelligent by incorporating widespread computing power, embedded systems, communication technologies, sensor networks, internet protocols, and various applications.

## 1.5. Industrial IoT (IIoT)

The machines used in IIoT operate autonomously, allowing them to decide on their energy efficiency and operating speed. Smart production machines can communicate with other objects or machines on the network directly or using the cloud computing system.

Since production settings are also made via the cloud, smart machines make the necessary updates via the cloud when they are connected to the cloud (Chen et al., 2012).

IIoT systems can respond quickly to situations such as raw material supply and demand changes. In addition, monitoring production simultaneously with data collected from machines provides significant advantages in terms of cost and competition. In addition, automated flexible production enables rapid adaptation to changing customer demands, preventing unnecessary stockholding and thus reducing costs (Snatkin et al., 2013).

The use of IoT systems in industry provides many advantages (Xu, 2011):

- Smart production systems with IoT technology instantly communicate with each other to control production and minimize human control.
- Machine malfunctions can be detected in advance and machine downtime can be reduced.
- Raw material shortages needed in production can be quickly detected and orders can be placed.
- Managers can instantly learn production and malfunction information about the factory from anywhere in the world and share this information with customers.

## 2. Smart Factories and Use of Smart Technologies

The term "smart" generally refers to the ability of computers to use data and replicate human-like capabilities by merging this data with human intelligence and practical skills. This definition encompasses technologies enhanced by AI, IoT, cloud computing, and big data. Additionally, the concept of smart systems may sometimes apply to systems that do not incorporate these specific technologies (Alter, 2020).

Smart service systems are designed to learn from both received and transmitted data, adapt in real-time, make informed decisions based on this data, and execute future plans. These systems achieve their goals through functions such as self-organization, self-correction, self-monitoring, and self-control. The realization of these functions is made possible by leveraging technologies related to perception, coordination, communication, actuation, control, and more (Medina-Borja, 2015). Smart Service Systems are service systems that can learn, make decisions and improve their future capabilities by considering the data received, transmitted or processed. Such service systems give importance to human beings so that aspects such as knowledge, skills and values are defined by the members of the system (Longo et al., 2019).

Smart systems use sensors related to traffic flow and public service systems in big cities and can make advanced predictions using this information. Additionally, IoT technology sends data over a network for communication between objects. This data is used to create environmentally friendly products and create smart cities using IoT technology (Chen et al., 2022).

## 2.1. Smart factories

Compared with traditional factories, smart factories reduce storage, increase resource utilization, reduce production errors, shorten the time required for customer-specific product production, enable faster production, and significantly reduce storage in the supply chain process (Wang and Zhou, 2015). In Industry 4.0, objects are revitalized through communication with one another, leading to the creation of smart systems powered by advancements in robotics, AI, augmented reality, and 3D printing. The primary objective is to establish smart factories capable of autonomously managing their production processes (Wang and Zhou, 2015). To achieve this, it is essential to integrate smart objects with big data analysis. Thus, small-volume and customized products can be produced profitably and efficiently in smart factories (Wang et al., 2016).

Smart factories are composed of intelligent machines and systems that identify business needs through sensors, communicate with remote production tools via the internet, and access essential production information from big data. Internet technology enhances communication and interaction among production tools within smart factories. All production resources—such as machines, sensors, and robots—are capable of participating in the automatic exchange of information and monitoring the control, maintenance, and management of the production process.

These processes will not only be managed by a decentralized system but will also be interconnected in an end-to-end system, meaning they are controlled in an interdependent manner (Jian et al., 2016).

Smart factories differ significantly from traditional factories in terms of production processes, production methods, resource utilization, inventory management, routing, decision-making processes, organizational structure, information systems, maintenance management, quality control, maintenance management, and information management (Wang et al., 2016). The ability to connect machines and objects with IoT technology to carry out production brings numerous advantages to businesses. In smart factories, products move independently through production processes and are always easily identifiable and findable. Smart factories make the increasing complexity of production processes manageable for employees and, at the same time, profitable and attractively sustainable in urban environments (Hofmann and Rüsch, 2017).

Smart factories utilize smart technologies not only in the production process but also across various other areas. These factories apply smart technologies in quality control, maintenance management, and SCM components, thereby enhancing the system's flexibility and efficiency. This section offers insights into the components of smart factories that leverage smart systems.

## 2.2. Use of smart technologies in quality control

Production and quality control processes are an inseparable whole. Quality control is a process that is mostly carried out by human hands. In order for manufacturing companies to achieve their goals, they need to produce their products in high quality. Human vision ability led to the development of image processing technology, which has recently found application across various fields. By applying deep learning models with image processing technology, it is possible to detect defects in quality control and prevent possible errors. Thus, by using image processing and deep learning technologies in quality control, it is ensured that faulty products are not sent for shipment. With the use of AI and smart technologies in production, production errors, equipment failures, and unexpected changes in inputs such as raw materials can be predicted, and production processes can thus be optimized (Pipiay et al., 2021).

By using smart technologies in quality management systems, businesses can improve quality control, catch mistakes early, and lower costs related to quality by preventing errors. Utilizing a quality management system driven by smart technologies enables businesses to uncover innovative applications of quality that support decision-making, providing them with significant competitive advantages over other companies (Yang et al., 2019).

## 2.3. Use of smart technologies in maintenance management

Industry 4.0 technologies enable real-time monitoring of machines and production lines. Sensors can gather various information about these machines or lines to assess whether maintenance is necessary. Predictive maintenance involves monitoring different components of the system through sensors to evaluate their remaining useful life and determine if maintenance is required. AI technologies significantly reduce equipment downtime, minimize interruptions, detect production errors early, and enhance supply chain operations by utilizing ML techniques based on historical data from manufacturers (Angelopoulos et al., 2020). Predictive maintenance can be defined as the process of estimating the duration of failure-free operation for equipment or machinery within a specified timeframe. Sensors on the equipment or machines collect data such as pressure, temperature, noise, and current draw to estimate the time they can operate without failure. Predictive maintenance integrates ML and IIoT technologies to predict when production equipment will require maintenance and to address issues that lead to failures (Carvalho et al., 2019).

The lack of fault detection during maintenance after a fault occurs, along with prolonged maintenance time and the potential for a single fault to trigger multiple faults, can reduce overall efficiency. However, increasing the frequency of regular maintenance before failures occur can increase total maintenance time and reduce efficiency. Predictive maintenance is used to determine when maintenance is needed at that exact point. The technologies brought by Industry 4.0 enable machines to be monitored in real time.

By simultaneously monitoring information about the machines with their sensors, we can determine the need for maintenance (Wang et al., 2016). This process increases the efficiency of the machines and supports the realization of failure-free production.

## 2.4. Use of smart technologies in supply chain management (SCM)

SCM integrates all business processes from suppliers to the end consumer, ensuring that the necessary services, products, and information are provided to all stakeholders and customers (Lambert & Cooper, 2000). The primary goal of SCM is to deliver the right product to the right customer on time, in the desired quality and quantity. An effective SCM is expected to result in reduced cycle times and lower inventory levels, thereby increasing productivity (Raoufi et al., 2024).

The primary goal in designing, analyzing, and modeling traditional supply chain (SC) is to optimize processes, beginning with the procurement of raw materials from suppliers and extending to the distribution of products to customers. In contrast, smart SC exhibit a more dynamic structure, allowing them to quickly adapt to market changes. Technologies such as IoT systems, intelligent transportation systems, 5G technology, and CPS are utilized in the supply process and transportation of raw materials or products. Additionally, the incorporation of 3D printers in production and assembly processes is a key element that facilitates SC operations (Correia et al., 2021). The use of 3D printers can influence SC in various ways, including enhancing production flexibility, accelerating product development, and minimizing material waste (Attaran, 2017).

To monitor changes in the SC process, it is essential to obtain up-to-date data. Digital twin (DT) technology enables the collection of information from various sources within the SC. This allows for the current performance of the SC to be monitored and facilitates necessary analysis and modeling. Employing DT technology enables companies to proactively address potential delays for products affected by traffic congestion or adverse weather conditions. Additionally, blockchain technology has recently been implemented to enhance SC effectiveness (Iddum et al., 2022).

Smart storage and inventory management, autonomous vehicles, robotic systems, blockchain, and AI-based technologies offer significant benefits to companies seeking to integrate SCM. Advantages such as decision support systems, accurate demand forecasting, real-time transportation tracking, just-in-time production, and efficient storage are among the many benefits that smart technologies contribute to the SC. Smart technologies utilized in SCM perform various functions, including monitoring and controlling operations, aiding decision-making processes, and storing and processing data. Additionally, they simulate production and supply processes. Stakeholders in the SC employ technologies such as RFID and barcode systems, mobile data, order processing, and electronic data integration to facilitate effective communication. These technologies are instrumental in executing activities like forecasting and inventory analysis, vehicle fleet tracking, production planning and oversight, route planning, international trade planning, demand forecasting, and enhancing efficiency within the SC (Iddum et al., 2022).

Digital technologies incorporate predictive analytics and automation capabilities to accelerate time to market, swiftly predict and address issues, reduce planning cycles, enhance decision-making, and deliver optimal service to all stakeholders. Smart SC assesses the stakeholders of the SC as a unified entity to ensure customer satisfaction and achieve substantial profits.

AI enables rapid responses to customer demand trends through data analysis, leading to improved quality and efficiency driven by feedback. This technology can oversee all phases of the SC. AI solutions offer numerous social and economic benefits for SC, including the ability to document a high volume of transactions that would otherwise require human effort and to streamline inventory and warehouse management processes (IBM, 2025). Moreover, leveraging technologies such as ML and AI for demand forecasting is crucial for generating accurate predictions. This framework provides immediate solutions to various issues that could adversely impact production and logistics activities. Some of these challenges include difficulties in supplier communication, disruptions in manufacturing processes, and challenges in meeting retailer demands and end-user needs. Additionally, by developing a flexible SC that integrates advanced solutions, organizations can effectively plan subsequent steps in the process, update demand forecasts to reflect changes, and easily adapt to evolving industry conditions.

## 2.5. Advantages of smart factories

The advantages provided by smart technologies used in smart factories are as follows:

1. High Competitiveness: Equipping machines with the internet and strong networks in smart factories enables businesses to be more efficient and more competitive worldwide.
2. Performance Efficiency: Smart factories produce goods faster and more effectively due to the simultaneous collection of data throughout the entire supply chain, the instant detection of emergencies and errors, the implementation of

interventions, and the improvement of processes based on the data collected from various objects.

3. High Productivity: Smart production processes enable simultaneous access to data within the SC This access to real-time information on products and materials needed by manufacturers and suppliers, along with rapid provisioning, enhances production efficiency.
4. Lower Costs: Smart factories enable a more accurate assessment of inventory needs, facilitate better recruitment decisions, minimize process variability, and reduce waste rates, resulting in lower costs compared to traditional factories.
5. Quality: Smart factories can detect the causes of errors more quickly and accurately with the technologies they use and can quickly take the necessary precautions against machines, people, or environmental elements that cause poor quality. This is effective in producing higher-quality products with lower waste rates.
6. Innovative Business Models: New services, processes, and business models can be developed by communicating with each other between different smart systems in smart factories.
7. Improvement in Working Conditions: Employees' working hours can become more flexible and independent as smart machines adjust processes according to the human work cycle.
8. More Flexibility: Smart technologies in smart factories enable more flexible production by adapting to system and process changes more quickly.
9. High Security: Carrying out activities with smart technologies in smart factories is effective in reducing employee errors and the number of work accidents (https://www2.deloitte.com/; https://www.datexcorp.com/)

## IOT AND INDUSTRY 4.0: NEW GENERATION PRODUCTION AND BUSINESS MODELS

IoT and smart technologies are widely used in many areas, especially in industry. In this section, the areas where smart technologies are used and their sample applications in industry and other areas are given.

### 1. Use of IoT and Industry 4.0: Advanced Production and Business Models

Smart factories are manufacturing settings that arose from the convergence of Industry 4.0 and the IoT. These factories are founded on three fundamental principles:

1. Vertical Integration: The amalgamation of production floor equipment with upper management systems, including ERP and MES (Manufacturing Execution Systems) (Wang, Torngren & Onori, 2015).
2. Horizontal Integration: Uninterrupted data exchange among SC, logistics, and customer relationship management.
3. Edge Computing: The deployment of data processing and analytical functions at the source (machine level) to diminish reliance on cloud infrastructure (Gubbi et al., 2013).

In this design, the IoT performs two essential roles:

- Machine Data Collection: Surveillance of parameters including temperature, pressure, and energy use via sensors.
- Autonomous Decision Making: Automated actions are executed in processes such as production velocity, quality assurance, or energy optimization through the analysis of gathered data using AI algorithms (Lee et al., 2015). At Bosch's Augsburg facility, IoT-based solutions enhanced energy efficiency by 25% through simulations that generated a digital twin for each component on the production line.

Likewise, Siemens' Amberg facility decreased the error rate per product from 12 to 1 in 1 million due to the use of IoT and AI. Siemens' digital twin technology has achieved a 45% reduction in production errors by utilizing simulations to create a virtual replica of a physical machine. (Hermann et al., 2016).

A significant contribution of IoT in the logistics business is the enhancement of transparency in SCM. Intelligent RFID tags and GPS tracking systems enhance inventory management while reducing time and costs in transportation procedures (Kamble et al., 2018). For instance, major logistics firms like Walmart and Amazon enhance their inventory management and react more swiftly to consumer requests with IoT-enabled SCM (Xu et al., 2018).

The waste of time and the environmental damage that it creates are both caused by traffic congestion in cities. Using sensors, global positioning systems (GPS), and AI algorithms, IoT-based intelligent traffic management systems optimize the flow of traffic (Javaid et al., 2018). According to Swathi et al. (2016), smart traffic lights, for instance, evaluate the amount of traffic congestion, change the duration of the signal, and make certain that vehicles burn less fuel. This results in an increase in energy efficiency as well as a reduction in the negative effects on the environment. The emergence of smart factories is revolutionizing conventional business models across four principal dimensions.

Product-service hybridization (servitization) denotes the transition from selling tangible things to service-oriented income models. For instance, Kaeser Kompressoren has implemented a pay-per-use model for its clientele by providing "compressed air services" rather than selling air compressors. IoT sensors tracked equipment performance, decreasing maintenance expenses by 30% and allowing the company to derive 40% of its revenue from service-oriented contracts (Xu et al., 2014). This model enhances manufacturers' consumer loyalty while preserving their competitive advantage.

The second shift is the advent of data-driven revenue models. The substantial data generated by IoT provides enterprises the opportunity to develop new revenue sources. An industrial machinery maker can evaluate its customers' production data and offer sector-specific benchmark reports. Gubbi et al. (2013) forecast that these data markets will augment the global economy by $13 trillion by 2030. Moreover, General Electric's Predix platform analyzes industrial IoT data to deliver predictive maintenance solutions to clients, yielding $1.2 billion in annual revenue from these services.

The third axis pertains to dynamic SCM. The integration of IoT and blockchain enhances transparency and adaptability inside supply networks. Philips diminished rotting risk by 89% through the utilization of IoT sensors to monitor temperature and humidity data in pharmaceutical shipping containers (Wang et al., 2015). Walmart similarly decreased traceability in the food SC to 2.2 seconds with a blockchain-based IoT system, resulting in a 30% reduction in recall costs.

The fourth transformation pertains to the advancement of workforce competencies. The demand for conventional unskilled labor is transitioning towards "digital hybrid" people possessing digital literacy and data analytics competencies. At BMW's Leipzig facility, employees utilize augmented reality (AR) glasses to oversee intricate assembly operations, resulting in a 17% enhancement in operational efficiency (Lee et al., 2015). Moreover, human-machine collaboration (cobots) has diminished error rates in production lines from 12% to 1.5%. This change also presents considerable problems. Cybersecurity threats represent the paramount vulnerability of smart manufacturing. In 2021, a ransomware assault on an automotive supplier's IoT network disrupted operations in 24 facilities, resulting in losses of $50 million (Xu et al., 2014). Moreover, smart factories utilize 40% more energy than conventional factories, which undermines sustainability objectives (Hermann et al., 2016).

Consequently, Industry 4.0 and IoT are revolutionizing the manufacturing landscape, compelling enterprises to implement a profound transition in their revenue models and supply chains. The advancement of cybersecurity and human resource competencies is as essential as expenditures in technology infrastructure.

## 2. The Role of Digital Twins and Autonomous Systems

Digital twin (DT) technology is a cutting-edge digital transformation instrument that facilitates the development of virtual representations of physical entities, systems, and processes. This technology enables firms to monitor, simulate, and optimize each phase of the product lifecycle inside a digital framework. Digital transformation systems encompass the amalgamation of sensor technologies, data analytics, AI, IoT, and sophisticated modeling methodologies.

This approach renders firms more data-driven, adaptable, and reliable in their operational and strategic decision-making processes (Tao et al., 2019). DT technology significantly enhances operational efficiency by monitoring and managing business processes within intricate systems in a digital context. In the manufacturing industry, DT applications enable the early detection of production line faults and facilitate more effective maintenance planning (Fuller et al., 2020). This diminishes production losses, decreases maintenance expenses, and maximizes resource utilization.

Nonetheless, DT systems extend beyond production. They are extensively utilized in fields including healthcare, transportation, energy, smart cities, and defense. In the energy sector, power plant performance may be monitored in real time, whilst in the healthcare sector, tailored treatment programs can be formulated through patient simulations. In smart cities, data including traffic density, infrastructure utilization, and energy consumption may be continuously examined to enhance urban management (Kritzinger et al., 2018).

In recent years, the significance of DT technology has escalated, particularly within the defense and aviation industries. The implementation of DT has become essential in various areas for managing intricate systems and guaranteeing operational continuity. DT systems enhance mission safety and offer a strategic edge to decision-makers through real-time situational awareness and remote monitoring and maintenance capabilities.

The predominant and efficacious application of DT technology is observed in the industry. In the context of Industry 4.0, automation, adaptability, and traceability have emerged as essential components of manufacturing processes. DTs enable the digital modeling of production lines, simulation of machine behavior, prediction of possible faults, and optimization of maintenance processes (Kritzinger et al., 2018). Siemens announced that it attained nearly 99% product quality in its production process by employing DT technologies at its Amberg facility in Germany. In the aerospace and defense industries, system complexity and mission safety are paramount. Consequently, DT technology is essential in these industries. General Electric has created DT systems that can forecast maintenance requirements and enhance flight safety by continuously monitoring jet engine performance (Fuller et al., 2020). Furthermore, firms like Lockheed Martin develop digital prototypes of fighter jets, expediting testing procedures, minimizing expenses, and enhancing technical precision.

In the energy sector, DT technology is employed for optimizing power plant efficiency, planning maintenance procedures, and balancing energy consumption. For instance, the energy conglomerate Shell uses DTs to simulate production processes and do risk assessments at oil refineries. Analyzing each turbine's operational data allows the DTs of wind turbines to optimize energy production (Tao et al., 2019). The potential of DTs in the healthcare sector is transformative for the personalization of patient treatment. Simulations conducted on digital models of human organs provide preemptive testing of treatment processes, facilitating more precise decision-making. Philips develops personalized DTs for cardiac patients and evaluates pre-operative scenarios. Furthermore, DTs enhance the upkeep and administration of medical gadgets used in hospitals (Barricelli et al., 2019).

DT technology presents significant promise for urban management. For instance, in the context of the "Virtual Singapore" initiative executed across the city, a comprehensive digital model of the entire city has been created, serving as a decision support system in various domains, including transportation planning and environmental impact assessment. This method guarantees the cost-effective management of infrastructure projects and facilitates the development of scenarios addressing hazards such as natural disasters (Kritzinger et al., 2018). Notwithstanding the significant advantages conferred by DT technology, many problems persist. Challenges such as data security, system integration, lack of standards, and substantial investment expenditures may impede companies during the technological shift (Fuller et al., 2020). The amalgamation of AI, big data, and 5G technologies will substantially aid in addressing these difficulties.

DT technology is a virtual representation of physical assets that facilitates the simulation and optimization of manufacturing processes (Rosen et al., 2015). Major corporations like Siemens and General Electric employ DTs in their manufacturing processes to proactively identify problems and enhance production efficiency (Grieves & Vickers, 2017). DTs are employed in autonomous production systems, enabling robots to assess their status. This facilitates system optimization autonomously and accelerates processes (Tao et al., 2018).

A consequence of this is that the technology of DTs plays a significant part in the optimization of corporate processes. The skills of real-time data analysis, simulation, and prediction that it possesses allow it to improve the efficiency of enterprises, lower their expenses, and boost their competitiveness. In spite of this, there are challenges that need to be conquered before this technology can become widely used. These challenges include data security, a lack of standards, and integration difficulties.

## RESULTS

In this section, the profound transformation brought about by digitalization and Industry 4.0 in company structures and strategies will be addressed; the opportunities offered by smart systems and IoT from logistics to customer services, the security and data privacy risks encountered in the implementation of these technologies, the challenges of standardization and technological compatibility, and the competitive advantages of IoT and smart systems will be briefly discussed under these headings.

### 1. Challenges and Opportunities

Digitalization and Industry 4.0 not only transform the technical department, but they also bring changes to the company's structure and strategy. As a result, new opportunities arise in logistics and customer service, which are essential in transport management, particularly in production, material handling, warehousing, and logistics. Smart systems and the IoT empower organizations to be resilient and adaptive, enabling them to become more flexible, efficient, innovative, and scalable in ever-changing markets.

The benefits derived from the effective management and operation of IoT systems and smart devices will continue to outweigh the challenges we face today. Although the issues are growing, solutions are also consistently improving. This ongoing progress is driven by technological advancements, including information systems and innovative processes, which enable comprehensive order monitoring from start to finish, daily movement of equipment for order preparation, and efficient procurement execution. Businesses will be able to make quicker and more informed decisions thanks to increasingly intelligent and interconnected systems. Moreover, this evolution will contribute to improved employee performance.

In summary, companies that seize the opportunities presented by IoT and smart systems will likely emerge as leaders in the global competitive landscape. A crucial factor for maintaining competitiveness is ensuring operational efficiency and implementing safety measures. IoT devices and elements must align with the evolving realities of businesses regarding data protection, integration of core mechanisms, and technological identification. On the other hand, companies face some challenges in using these technologies. These challenges will be discussed in this section.

### 2. Security and Data Privacy Risks

The IoT and Industry 4.0 technologies offer significant innovations across a wide range of areas, from manufacturing processes to smart city systems. IoT devices can make business processes more efficient by generating large amounts of data and sharing this data in real-time, but this data flow can become vulnerable to malicious attacks. Especially IIoT systems, which are used in critical processes such as production lines, energy management, and logistics operations, exhibit a more sensitive structure in terms of cybersecurity risks (Sadeghi et al.,2015). One of the most common threats encountered in IoT systems is unauthorized access and data breaches. In the research conducted by Chen and Bridges (2017), it was noted that many IoT devices are easily targeted due to their lack of embedded security mechanisms. The Mirai Botnet attack is one of the most well-known examples of such threats; it targeted millions of IoT devices and caused major system collapses globally (Chen & Bridges, 2017). IoT components in industrial systems are also exposed to similar attacks, causing production lines to halt and resulting in significant economic losses.

IoT systems typically collect user data continuously, storing and processing it in cloud environments.

However, a lack of proper data encryption methods or their incorrect implementation can result in the leakage of personal information. Seneviratne et al. (2017) highlight that IoT-based devices in the healthcare sector exhibit significant deficiencies in safeguarding patient information, allowing sensitive data to be transmitted to unauthorized third parties. These IoT systems possess security vulnerabilities that pose risks not only to individual users but also to corporate entities and national security. Herterich et al. (2015) note that the dependency on IoT in factories and smart city infrastructures exacerbates the exploitation of these vulnerabilities. Industrial espionage, unauthorized data access, and malicious software can lead to large-scale data breaches via IoT devices. When critical infrastructures—such as electric grids, water systems, and transportation networks—are managed by IoT systems, attacks on these systems represent a substantial threat that could result in national security issues.

To minimize these emerging risks, the development of encryption algorithms, secure authentication systems, and regular security updates is of great importance. Two-factor authentication systems ensure that IoT devices are protected against unauthorized access, while decentralized security solutions like blockchain technology can also enhance data security. Additionally, AI-powered cybersecurity solutions detect abnormal network activities and offer preventive measures against attacks.

In conclusion, while IoT and Industry 4.0 technologies provide efficiency and automation advantages in the business world, the security vulnerabilities and data privacy threats posed by these systems should not be overlooked. Advanced security protocols, legal regulations, and awareness-raising training play a critical role in ensuring the security of IoT systems. Until the mentioned security developments occur, tightening existing security measures and continuously monitoring vulnerabilities are of great importance for the widespread adoption of IoT as a reliable technology.

## 3. Standardization and Technological Compatibility Challenges

The IoT and Industry 4.0 offer significant innovations in the digital transformation process, and standardization and technological compatibility have become major necessities for the successful global implementation of these technologies. IoT systems and smart factories developed by different manufacturers often operate with their own proprietary software, hardware, and protocols, leading to a lack of interoperability between these systems. Saleem et al. (2018) emphasize that the lack of standards in IoT-based systems prevents devices and software from working seamlessly together and causes significant issues in big data management.

Next-generation smart devices that need to be integrated into traditional production systems differ in terms of data formats, communication protocols, and system architectures. Smart factories and CPS largely utilize sensors, cloud computing, and AI-supported analytical systems. However, the development of these components by different manufacturers makes seamless integration of the systems difficult. Tariq et al. (2019) state that due to the lack of interoperability between different IoT platforms, there can be significant delays in data sharing and management.

Technological compatibility challenges also pose a significant problem on a global scale. This issue makes it difficult for international companies to adapt the technologies they develop in one country to other markets. Many industry standards are still in the development phase today, and a fully global standard for IoT systems has not yet been established. Jiang and colleagues (2018) state that this situation could delay the Industry 4.0 transformation process and lead to inefficiencies in global production networks. To overcome these challenges, international standard organizations (such as ISO, IEEE, ETSI) and industry leaders are collaborating to develop global IoT standards. Open-source IoT platforms and universal data formats are seen as one of the most important strategies for harmonizing different systems.

AI-powered data transformation tools and standardized API (Application Programming Interface) solutions are among the innovative approaches that can help overcome these inconsistencies.

Ultimately, the lack of standardization and technological compatibility poses one of the most significant challenges to the global adoption of IoT and Industry 4.0. It is essential to integrate systems created by various manufacturers, align them with international data security standards, and establish universal communication protocols. When these processes are effectively managed, IoT-based systems can be implemented more swiftly and reliably, thereby accelerating the transformation process of Industry 4.0.

## 4. Competitive Advantages of IoT and Smart Systems

IoT and smart systems are among the critical technologies that enable businesses to gain a competitive advantage in the digital transformation process.

In today's rapidly changing industrial environment, it is of great importance for businesses not only to increase productivity but also to make their processes more efficient and flexible. IoT-supported smart systems are widely used in a broad range of areas, from production to supply chain management, customer experience to decision support systems (Hou et al., 2023).

IoT devices analyze large datasets from sensors, providing businesses with real-time data and helping them make more informed decisions. For example, smart systems in production lines can detect machine stoppages in advance, optimize maintenance processes, and minimize production interruptions (Yang et al., 2019). This situation increases operational efficiency while providing financial advantages to businesses by reducing costs. In addition, compared to traditional production models, IoT-supported smart factories can respond more quickly to customer demands and dynamically alter production processes (Hou et al., 2023).

The advantages provided in supply chain management and logistics processes are also one of the competitive benefits that IoT offers to businesses. Smart supply chain management systems can optimize inventory levels, reduce delays in logistics processes, and enhance efficiency by enabling more transparent data sharing among supply chain stakeholders (Yang et al., 2019). Especially, with the integration of blockchain and IoT, reliability in the supply chain is increasing and fraud is being prevented.

In conclusion, the application of IoT and smart systems in the business world has become an indispensable topic for companies that want to make a difference. Advantages arising in areas such as real-time data acquisition, flexibility, customer-centric production, supply chain management, and sustainability make the integration of these technologies into business models indispensable. To fully benefit from these advantages, businesses need to invest in technological compliance processes and prioritize issues such as security and data privacy.

## CONCLUSION

In this book chapter, the use of smart systems and IoT technologies for business optimization in the field of engineering has been comprehensively addressed. Especially in the context of digital transformation and Industry 4.0, the importance of these technologies has been emphasized, highlighting how they create an environment that enables the development of smarter, more autonomous, and innovative systems that change the traditional structures of today's industrial production and business processes. Smart systems, by using advanced technologies such as big data analytics, AI, ML, and cloud computing, significantly enhance the operational efficiency of businesses. Thanks to these systems, production errors are minimized, operational costs are reduced, production times are shortened, and more effective decision-making mechanisms are created. The use of real-time data analysis and predictive analytics methods provides businesses with unprecedented accuracy and flexibility.

The IoT forms the essential infrastructure necessary for all these smart systems to operate effectively. Thanks to IoT, the continuity of communication between machine-to-machine and machine-to-human is ensured, making it possible to monitor systems and processes in real-time. This situation enables predictive maintenance applications, inventory optimization, improvement of logistics processes, and more efficient energy management. The real-time and detailed data flow provided by IoT devices increases transparency in supply chains, enabling companies to quickly adapt to market demands and sudden changes.

The notion of the intelligent factory, most notably in the section, points out issues solved by smart systems and IoT technologies in application. For instance, through virtual twin technologies, enterprises have the opportunity to simulate and perfect their production processes in a virtual environment prior to carrying them out physically. This in turn facilitates the minimization of risks, the enhancement of production quality, and the provision of innovation process at a faster pace. The advantages granted to leading companies such as Siemens and Bosch in this field clearly indicate the good and real impacts of technology on businesses. Alongside all these advantages, there are also challenges encountered in the integration of smart systems and IoT into business processes. The most important of these are security and data privacy risks. The size and nature of the data collected by IoT systems can create a vulnerable environment against cyber attacks. Cybersecurity incidents like the Mirai botnet attack clearly highlight the importance of establishing strong security protocols against such risks.

Therefore, it is important for businesses to invest in methods such as AI-based security solutions, blockchain technologies, and advanced authentication mechanisms. Another area of difficulty is the lack of standardization in IoT and smart systems and the issues of technological compatibility. The use of their own software and hardware systems by different manufacturers prevents the systems from working seamlessly with each other. This situation complicates integration processes, increases costs, and restricts innovation. Therefore, establishing global standards with the participation of international organizations such as ISO, IEEE, ETSI, and industry stakeholders is of critical importance. Looking to the future, it is anticipated that companies effectively utilizing the potential of IoT and smart systems will gain significant competitive advantages. These advantages can be listed as faster decision-making ability, more efficient cost management, agility, customized production capacity, and sustainability. Especially in the dimension of sustainability, practices such as energy management, more efficient use of resources, and reduction of carbon footprint will help companies fulfill their environmental responsibilities and strengthen their brand image.

Finally, some strategic recommendations are offered to businesses to make the most effective use of these technologies: Companies should not limit their digital transformation processes to just technical innovations; they should also shape their corporate cultures and security understandings according to the digital transformation vision. Investments in IoT and smart systems should be directly aligned with the overall strategic goals of the companies and should serve clear business objectives. Training programs should be organized to enhance employees' digital skills and analytical competencies, and a culture of human-machine collaboration should be promoted. By actively participating in organizations that contribute to the establishment of international standards, support should be provided for solving global compatibility and integration issues.

In conclusion, Industry 4.0 and IoT-based smart systems are emerging as indispensable technologies for future businesses. It is clear that investments in these technologies, when supported by the right strategy and corporate transformation, will enhance companies' competitive power and better prepare them for future uncertainties. In this context, future research aimed at overcoming existing problems will develop new solutions and innovative applications, enabling technology to reach and be adopted by a wider audience.

# REFERENCES


Alter, S. (2020). Making sense of smartness in the context of smart devices and smart systems. *Information Systems Frontiers*, *22*(2), 381–393. DOI: 10.1007/s10796-019-09919-9

Angelopoulos, A., Michailidis, E. T., Nomikos, N., Trakadas, P., Hatziefremidis, A., Voliotis, S., & Zahariadis, T. (2020). Tackling faults in the Industry 4.0 era—A survey of machine-learning solutions and key aspects. *Sensors (Basel)*, *20*(1), 109. DOI: 10.3390/s20010109 PMID: 31878065

Attaran, M. (2017). Additive Manufacturing: The most promising technology to alter the supply chain and logistics. *Journal of Service Science and Management*, *10*(3), 189–206. DOI: 10.4236/jssm.2017.103017

Atzori, L., Lera, A., & Morabito, G. (2010). Internet of things: A survey. *Computer Networks*, *54*(15), 2787–2805. DOI: 10.1016/j.comnet.2010.05.010

Barricelli, B. R., Casiraghi, E., & Fogli, D. (2019). A survey on digital twin: Definitions, characteristics, applications, and design implications. *IEEE Access : Practical Innovations, Open Solutions*, *7*, 167653–167671. DOI: 10.1109/ACCESS.2019.2953499

Belli, L., Davoli, L., Medioli, A., Marchini, P. L., & Ferrari, G. (2019). Toward Industry 4.0 with IoT: Optimizing business processes in an evolving manufacturing factory. *Frontiers in ICT (Lausanne, Switzerland)*, *6*, 17. Advance online publication. DOI: 10.3389/fict.2019.00017

Carvalho, T. P., Soares, F. A., Vita, R., Francisco, R. D. P., Basto, J. P., & Alcalá, S. G. (2019). A systematic literature review of machine learning methods applied to predictive maintenance. *Computers & Industrial Engineering*, *137*, 106024. DOI: 10.1016/j.cie.2019.106024

Chen, M., Wan, J., & Li, F. (2012). Machine-to-machine communications: Architectures, standards and applications. *KSII Transactions on Internet and Information Systems*, *2*, 480–497. DOI: 10.3837/tiis.2012.02.002

Chen, Q., & Bridges, R. A. (2017). Automated behavioral analysis of malware: A case study of WannaCry ransomware. In *2017 16th IEEE International Conference on Machine Learning and Applications (ICMLA)* (pp. 454–460). https://doi.org/ DOI: 10.1109/ICMLA.2017.0-119

Chen, Z., Sivaparthipan, C. B., & Muthu, B. (2022). IoT based smart and intelligent city energy optimization. *Sustainable Energy Technologies and Assessments, 49*, 101724. DOI: 10.1016/j.seta.2021.101724

Correia, D. M., Teixeira, L., & Marques, J. L. (2021). Smart supply chain management: The 5W1H open and collaborative framework. In *2021 IEEE 8th International Conference on Industrial Engineering and Applications (ICIEA)* (pp. 401–405). https://doi.org/DOI: 10.1109/ICIEA52957.2021.9436817

Datexcorp. (2025, March). Inventory Management. https://www.datexcorp.com/

Deloitte. (2025, March). https://www2.deloitte.com/

Fuller, A., Fan, Z., Day, C., & Barlow, C. (2020). Digital twin: Enabling technologies, challenges and open research. *IEEE Access : Practical Innovations, Open Solutions, 8*, 108952–108971. DOI: 10.1109/ACCESS.2020.2998358

Grieves, M., & Vickers, J. (2017). Digital twin: Mitigating unpredictable, undesirable emergent behavior in complex systems. In *Transdisciplinary Perspectives on Complex Systems* (pp. 85–113). https://doi.org/DOI: 10.1007/978-3-319-38756-7_4

Gubbi, J., Buyya, R., Marusic, S., & Palaniswami, M. (2013). Internet of Things (IoT): A vision, architectural elements, and future directions. *Future Generation Computer Systems, 29*(7), 1645–1660. DOI: 10.1016/j.future.2013.01.010

Gündüz, M. Z., & Daş, R. (2018). Nesnelerin interneti: Gelişimi, bileşenleri ve uygulama alanları. *Pamukkale Üniversitesi Mühendislik Bilimleri Dergisi, 24*(2), 327–335. DOI: 10.5505/pajes.2017.89106

Hermann, M., Pentek, T., & Otto, B. (2016, January). Design principles for Industrie 4.0 scenarios. In *2016 49th Hawaii International Conference on System Sciences (HICSS)* (pp. 3928–3937). IEEE. https://doi.org/DOI: 10.1109/HICSS.2016.488

Herterich, M. M., Uebernickel, F., & Brenner, W. (2015). The impact of cyber-physical systems on industrial services in manufacturing. *Procedia CIRP, 30*, 323–328. DOI: 10.1016/j.procir.2015.02.110

Hofmann, E., & Rüsch, M. (2017). Industry 4.0 and the current status as well as future prospects on logistics. *Computers in Industry, 89*, 23–34. DOI: 10.1016/j.compind.2017.04.002

Hou, K. M., Diao, X., Shi, H., Ding, H., Zhou, H., & De Vaulx, C. (2023). Trends and challenges in AIoT/IIoT/IoT implementation. *Sensors (Basel), 23*(11), 5074. DOI: 10.3390/s23115074 PMID: 37299800

IBM. (2025, March). The smarter supply chain of future. https://www.ibm.com/downloads/cas/78NWEBYM

Iddum, T. S., Sahu, S., Sujatha, R., & Kumar, C. V. R. (2022). Ethereum blockchain based logistics application for smart supply chain management (pp. 1–7). https://doi.org/DOI: 10.1109/GlobConPT57482.2022.9938171

Javaid, S., Sufian, A., Pervaiz, S., & Tanveer, M. (2018, February). Smart traffic management system using Internet of Things. In *2018 20th International Conference on Advanced Communication Technology (ICACT)* (pp. 393–398). IEEE. https://doi.org/DOI: 10.23919/ICACT.2018.8323770

Jian, Q., Ying, L., & Grosvenor, R. (2016). A categorical framework of manufacturing for Industry 4.0 and beyond. *Procedia CIRP*, *52*, 173–178. DOI: 10.1016/j.procir.2016.08.005

Jiang, J. R. (2018). An improved cyber-physical systems architecture for Industry 4.0 smart factories. *Advances in Mechanical Engineering*, *10*(6), 1687814018784192. DOI: 10.1177/1687814018784192

Kritzinger, W., Karner, M., Traar, G., Henjes, J., & Sihn, W. (2018). Digital twin in manufacturing: A categorical literature review and classification. *IFAC-PapersOnLine*, *51*(11), 1016–1022. DOI: 10.1016/j.ifacol.2018.08.474

Lambert, D. M., & Cooper, M. C. (2000). Issues in supply chain management. *Industrial Marketing Management*, *29*(1), 65–83. DOI: 10.1016/S0019-8501(99)00113-3

Lee, H., & Özer, Ö. (2007). Unlocking the value of RFID. *Production and Operations Management*, *16*(1), 40–64. DOI: 10.1111/j.1937-5956.2007.tb00165.x

Lee, J., Bagheri, B., & Kao, H. A. (2015). A cyber-physical systems architecture for Industry 4.0-based manufacturing systems. *Manufacturing Letters*, *3*, 18–23. DOI: 10.1016/j.mfglet.2014.12.001

Longo, F., Nicoletti, L., & Padovano, A. (2019). Ubiquitous knowledge empowers the smart factory: The impacts of a service-oriented digital twin on enterprises' performance. *Annual Reviews in Control*, *47*, 221–236. DOI: 10.1016/j.arcontrol.2019.01.001

Lu, Y. (2017). Industry 4.0: A survey on technologies, applications and open research issues. *Journal of Industrial Information Integration*, *6*, 1–10. DOI: 10.1016/j.jii.2017.04.005

Medina-Borja, A. (2015). Editorial column—Smart things as service providers: A call for convergence of disciplines to build a research agenda for the service systems of the future. *Service Science*, *7*(1), ii–iv. DOI: 10.1287/serv.2014.0090

Menon, A., He, Y., Bahirat, P., & Knijnenburg, B. P. (2018). A data-driven approach to designing for privacy in household IoT. *ACM Transactions on Interactive Intelligent Systems, 1*(1). *ACM Transactions on Interactive Intelligent Systems*, *1*(1), 1–47. DOI: 10.1145/3241378

Mrugalska, B., & Wyrwicka, M. K. (2017). Towards lean production in Industry 4.0. *Procedia Engineering*, *182*, 466–473. DOI: 10.1016/j.proeng.2017.03.135

Pipiay, G. T., Chernenkaya, L. V., & Mager, V. E. (2021, January). Quality indicators of instrumentation products according to the "Quality 4.0" concept. In *IEEE Conference of Russian Young Researchers in Electrical and Electronic Engineering (ElConRus)* (pp. 1032–1036). https://doi.org/DOI: 10.1109/ElConRus51938.2021.9396535

Raoufi, K., Sutherland, J. W., Zhao, F., Clarens, A. F., Rickli, J. L., Fan, Z., Huang, H., Wang, Y., Lee, W. J., Mathur, N., Triebe, M. J., Desabathina, S. S., & Haapala, K. R. (2024). Current state and emerging trends in advanced manufacturing: Smart systems. *International Journal of Advanced Manufacturing Technology*, *134*(7), 3031–3050. DOI: 10.1007/s00170-024-14279-z

Rosen, R., Wichert, G. V., Lo, G., & Bettenhausen, K. D. (2015). About the importance of autonomy and digital twins for the future of manufacturing. *IFAC-PapersOnLine*, *48*(3), 567–572. DOI: 10.1016/j.ifacol.2015.06.141

Sadeghi, A. R., Wachsmann, C., & Waidner, M. (2015). Security and privacy challenges in industrial Internet of Things. In *Proceedings of the 52nd Annual Design Automation Conference* (pp. 1–6). https://doi.org/DOI: 10.1145/2744769.2747942

Saleem, J., Hammoudeh, M., Raza, U., Adebisi, B., & Ande, R. (2018). IoT standardisation: Challenges, perspectives and solution. In *Proceedings of the 2nd International Conference on Future Networks and Distributed Systems* (pp. 1–9). https://doi.org/DOI: 10.1145/3231053.3231103

Sarker, I. H. (2022). AI-based modeling: Techniques, applications and research issues towards automation, intelligent and smart systems. *SN Computer Science*, *3*(2), 158. DOI: 10.1007/s42979-022-01043-x PMID: 35194580

Sarker, I. H., Kayes, A. S. M., Badsha, S., Alqahtani, H., Watters, P., & Ng, A. (2020). Cybersecurity data science: An overview from machine learning perspective. *Journal of Big Data*, *7*(1), 1–29. DOI: 10.1186/s40537-020-00318-5

Seneviratne, S., Hu, Y., Nguyen, T., Lan, G., Khalifa, S., Thilakarathna, K., Hassan, M., & Seneviratne, A. (2017). A survey of wearable devices and challenges. *IEEE Communications Surveys and Tutorials*, *19*(4), 2573–2620. DOI: 10.1109/COMST.2017.2731979

Shukurova, N. K. (2025). Economic efficiency and innovative development strategies in the service sector. *EduVision: Journal of Innovations in Pedagogy and Educational Advancements, 1*(2), 112–119. https://brightmindpublishing.com/index.php/ev/article/view/97

Snatkin, A., Karjust, K., Majaki, J., Aruvali, T., & Eiskop, T. (2013). Real time production tracking system in SME. *Estonian Journal of Engineering*, *19*(1), 32–75. DOI: 10.3176/eng.2013.1.06

Tao, F., Zhang, H., Liu, A., & Nee, A. Y. C. (2019). Digital twin in industry: State-of-the-art. *IEEE Transactions on Industrial Informatics*, *15*(4), 2405–2415. DOI: 10.1109/TII.2018.2873186

Tariq, N., Asim, M., Al-Obeidat, F., Zubair Farooqi, M., Baker, T., Hammoudeh, M., & Ghafir, I. (2019). The security of big data in fog-enabled IoT applications including blockchain: A survey. *Sensors (Basel)*, *19*(8), 1788. DOI: 10.3390/s19081788 PMID: 31013993

Thames, L., & Schaefer, D. (2016). Software defined cloud manufacturing for Industry 4.0. *Procedia CIRP*, *52*, 12–17. DOI: 10.1016/j.procir.2016.07.041

Timm, I. J., & Lorig, F. (2015). Logistics 4.0: A challenge for simulation. In *2015 Winter Simulation Conference (WSC)* (pp. 3118–3119). IEEE. https://doi.org/DOI: 10.1109/WSC.2015.7408428

Wang, S., Torngren, M., & Onori, M. (2015). Current status and advancement of cyber-physical systems in manufacturing. *Journal of Manufacturing Systems*, *37*, 517–527. DOI: 10.1016/j.jmsy.2015.04.008

Wang, S., Wan, J., Zhang, D., Li, D., & Zhang, C. (2016). Towards smart factory for Industry 4.0: A self-organized multi-agent system with big data based feedback and coordination. *Computer Networks*, *101*, 158–168. DOI: 10.1016/j.comnet.2015.12.017

Xu, L. D. (2011). Enterprise systems: State-of-the-art and future trends. *IEEE Transactions on Industrial Informatics*, *7*(4), 630–641. DOI: 10.1109/TII.2011.2167156

Xu, L. D., He, W., & Li, S. (2014). Internet of Things in industries: A survey. *IEEE Transactions on Industrial Informatics*, *10*(4), 2233–2243. DOI: 10.1109/TII.2014.2300753

Xu, L. D., Xu, E. L., & Li, L. (2018). Industry 4.0: State of the art and future trends. *International Journal of Production Research, 56*(8), 2941–2962. DOI: 10.1080/00207543.2018.1444806

Yang, H., Kumara, S., Bukkapatnam, S. T. S., & Tsung, F. (2019). The Internet of Things for smart manufacturing: A review. *IISE Transactions, 51*(11), 1190–1216. DOI: 10.1080/24725854.2018.1555383

Yang, J. P., Wang, W. L., & Zhou, S. K. (2019). A design of integrated quality management system based on artificial intelligence (AI) technology. *Destech Transactions on Computer Science and Engineering*, 15–20.

Zahid, A., Leclaire, P., Hammadi, L., Costa-Affonso, R., & El Ballouti, A. (2025). Exploring the potential of Industry 4.0 in manufacturing and supply chain systems: Insights and emerging trends from bibliometric analysis. *Supply Chain Analytics. Article, 100108*, 100108. Advance online publication. DOI: 10.1016/j.sca.2025.100108

## ADDITIONAL READING

Bandyopadhyay, D., & Sen, J. (2011). Internet of Things: Applications and challenges in technology and standardization. *Wireless Personal Communications, 58*(1), 49–69. DOI: 10.1007/s11277-011-0288-5

Cañas, H., Mula, J., Díaz-Madroñero, M., & Campuzano-Bolarín, F. (2021). Implementing Industry 4.0 principles. *Computers & Industrial Engineering, 158*, 107379. DOI: 10.1016/j.cie.2021.107379

Gubán, M., & Kovács, G. (2017). Industry 4.0 conception. *Acta Technica Corviniensis – Bulletin of Engineering, 10*(1), 111.

Rahman, H., & Hussain, M. I. (2020). A comprehensive survey on semantic interoperability for Internet of Things: State-of-the-art and research challenges. *Transactions on Emerging Telecommunications Technologies, 31*(12), e3902. DOI: 10.1002/ett.3902

# Chapter 12
# Digital Innovation as the Backbone of Developing Economies: A Study

**T. Nithya**
https://orcid.org/0000-0001-9738-9910
*Velalar College of Engineering and Technology, India*

**V. Chandrasekaran**
*Velalar College of Engineering and Technology, India*

**S. Mahendrakumar**
https://orcid.org/0000-0001-6966-2545
*Velalar College of Engineering and Technology, India*

**N. V. Keerthana**
https://orcid.org/0000-0001-8878-2083
*Velalar College of Engineering and Technology, India*

**A. Priya Keerthana**
*Velalar College of Engineering and Technology, India*

## ABSTRACT

*Digital innovation has emerged as a transformative force driving the growth and development of economies worldwide, particularly in developing nations. By leveraging technology, these countries can overcome structural challenges, improve productivity, and achieve sustainable economic progress. In an era defined by rapid technological advancements, digital tools and platforms are playing a pivotal role*

DOI: 10.4018/979-8-3373-1534-8.ch012

*in fostering economic inclusion, enhancing public services, and creating opportunities for individuals and businesses alike. Developing economies often face hurdles such as limited infrastructure, inefficient systems, and high levels of poverty. Digital innovation offers solutions to these issues by introducing cost-effective, scalable technologies that address local challenges while fostering global competitiveness. Mobile connectivity, for instance, has revolutionized access to financial services in regions where traditional banking systems are underdeveloped.*

## INTRODUCTION

Economic Inclusion: A significant catalyst for change in the economies of developing countries is the area of digital innovation. Developing economies tend to use the benefits of available digital technologies to overcome their development obstacles and propel faster socioeconomic changes. This is coupled with the fact that most developing countries do not have the culture of creating and providing the required data set that measuring organizations can use to measure their transition to a digitized economy (Oloyede, Faruk, Noma, Tebepah, & Nwaulune, 2023).The use of digital innovation can be seen in areas such as poverty alleviation or the use of digital health platforms to enhance healthcare service delivery. The term Digital Economy was invented by Tapscot Don, an American businessman, and fell in its birth year between the definitions of 'digital economy' have been changing. In the beginning, the definition and related analyses focused on Internet adoption, use, and several related economic implications The definition of Digital Economy has changed over time to take account of the shift in technology in terms of how people are engaging with each other for higher social welfare, individual interacting with businesses, companies interacting (Oloyede et al., 2023). Digital innovation has a boundless impact, as its utilization expands, so does it elicit change in developing economies, making it global interconnectedness' strongest weapon for change (Dana, Salamzadeh, Hadizadeh, Heydari, & Shamsoddin, 2022). We will also look into the different sectors that apply digital innovation and the problems that should be solved in order to reap maximum benefit in the future.

## A REVIEW - DIGITAL TECHNOLOGIES AND TRANSFORMATIONS

Digital technologies drive transformative changes by enhancing innovation, efficiency, and resilience across ecosystems. Figure 1 illustrates the key factors influencing digital innovation ecosystem resilience, including digital technology application, human resources, R&D investment, inter-provincial competitive pressure, and innovation environment. These elements operate across organizational and environmental levels to strengthen ecosystem adaptability and sustainability.

*Figure 1. Digital Innovation Ecosystem Resilience*

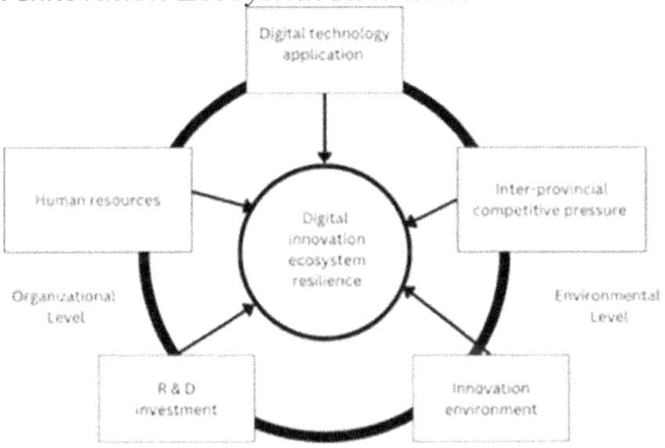


## CONTRIBUTION OF DIGITAL INNOVATION TO ECONOMIES AND BUSINESSES

Digital innovation assists in the development of economies and is extremely useful in overcoming existing traditional barriers, increasing productivity, and creating new opportunities for development (Steiber, Alänge, Ghosh, & Goncalves, 2021). Under technology, digitization and digital transformation are all driven by forces that fuel the techno-centric world: exponential growth of computing power, increasing ubiquitous connectivity, and big data. The third piece of the challenge of any meta-strategy for digital transformation is its complicated strategic implications. Digital transformation can alter the potential for value creation and capture (Furr, Ozcan, & Eisenhardt, 2022). With the purchase and availability of digital technologies, people can achieve efficiency and new standards of accessible productivity within developing nations. The importance of innovation in its limitless form goes beyond imagination; a few examples of importance can be seen in the role of innovation in the alleviation of poverty or the improvement in the quality of life of the population in a country. Increased economic transformation valuation – innovation plays an important role in economic growth. It shall apply toward an increase in GDP.

- Various forms of innovation allow for the improvement of existing processes. Businesses use big data analytics to make informed decisions and better understand consumer behaviour. E-commerce platforms connect sellers and buyers using digital marketplaces to stimulate trade and create new revenue streams.
- Digital innovation helps in the emergence of new industries such as Fintech, edtech, and health tech, which diversify economies and create job opportunities.

In Kenya, mobile payment solutions like M-Pesa have revolutionized financial transactions and created jobs in app development, highlighting the features and challenges of the digital economy (Bennett & McWhorter, 2021). Key features include the rise in e-commerce, digital services, economic growth, cybersecurity, and customer service. Cloud-based solutions and Internet of Things (IoT) integration lower infrastructure and maintenance costs, making advanced technologies more accessible for resource-constrained SMEs (Ali & Hajjar, 2024). Digital content creation platforms like YouTube and TikTok allow users to earn their salary through digital media. Digital platforms like Uber allow workers and entrepreneurs to adopt flexible employment opportunities. Businesses can expand digitally, which enables them to enter the international market. This is important because it helps to increase trade and attract foreign investment. India has become a world power in software and IT services development. In Indonesia, there occurred a major transformation occurred in the digital world which transforming all businesses into a virtual community (Aminah & Saksono, 2021).

## FINANCIAL INCLUSION AND FINTECH REVOLUTION

Especially in emerging economies, financial inclusion is an important aspect of socio-economic advancement because much of the population is excluded from the mainstream banking system. An important strategic implication of digital transformation is to motivate establishment or membership in a platform as a desirable strategy for global companies  Establishing vs. joining a platform also necessitates consideration of trade-offs between short- and long-term results. But for small and medium enterprises, partnerships with platforms like the Mastercard Farmers Network in Africa or Pinduoduo in China are net positive over the short and long term because they allow access to tools or markets that were previously inaccessible (Furr et al., 2022). The Fintech revolution, which is powered by digital innovation changed how financial services are accessed and provides solutions to increase access and affordability. By adoption of mobile money, blockchain, and artificial intelligence, fintech has changed the economic and financial landscape for the better.

## HEALTHCARE'S ADAPTATION OF DIGITAL TRANSFORMATION

- The use of new technologies in healthcare, also known as Digital Transformation in Healthcare, has improved the entire healthcare sector's service delivery and communication systems. The introduction of telemedicine and remote healthcare services indeed stands out as revolutionary.
- Telemedicine improves availability, especially for patients in remote, rural regions or areas that do not have adequate healthcare facilities.
- This innovation has also been proven to reduce costs while improving health care service delivery for both patients and service providers. In addition, with Integrated and pervasive health information systems, medical data is collected, stored, and processed in a completely different way.

These systems have revolutionized patient information management. They support real-time patient information retrieval, disease prediction, and patient-specific treatment. This detail enables healthcare professionals to make informed decisions, providing them with tangible evidence on which to base their calculations and outcomes (Danladi, Prasad, Modibbo, Ahmadi, & Ghasemi, 2023). Additionally, mHealth applications and wearable health technologies have enabled individuals to self-monitor using smartphones, watches, and fitness smart devices. These innovations allow valuable data to be captured concerning a person's key health parameters, like heart rate and physical exertion.

### Education and Skill Development through Digital Platforms

Education and Skill Development via Digital Platforms has changed the way we learn by using technology to enhance the accessibility, inclusiveness, and flexibility of education. E-learning providers such as Coursera, Udemy, and Khan Academy have transformed education, making it accessible to people all over the world, irrespective of their location or financial situation. Such self-paced learning systems are designed for people who are professionals, students living in far-off regions, and those suffering from disabilities. Concerning various reasons behind the conversion and e-government adaptation of the government; The shift toward e-government is a sign of a broader trend to reform the public sector, which has built up in numerous nations in ongoing years, which has been driven in general by the desires of residents the world over, who are putting new requests on governments.

Some requests are given for efficiency, transparency, and overall better performance, and some are trained by Innovative Wave (Aminah & Saksono, 2021). Moreover, digital literacy programs serve the underprivileged communities by providing them with the required skills to succeed in the digital world.

Governments and institutions around the globe have started programs to provide essential digital education, including how to use computers and the internet, to less privileged communities, thereby reducing the gap in technology use and promoting socio-economic inclusion (Kraus et al., 2021). A further important element of the educational digital ecosystem is the contribution of Massive Open Online Courses (MOOCs) to skill development. MOOCs have been started by well-known colleges such as MIT, Harvard, and Stanford, which offer their courses without charge or at a minimal cost. Digital transformation in the government sector is being complicated by issues such as software integration and upgrades, network interoperability, real-time processes, application synchronization, and security. Inter-organisational information integration has emerged as a key enabler of e-government. Crossing traditional government lines to integrate and share information requires interactions among many players (Schneider & Kokshagina, 2021). These courses cover a wide range of subjects, including artificial intelligence. These courses often include certifications that enhance employability and career advancement opportunities. MOOCs also help learners stay on top of emerging trends and technologies, which is especially important in fields such as IT, healthcare, and engineering, where you can never have enough skills. Interactive elements such as forums, peer reviews, and live sessions foster a collaborative learning environment that enriches the education process.

In general, they increasingly make education more accessible and inclusive, develop skills and knowledge, and promote lifelong learning. The transformation is what drives a part of personal growth, ability to maximize potential, and employability, and is a devastatingly key (yet neglected) aspect of global economic development.

## DIGITAL INFRASTRUCTURE AS A CATALYST FOR DEVELOPMENT

Digital Infrastructure as a Catalyst for Development plays a pivotal role in driving socio-economic growth by enabling widespread access to technology and connectivity. At the heart of this transformation is the importance of internet penetration and connectivity, which act as foundational elements for modern economies.

- For developing regions, increased connectivity reduces the digital divide, empowering individuals and businesses to engage in the global economy.

Countries with high internet penetration often see significant improvements in GDP, innovation, and social inclusion, making it essential for governments and organizations to invest in broadband expansion and affordable data plans.

- The advent of 5G technology further accelerates this transformation by providing faster speeds, lower latency, and greater connectivity capacity.
- With 5G, innovations like smart cities, autonomous vehicles, telemedicine, and IoT devices become more viable, fostering advancements in various sectors. Affordable devices, such as budget-friendly smartphones and tablets, complement this progress by ensuring that even low-income populations can access digital resources.
- Public-private partnerships (PPPs) play a critical role in building robust digital infrastructure. Governments, in collaboration with private sector stakeholders, are investing in projects to expand broadband networks, establish data centers, and improve rural connectivity. This denotes that organizational, environmental, and technological factors significantly affect the ability of SMEs to digitalize their business. These results are a perfect symptom of the influence of organizational, environmental, and technological factors on SMEs' ability to digitalize their process and activities. These findings are in line with the earlier results and findings in the literature that a sufficient level of technology resources (Alraja, Hussein, & Ahmed, 2021).
- For instance, collaborations like India's BharatNet project aim to provide high-speed internet to rural areas, enabling access to education, healthcare, and e-governance. Private companies bring expertise, technology, and funding, while governments provide policy support and regulatory frameworks to ensure equitable distribution of resources.
- These partnerships also foster innovation by combining public accountability with private efficiency, resulting in cost-effective and sustainable solutions.

Moreover, digital infrastructure lays the foundation for inclusive economic growth. It supports small and medium-sized enterprises (SMEs) by giving them access to e-commerce platforms, digital payments, and online marketplaces, enabling them to scale their businesses. Enhanced connectivity also creates opportunities for remote work and online learning, which became especially critical during the COVID-19 pandemic (Ragazou, Passas, & Sklavos, 2022).

In addition to economic benefits, digital infrastructure contributes to social development by improving access to essential services. For example, telemedicine platforms reduce healthcare disparities by connecting remote areas with medical professionals, while e-governance initiatives streamline public service delivery, enhancing transparency and accountability. Furthermore, digital inclusion promotes gender equality by providing women with access to education, employment opportunities, and entrepreneurship resources (Bansal, Panchal, Jabeen, Mangla, & Singh, 2023). In conclusion, digital infrastructure acts as a catalyst for development by fostering economic growth, reducing inequalities, and enhancing quality of life. By prioritizing internet penetration, leveraging 5G technology, ensuring affordability, and promoting public-private partnerships, nations can create a more connected and inclusive future where digital transformation drives sustainable progress.

## DIGITAL INNOVATION IN AGRICULTURE AND SMART AGRICULTURE

Changing the Way Farmers Harvest. The most significant part of this transformation is the adoption of IoT and Artificial Intelligence (AI) technology in precision farming, helping farmers improve asset utilization and increase crop output. IoT-enabled devices, including drones and sensors, help monitor soil condition, humidity, temperature, and the health of crops in real-time by being deployed in fields. These devices gather a great deal of information about the environment, and AI algorithms processing the data deliver insights, like when to irrigate, fertilize, or harvest crops. 4IR technologies, such as precision agriculture and robotics, can optimize resource use, reduce waste, and improve crop yields, leading to increased food production. 4IR can promote sustainable agricultural practices by reducing reliance on pesticides and fertilizers, conserving water resources, and minimizing environmental impact (Uwaoma et al., 2023). This is both reducing waste and reducing environmental damage. Also, farmers' capacity to meet global food demand in a resource-efficient manner is that AI-driven utilities can predict pest infestations or diseases and advise farmers to take remedial action to avoid losing crops and money.

One of the most notable developments is the emergence of mobile apps that deliver vital information to farmers, such as the weather, market prices, and agricultural methods to follow (Dayıoğlu & Turker, 2021). These apps empower farmers, especially in rural areas, to make informed decisions and plan their activities efficiently. Accurate weather predictions help them determine optimal planting and harvesting times, reducing losses due to unforeseen weather events. This is both reducing waste and reducing environmental damage. Also, farmers' capacity to meet global food demand in a resource-efficient manner is that AI-driven utilities can predict pest infestations or diseases and advise farmers to take remedial action to avoid losing crops and money. One of the most notable developments is the emergence of mobile apps that deliver vital information to farmers, such as the weather, market prices, and agricultural methods to follow.

These technologies are most commonly referred to as the IoT, which has offered new opportunities for precision farming (Thuan et al., 2022). IoT-powered smart irrigation systems can make real-time decisions on the use of water gained from weather data and soil conditions that help conserve water and money. Likewise, AI-enabled robots are being used for activities such as sowing, weeding, and harvesting, mitigating labor shortages as well as improving efficiency. In conclusion, digital innovations in farming, driven by IoT, AI, mobile applications, and blockchain, are reshaping the agricultural sector. These technologies not only enhance productivity and profitability for farmers but also promote sustainability and transparency across the supply chain.

In addition to these core technologies, drones and satellite imagery are increasingly being used in modern farming to map fields, monitor crop growth, and identify areas needing attention. Smart irrigation systems powered by IoT can adjust water usage automatically based on real-time weather data and soil conditions, conserving water and reducing costs. Similarly, AI-powered robotics is being deployed for tasks like planting, weeding, and harvesting, addressing labor shortages and increasing efficiency (Maiurova et al., 2022).

## E-GOVERNANCE AND PUBLIC SERVICE DELIVERY

E-Governance and Public Service Delivery have emerged as transformative approaches to modern governance, leveraging digital platforms and advanced technologies to enhance efficiency, transparency, and citizen engagement (Kurniawan, Othman, Hwang, & Gikas, 2022). Central to this transformation is the adoption of digital platforms for transparent governance, which simplify the delivery of public services and improve accountability. E-governance initiatives, such as online portals and mobile applications, enable citizens to access services like paying taxes, applying for government schemes, or obtaining essential documents without the need for physical visits. These platforms reduce bureaucratic delays, eliminate corruption by minimizing human intervention, and promote transparency through real-time tracking of applications. For instance, platforms like India's DigiLocker allow citizens to securely store and share digital versions of government-issued documents, while Estonia's e-Residency program enables seamless interaction with public services online, setting a benchmark for digital governance worldwide.

The concept of smart cities further exemplifies the integration of digital innovation into urban planning and public service delivery (Felicetti, Corvello, & Ammirato, 2024). Smart cities use technology-driven solutions to address challenges related to urbanization, such as traffic congestion, waste management, and energy consumption. IoT-enabled sensors and networks collect data from various city systems, providing real-time insights for efficient management of resources. For example, smart traffic management systems analyze traffic flow and adjust signal timings dynamically to reduce congestion, while smart grids optimize energy distribution to minimize waste (Khan, Piprani, & Yu, 2022). In waste management, sensor-equipped bins notify authorities when they need to be emptied, ensuring timely disposal and cleaner urban spaces. Additionally, smart cities focus on citizen well-being by incorporating green spaces, intelligent public transport systems, and sustainable housing projects. Initiatives like Singapore's Smart Nation and Barcelona's smart city project demonstrate how technology can create livable, sustainable, and efficient urban environments.

Another critical component of e-governance is the role of AI and data analytics in policymaking. AI enables governments to process vast amounts of data quickly, identify patterns, and derive actionable insights that inform policy decisions. Data-driven policymaking enhances efficiency by addressing specific challenges with targeted solutions. For example, predictive analytics can forecast the impact of proposed policies, helping policymakers make informed choices. AI is also being used to analyze public feedback collected through digital platforms, ensuring citizen-centric governance. In healthcare, AI algorithms analyze disease trends to allocate resources effectively, while in education, data analytics identify gaps in access or quality, enabling targeted interventions. Moreover, AI-powered chatbots and virtual assistants are transforming citizen services by providing instant responses to queries, reducing the burden on traditional call centers.

E-governance also plays a significant role in disaster management and public safety. AI and data analytics are used to predict natural disasters like floods, earthquakes, or cyclones, enabling timely evacuation and resource deployment (Nugraha, Setiawan, Nathan, & Fekete-Farkas, 2022). Digital platforms streamline the coordination of relief efforts, ensuring that aid reaches affected communities efficiently. Additionally, surveillance technologies, including AI-powered facial recognition systems, enhance public safety by monitoring crowded areas and identifying potential threats (Ghosh, Hughes, Hodgkinson, & Hughes, 2022).

However, the success of e-governance relies heavily on digital inclusion. Ensuring that all citizens, especially those in rural and marginalized communities, have access to the internet and digital literacy programs is essential for equitable service delivery. Governments worldwide are working on bridging the digital divide by investing in broadband infrastructure, affordable devices, and awareness campaigns.

## WOMEN EMPOWERMENT THROUGH DIGITAL PLATFORMS

Women's Empowerment through Digital Platforms has become a cornerstone of fostering gender equality and providing opportunities for women to thrive in the digital age. By leveraging technology, digital platforms have emerged as powerful tools to bridge the gender gap, enable entrepreneurship, and showcase inspiring success stories. These platforms empower women not only to participate in the workforce but also to lead initiatives, gain financial independence, and contribute to societal progress.

One of the most significant ways digital platforms empower women is by bridging the gender gap through digital literacy. Access to technology alone is not enough; equipping women with the skills and confidence to use digital tools is essential for them to reap the benefits of the digital revolution (Du, Shen, Song, & Zhang, 2023). Digital literacy programs specifically tailored for women are breaking barriers to education and employment, particularly in underserved communities. For instance, initiatives like Google's Internet Saathi in India have trained millions of rural women to use smartphones and the internet, enabling them to access information, connect with markets, and explore new opportunities. Digital literacy helps women understand and access government schemes, healthcare services, and e-learning platforms, empowering them to make informed decisions and improve their quality of life. By reducing the digital divide, these programs also promote inclusivity, ensuring that women are not left behind in an increasingly tech-driven world.

Digital platforms also play a pivotal role in enabling women entrepreneurs and workers, providing them with tools to start, grow, and sustain their businesses (Bai, Qi, Li, & Xu, 2023). E-commerce platforms, such as Etsy, allow women to market their products globally, eliminating the constraints of local markets. These platforms offer user-friendly interfaces, affordable listing fees, and marketing tools, empowering women to showcase their creativity and craftsmanship. Similarly, gig economy platforms like Upwork and Freelancer enable women to work from home, balancing professional and personal responsibilities while earning a steady income. These platforms are especially transformative for women in conservative or rural societies where traditional roles often limit their mobility and opportunities for employment.

In addition to e-commerce and freelancing platforms, microfinance and financial technology (fintech) platforms are critical enablers of women's empowerment. Platforms like GrameenBank and Kiva provide women entrepreneurs with access to microloans, enabling them to invest in small businesses or pursue skill development. Digital payment systems such as eSewa in Nepal simplify financial transactions, allowing women to save, invest, and manage their finances independently. Access to financial services not only boosts women's economic participation but also enhances their decision-making power within households and communities.

## RESEARCH METHODOLOGY

In order to investigate how digital innovation supports economic development in underdeveloped countries, the study employs a descriptive and exploratory research design. It aims to outline current digital projects and investigate their wider socioeconomic effects. The study applies thematic analysis to code qualitative data and identify patterns related to:

- Digital inclusion
- Technology adoption challenges
- Socioeconomic impacts (job creation, poverty reduction)
- Institutional responses (regulations, investments)

This study uses a descriptive and exploratory research style to investigate how digital innovation propels economic growth in developing nations. The study mostly uses secondary data that was gathered from reliable sources, including government policy documents, academic journals, UNDP publications, and World Bank reports. The influence of digital innovation on a variety of industries, including banking, healthcare, education, and government, is evaluated through comparative analysis. To find patterns and trends about digital adoption, inclusion, and growth, thematic analysis is utilized. Clear data representation is achieved through the use of fundamental statistical techniques and visualization platforms like Tableau and Excel.

## CHALLENGES AND FUTURE PROSPECTS

Adopting digital innovation is transforming industries and societies, yet it is not without its challenges. One of the most prominent barriers is the digital divide and unequal access to technology. In many parts of the world, access to the internet and digital devices remains a privilege rather than a universal right. Figure 2 illustrates a structured approach to driving strategic transformation within an organization. It begins with defining a clear vision and setting specific goals, which serve as the foundation for all future actions. Next, the current landscape is assessed to identify existing capabilities, challenges, and opportunities. Based on this assessment, key initiatives are prioritized and selected to align with the overarching vision. An implementation plan is then developed to guide execution. Finally, the process emphasizes the importance of embracing continuous learning and iteration to adapt to changing circumstances and ensure ongoing improvement.

Another significant challenge is the increasing prevalence of highlighting the features and challenges of the digital economy. Key features include the rise in e-commerce, digital services, economic growth, cybersecurity threats, and privacy concerns. As organizations and individuals become more reliant on digital platforms, they are exposed to risks such as data breaches, ransomware attacks, and identity theft (Kulandaivel & Kulandaivel, 2025). Cybercriminals are constantly evolving their methods, targeting vulnerabilities in systems to exploit sensitive data. For individuals, concerns over data privacy often deter them from adopting digital solutions, particularly in regions with weak data protection regulations. For businesses, cyber attacks can lead to financial losses, reputational damage, and loss of consumer trust. Governments and organizations must prioritize highlighting the features and challenges of the digital economy. Key features include the rise in e-commerce, digital services, economic growth, cybersecurity by implementing robust protection measures, raising awareness about safe online practices, and enforcing stringent data privacy laws. Trust in digital systems is essential for widespread adoption, and without addressing these risks, the potential of digital innovation remains limited.

The lack of digital skills and resistance to change also hinders the adoption of digital innovation. Many individuals, especially in older age groups or less educated populations, struggle to adapt to new technologies due to a lack of familiarity or confidence in using digital tools. This digital illiteracy prevents them from fully participating in modern economies or accessing vital services. For businesses, resistance to change among employees and management can slow the integration of digital technologies. People are often hesitant to adopt innovations due to fear of job displacement, disruption to established workflows, or simply the challenges of learning new systems. Overcoming this barrier requires comprehensive digital literacy programs, targeted training initiatives, and change management strategies that emphasize the benefits of digital transformation while addressing fears and misconceptions.

Furthermore, cultural and organizational inertia can prevent progress in adopting digital innovation. In many cases, traditional industries or bureaucratic systems are slow to embrace change, clinging to outdated methods instead of investing in modernization. Resistance to change is often fuelled by a lack of understanding of the long-term value of digital transformation or an unwillingness to allocate resources toward uncertain returns.

*Figure 2. Steps for Digital Innovation*

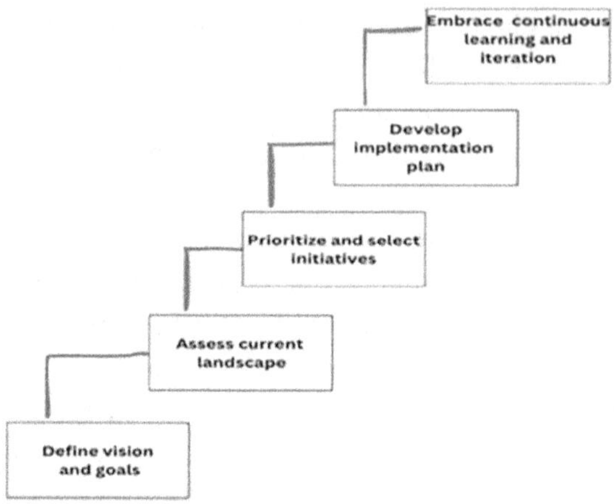


## FUTURE PROSPECTS AND POLICY RECOMMENDATIONS

The future of digital innovation lies in creating ecosystems that foster growth, inclusivity, and sustainability, with governments playing a pivotal role in shaping these landscapes. Governments must act as enablers, designing policies and frameworks that encourage innovation while ensuring equitable access to its benefits. By investing in research and development (R&D), offering incentives for startups, and creating infrastructure for digital connectivity, governments can nurture environments conducive to technological advancement. Policies that encourage Public-Private Partnerships (PPPs) are essential, as they combine the innovation and agility of the private sector with the resources and regulatory support of the public sector. Governments also need to focus on education and skill-building programs to create a digitally literate workforce capable of driving and sustaining technological progress. These efforts should include bridging the digital divide by ensuring that rural and marginalized communities have access to the tools and training necessary to participate in the digital economy.

The importance of international collaboration and investment cannot be overstated in realizing a digitally empowered future. No country can achieve technological supremacy in isolation; collaboration between nations is essential to address global challenges, such as highlighting the features and challenges of the digital economy. Key features include the rise in e-commerce, digital services, economic growth, cybersecurity threats, climate change, and health crises. International organizations and trade agreements can help standardize policies, promote knowledge sharing, and create frameworks for cross-border cooperation. Collaborative initiatives, such as the European Union's Digital Agenda or global efforts to regulate AI ethics, demonstrate how countries can work together to foster innovation responsibly. Additionally, international investments in technology transfer and infrastructure development can help underdeveloped regions catch up with global advancements. In network visualization, there are text labels, circles, connections, and color areas. By default, items are represented by labels and circles (Baskar, Venkatesan, & Sangeetha, 2020). Developed nations have a responsibility to assist developing countries by providing funding, expertise, and resources, ensuring that digital innovation benefits all, rather than widening global inequalities.

A vision for a digitally inclusive global economy must prioritize access, equity, and sustainability. The concept of a digitally inclusive economy extends beyond providing internet access; it involves creating an environment where all individuals and businesses, regardless of geography, gender, or socioeconomic status, can participate meaningfully in the digital revolution. Achieving this requires a multi-faceted approach. In addition, compatible beliefs and practices with the used technology resources increase the SMEs' ability for digital transformation. Furthermore, the industry in which SMEs operates influences the SMEs' ability to transform into digital technologies (Alraja et al., 2021). First, policymakers must focus on affordability by reducing the costs of devices, data, and services. Programs that subsidize technology for low-income households or offer affordable internet packages can bring more people into the digital fold. Second, inclusion must address gender disparities, as women in many parts of the world face unique barriers to accessing and using technology. Initiatives that empower women through digital literacy, entrepreneurship, and workforce participation are critical for achieving a balanced digital economy.

Moreover, sustainability must be central to future digital strategies. The availability of cheap and increasingly skilled labour in developing economies, and the rapid and systematic liberalisation of the trade regime, was complemented by innovations in transport and communications infrastructure. Containerisation in the shipping and logistics sectors significantly reduced the costs of shipping products across the globe, and advances in communications technologies made it less costly to organise and control global supply chains (Kaplinsky & Kraemer-Mbula, 2022). As digital technologies expand, so does their environmental impact, from energy consumption to e-waste generation. Governments and corporations need to invest in green technologies, such as energy-efficient data centers, renewable-powered infrastructure, and sustainable manufacturing practices. Circular economy models for e-waste recycling and reuse must also be prioritized to minimize environmental harm while maximizing resource efficiency.

The ethical and social dimensions of digital innovation also demand attention. Technologies such as AI, blockchain, and IoT offer transformative potential but also pose risks related to data privacy, algorithmic bias, and job displacement. These results are a perfect symptom of the influence of organizational, environmental, and technological factors on SMEs' ability to digitalize their process and activities. These findings are in line with the earlier results and findings in the literature that a sufficient level of technology resources used by SMEs helps those enterprises to digitalize their processes (Pan, Xie, Wang, & Ma, 2022). Policymakers must establish ethical guidelines and regulatory frameworks to ensure that innovation respects human rights, promotes equity, and minimizes unintended consequences. Encouraging transparency, accountability, and inclusivity in technological development can help build public trust and maximize societal benefits.

In the long term, a digitally inclusive global economy should strive to empower every individual to contribute to and benefit from digital innovation. Governments and international bodies must collaborate to create policies that support entrepreneurship, ensure fair competition, and reduce trade barriers for digital goods and services. Education systems must evolve to equip students with the skills required for emerging fields such as AI, robotics, and quantum computing. Public awareness campaigns can help demystify technology and encourage adoption among hesitant populations.

## APPLICATION OF DIGITAL TECHNOLOGY

## E-GOVERNANCE

E-Governance refers to the application of digital technology in government administration to improve efficiency, transparency, and accessibility. It enables seamless interaction between the government, citizens, businesses, and other stakeholders through digital platforms, ensuring that public services are delivered in a timely and efficient manner. As an important concept in economics, total factor productivity (TFP) measures the impact of technological progress on production. It refers to the efficiency of production activities over a certain period and is mainly used to indicate the partial contribution of output that cannot be explained by the growth of capital and labour factors (Pan et al., 2022). E-Governance has transformed the traditional bureaucratic system, reducing paperwork, delays, and corruption, thereby making governance more effective and citizen-friendly.

E-Governance can be classified into four major categories: Government-to-Citizen (G2C), Government-to-Business (G2B), Government-to-Government (G2G), and Government-to-Employee (G2E). G2C services aim to make government services more accessible to the public through online platforms. These include digital portals for tax payments, obtaining birth and death certificates, applying for passports, and accessing e-healthcare services. Online grievance redressal systems allow citizens to report issues and track the status of their complaints, promoting accountability. G2B services facilitate interactions between the government and businesses, enabling online business registrations, licensing, and compliance with regulatory frameworks. Lastly, G2E services involve digitizing various government employee-related activities, such as payroll management, digital attendance, e-learning platforms for training, and online leave applications.

The implementation of E-Governance offers numerous benefits. One of the most significant advantages is transparency and accountability, as digital platforms reduce human intervention, thereby minimizing opportunities for corruption. Automation in public service delivery ensures that processes are conducted fairly and without bias. Another key benefit is efficiency and cost-effectiveness—by digitizing government processes, tasks that once took weeks or months can now be completed within hours or even minutes. This not only saves time but also reduces administrative costs associated with paperwork and manual processing. Additionally, E-Governance enhances public participation by providing citizens with easy access to government information, allowing them to engage in policymaking through online feedback mechanisms and digital surveys. This promotes a more inclusive and democratic governance system.

Furthermore, E-Governance bridges the digital divide by making government services accessible to people in remote and rural areas. Initiatives such as Common Service Centers (CSCs) provide internet access to underserved populations, enabling them to avail government services without traveling long distances. Digital transformation in SME business is a company's complex, multi-faceted activity of integrating digital technologies in all functional areas, processes, and activities to fundamentally change the way businesses deliver value to customers (Felicetti et al., 2024). The integration of technologies like AI, blockchain, and cloud computing further strengthens the security, efficiency, and reliability of digital governance systems. However, challenges such as digital literacy, cybersecurity threats, and inadequate infrastructure must be addressed to ensure the successful implementation of E-Governance.

E-Governance is a powerful tool that transforms governance by making it more transparent, efficient, and citizen-centric. By leveraging technology, governments can enhance public service delivery, improve administrative processes, and foster economic growth. As digital transformation continues to evolve, E-Governance will play a crucial role in shaping the future of governance, ensuring that public services are more accessible, inclusive, and effective.

## THE ROLE OF 5G AND IOT IN TRANSFORMING DEVELOPING ECONOMIES

The advent of 5G and the IoT is revolutionizing industries. For example, utilizing analytics tools available on e-commerce platforms or social media, MSMEs can track purchasing patterns, evaluate the effectiveness of promotions, and identify emerging market trends and economies worldwide, especially in developing nations (Hendrawan, Chatra, Iman, Hidayatullah, & Suprayitno, 2024). These advanced technologies offer high-speed connectivity, ultra-low latency, and real-time data processing, enabling governments, businesses, and individuals to improve efficiency, productivity, and innovation.

## UNDERSTANDING 5G AND IOT

5G technology provides faster internet speeds, lower latency, and the ability to connect millions of devices simultaneously. IoT, on the other hand, refers to the network of interconnected devices—ranging from smartphones and sensors to smart appliances and industrial machines—that collect and exchange data over the internet. When combined, 5G and IoT create a digital ecosystem where devices communicate efficiently, leading to increased automation, real-time decision-making, and improved resource management.

## IMPACT OF 5G AND IOT ON DEVELOPING ECONOMIES

### BOOSTING ECONOMIC GROWTH AND DIGITAL TRANSFORMATION

5G and IoT play a crucial role in accelerating economic development by supporting digital businesses, e-commerce, and fintech services. High-speed connectivity enables startups and small businesses to operate efficiently, reach a global market, and provide better customer experiences. Additionally, IoT-powered automation in industries such as agriculture, manufacturing, and logistics improves productivity and reduces costs, helping businesses grow and compete in the global market.

### ADVANCING SMART AGRICULTURE AND FOOD SECURITY

Agriculture remains the backbone of many developing economies. 5G and IoT revolutionize farming through smart agriculture, where farmers use IoT sensors to monitor soil moisture, temperature, and weather conditions. These sensors provide real-time data, enabling farmers to optimize irrigation, reduce water wastage, and increase crop yields. Drones equipped with IoT technology can survey large farmlands, detect pest infestations, and apply fertilizers more efficiently. By improving agricultural productivity, these technologies help ensure food security and rural development.

## TRANSFORMING HEALTHCARE SERVICES

Access to quality healthcare is a major challenge in developing countries, especially in remote areas. 5 G-powered telemedicine allows doctors to conduct virtual consultations with patients, reducing the need for travel. IoT-enabled medical devices, such as smart wearables and remote monitoring tools, help track patients' health conditions and send alerts in case of emergencies. Smart hospitals equipped with IoT systems can automate patient records, improve diagnostics, and enhance treatment plans, leading to better healthcare outcomes.

## ENHANCING SMART CITIES AND INFRASTRUCTURE

Developing nations face significant urbanization challenges, including traffic congestion, pollution, and inefficient public services. 5G and IoT drive smart city initiatives by enabling intelligent traffic management, real-time surveillance, and automated waste management systems. Smart lighting, energy-efficient buildings, and connected public transport systems enhance sustainability and improve urban living conditions. These advancements contribute to better governance and infrastructure development.

## IMPROVING FINANCIAL INCLUSION AND DIGITAL PAYMENTS

Financial services are essential for economic growth, and 5G and IoT enable secure and efficient digital payments, particularly in areas with limited banking infrastructure. IoT-powered point-of-sale (POS) systems, mobile wallets, and biometric authentication enhance financial accessibility, allowing more people to participate in the formal economy. This promotes financial inclusion, encourages entrepreneurship, and facilitates seamless transactions, driving overall economic growth.

## CHALLENGES AND THE WAY FORWARD

Figure 3 highlights the features and challenges of the digital economy. Key features include the rise in e-commerce, digital services, economic growth, and improved public services. However, it faces challenges like highlighting the features and challenges of the digital economy. Key features include the rise in e-commerce, digital services, economic growth, cybersecurity issues, job disruptions, and high infrastructure demands. Addressing these challenges is essential for building a secure and inclusive digital future.

*Figure 3. Challenges and features of the Digital Economy*

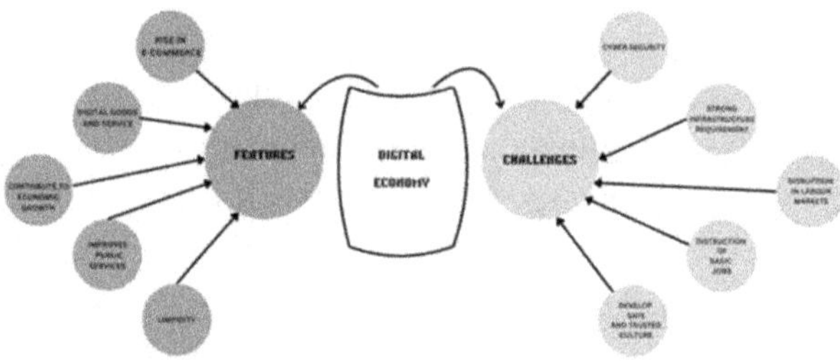


## CHALLENGES AND SOLUTIONS FOR EMERGING DIGITAL ECONOMIES

The rapid digital transformation in emerging economies has brought immense opportunities for economic growth, innovation, and improved public services. However, with increased reliance on digital technologies, these economies also face significant cybersecurity challenges. Utilizing research data on how each part of the system functions and reacts to the environment, and supported by data provided by sensors in the physical world, it can be analyzed to simulate real conditions and to respond to operational changes. Implies digital twins in key areas of industry, such as electricity, by enabling the control of energy resources distributed over the network with the appropriate software to analyze and visualize data (Strielkowski, Rausser, & Kuzmin, 2022). As businesses, governments, and individuals become more connected, cyber threats such as data breaches, ransomware attacks, identity theft, and financial fraud are becoming more prevalent.

## LACK OF ROBUST CYBERSECURITY INFRASTRUCTURE

Many emerging economies do not have well-developed cybersecurity frameworks, making them vulnerable to cyberattacks. Limited investment in security infrastructure, outdated software, and weak security protocols create an environment where cybercriminals can exploit vulnerabilities easily.

## RISE IN CYBERCRIME AND DATA BREACHES

As digital financial transactions, e-commerce, and cloud-based services grow, cybercriminals target sensitive personal and financial data. Phishing attacks, malware infections, and ransomware incidents have increased, leading to massive data breaches that compromise individuals' privacy and corporate security.

## WEAK REGULATORY FRAMEWORKS AND LAW ENFORCEMENT

Cyber laws in many developing countries are either outdated or poorly enforced. A lack of strict regulations and law enforcement agencies with cybersecurity expertise allows criminals to operate with minimal consequences. Additionally, there is limited international cooperation in tackling cross-border cyber threats.

## INCREASING THREAT OF CYBERATTACKS AND STATE-SPONSORED ATTACKS

Emerging digital economies are often targeted by cyberterrorists and state-sponsored hackers who aim to disrupt national infrastructure, financial systems, and government services. These attacks can lead to economic instability and pose national security threats.

## CHALLENGES IN SECURING IOT AND CLOUD COMPUTING

With the rise of IoT devices and cloud services, ensuring data security has become a major challenge. Many IoT devices lack strong security mechanisms, making them easy entry points for hackers. Similarly, cloud platforms that store large amounts of sensitive data are attractive targets for cyberattacks.

# CYBERSECURITY SOLUTIONS FOR EMERGING DIGITAL ECONOMIES

## STRENGTHENING CYBERSECURITY INFRASTRUCTURE

Governments and businesses must invest in modern cybersecurity technologies, such as firewalls, intrusion detection systems, encryption, and multi-factor authentication. Deploying AI-powered threat detection systems can help identify and mitigate cyber threats in real-time.

## IMPLEMENTING STRONG CYBERSECURITY POLICIES AND REGULATIONS

To combat cybercrime, governments should establish comprehensive cybersecurity laws that require businesses and institutions to follow strict security protocols. Regulations should include data protection laws, compliance standards, and cybersecurity reporting mechanisms to enhance accountability and transparency.

## RAISING CYBERSECURITY AWARENESS AND DIGITAL LITERACY

Educating individuals and organizations about cybersecurity risks is crucial. Governments, businesses, and educational institutions should conduct awareness campaigns, cybersecurity training programs, and workshops to teach safe online practices, password management, and threat identification.

## ENHANCING PUBLIC-PRIVATE COLLABORATION

Cybersecurity is a shared responsibility, and collaboration between the government, private sector, and international organizations is essential. Governments can work with cybersecurity firms, telecom companies, and financial institutions to establish national cybersecurity task forces that monitor threats and respond to incidents effectively.

## SECURING IOT DEVICES AND CLOUD PLATFORMS

The increasing adoption of IoT devices and cloud computing in emerging digital economies has brought numerous benefits, but it has also introduced significant cybersecurity risks. IoT devices, such as smart home appliances, industrial sensors, and healthcare monitors, often lack robust security mechanisms, making them easy targets for cybercriminals. The current methods of data infrastructure for industrial energy saving, which are being investigated here, show the potential for a precise and effective digital twin infrastructure in the industry. The digital twin technology in the energy industry enables the development and maintenance of intelligent networks with high-tech sensors, machine learning models and improved performance monitoring (Ragazou et al., 2022). One of the primary challenges is the absence of standardized security frameworks, leading to inconsistent protection across different devices and networks. Many IoT devices use weak or default passwords, allowing attackers to gain unauthorized access. Additionally, data transmitted between IoT devices and cloud servers is often unencrypted, making it vulnerable to interception and manipulation. Another major issue is the lack of regular firmware and software updates, which leaves devices exposed to known security vulnerabilities that hackers can exploit.

To address these security challenges, strong authentication mechanisms must be implemented, including Multi-Factor Authentication (MFA) and unique, complex passwords for each device. Role-based access control (RBAC) should also be enforced to ensure that only authorized users can control IoT devices. Additionally, end-to-end encryption should be applied to secure data transmission between IoT devices and cloud platforms, preventing unauthorized access and data breaches. Regular firmware and software updates are essential to patch vulnerabilities and enhance security. Manufacturers should integrate automatic update mechanisms to ensure devices receive the latest security patches. Cloud security is equally important, as cloud platforms store and process vast amounts of sensitive information. Implementing strong encryption protocols, access control policies, and intrusion detection systems can help protect cloud data from cyber threats. Organizations should also conduct regular security audits and penetration testing to identify potential vulnerabilities. Furthermore, governments and industry leaders should work together to establish standardized security protocols and regulations for IoT and cloud computing to create a safer digital ecosystem. By implementing these measures, emerging economies can protect their digital infrastructure, ensure data privacy, and foster trust in IoT and cloud-based services.

## PROMOTING CYBERSECURITY RESEARCH AND INNOVATION

In the rapidly evolving digital landscape, cybersecurity research and innovation play a crucial role in protecting emerging digital economies from growing cyber threats. As cybercriminals develop more sophisticated attack methods, traditional security measures often become inadequate, making continuous research essential for staying ahead of potential risks. Exchanges with individuals of varying ages, genders, professions, and personalities might provide entrepreneurs with valuable input for their businesses. For instance, by interacting with family members, entrepreneurs might receive new ideas for product improvements and services or receive valuable feedback on their business ideas (Soluk, Kammerlander, & Darwin, 2021). Investing in cybersecurity research enables the development of advanced security technologies such as AI-driven threat detection, blockchain-based authentication, and quantum-resistant encryption methods. Governments, academic institutions, and private organizations must collaborate to establish dedicated cybersecurity research centers, innovation hubs, and incubators that focus on identifying vulnerabilities and designing cutting-edge security solutions. Additionally, funding and grants should be provided to researchers and startups working on cybersecurity advancements, ensuring that financial constraints do not hinder innovation.

Encouraging collaboration between cybersecurity experts, policymakers, and industry leaders is also vital for addressing complex security challenges effectively. Public-private partnerships can facilitate knowledge sharing, joint development of security frameworks, and implementation of best practices across industries. Furthermore, organizing cybersecurity competitions, ethical hacking events, and bug bounty programs can motivate young talents and professionals to contribute innovative ideas to strengthen digital security. Education and skill development also play a significant role in promoting cybersecurity research. Universities should integrate specialized cybersecurity courses, hands-on training, and real-world cybersecurity simulations to equip students with the necessary expertise to combat emerging threats. Additionally, cybersecurity awareness campaigns should be conducted to educate businesses and individuals on the latest security trends and preventive measures. By fostering a culture of continuous learning, research, and technological advancements, emerging digital economies can build a resilient cybersecurity infrastructure that safeguards critical data, prevents cyberattacks, and ensures long-term digital growth.

## DEVELOPING INCIDENT RESPONSE AND RECOVERY PLANS

To strengthen cybersecurity resilience, organizations should create detailed response protocols for different types of cyber incidents, such as malware infections, data breaches, ransomware attacks, and Distributed Denial of Service (DDoS) attacks. These research findings have important policy implications and theoretical references for countries worldwide to widely promote the construction of 5G technology and combine different scenarios to promote the technology (Jiang & Xu, 2023).These protocols should outline immediate containment measures, such as isolating affected systems, blocking malicious traffic, and revoking compromised credentials to prevent further damage. Additionally, communication strategies should be included in the response plan to ensure timely notifications to stakeholders, employees, customers, and regulatory bodies about security incidents while maintaining transparency and trust. A critical aspect of incident response is forensic investigation, where security teams analyze attack vectors, identify vulnerabilities, and gather evidence to understand how the breach occurred. This helps organizations not only recover from incidents but also strengthen their security infrastructure to prevent future attacks.

Once the immediate threat is contained, a comprehensive recovery plan must be executed to restore affected systems, applications, and data. Organizations should maintain regular data backups and implement secure restoration procedures to minimize downtime and data loss. we incorporate into our conceptual framework the interactions between perceived usefulness and aggregate barriers to uptake of innovations for the circular plastic economy (Kolade et al., 2022). Cyber resilience can be further enhanced through post-incident analysis, where organizations evaluate their response effectiveness, identify gaps, and update security policies accordingly.

Conducting regular cybersecurity drills and simulations ensures that incident response teams remain well-prepared to handle real-world cyber threats efficiently. Moreover, organizations should stay updated with emerging cybersecurity trends and regulatory compliance requirements to continuously improve their response and recovery strategies. By investing in a well-structured incident response and recovery plan, businesses and governments can mitigate cybersecurity risks, protect sensitive data, and maintain trust in the digital ecosystem, ultimately ensuring long-term economic and technological stability.

## CHALLENGES IN SECURING IOT DEVICES: LACK OF STANDARDIZED SECURITY FRAMEWORKS

Many IoT manufacturers focus more on functionality and cost-efficiency rather than security. As a result, there are no universal security standards for IoT devices, making it difficult to enforce consistent protection across different devices and networks.

### WEAK AUTHENTICATION AND AUTHORIZATION MECHANISMS

Many IoT devices have weak or default passwords, making them easy targets for hackers. Cybercriminals often exploit unsecured devices to gain unauthorized access to personal data, manipulate industrial control systems, or launch large-scale Distributed Denial of Service (DDoS) attacks.

### UNENCRYPTED DATA TRANSMISSION

IoT devices frequently communicate over networks, but many do not use encryption. This leaves sensitive data vulnerable to interception by attackers who can modify, steal, or exploit it for malicious purposes.

## RESULTS

Figure 4 is the graphical representation of the data in Table I. Kenya has the largest adult population and the most mobile money users, and due to this fact, the contributes to the GDP. Based on the observation of Table I, the adult population directly correlates with mobile money users and contribution to GDP. It is very important to know that digital innovation will always correspond with the adult or youth population because most of the working professionals are in that range of age and they are very creative and there are a lot of digital innovations that have been innovated by the adults and that is the relationship between adult population and the innovations and at the same time the target is also adults. Table 1 shows mobile money adoption in four African countries. Kenya leads in all factors with 38.2 million users, which consists of 80% of adults and a 6% GDP contribution at the same time, Nigeria lags with 40% adoption despite its large population, reflecting untapped potential.

*Table 1. Mobile Money Adoption*

| Country | Mobile Money Users (Millions) | % Adult Population | Contribution to GDP (%) |
|---|---|---|---|
| Kenya | 38.2 | 80 | 6.0 |
| Tanzania | 21.5 | 65 | 3.8 |
| Ghana | 15.3 | 55 | 2.5 |
| Nigeria | 12.7 | 40 | 1.9 |

*Figure 4. Mobile Money Contribution to GDP*

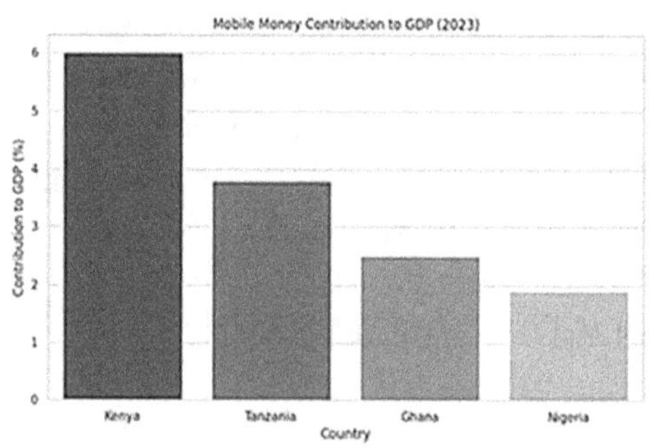


Figure 5 shows the graphical representation for the same. Both of the companies grew steadily, and we can see from the data that Mercado Libre reached 10.2B in 2023, and this reflects the rising internet penetration and customer trust in the online platforms. Mercado Libre's steeper growth reflects Latin America's larger market size, and Jumia's slower but consistent growth shows Africa's emerging E-commerce potential. Table 2 shows the revenue growth of Jumia and Mercado Libre and which are located in Africa and Latin America.

*Table 2. E-Commerce Growth*

| Year | Jumia Revenue (Million) | Mercado Libre Revenue (Billion) |
|---|---|---|
| 2020 | 160 | 3.9 |
| 2021 | 198 | 6.0 |
| 2022 | 228 | 8.5 |
| 2023 | 225 | 10.2 |

*Figure 5. E-Commerce Revenue Growth*

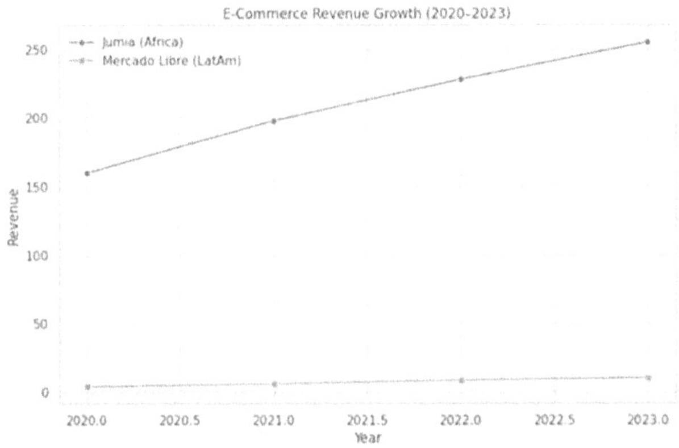


Figure 6 shows the graph representation for the same. This highlights the lack of adaptation and the untapped potential in this area. Figure 3 highlights the gap between adopters and non-adopters. Closing this gap could significantly boost agricultural productivity and rural incomes. Table 3 shows that 15.2 million Indian farmers use digital tools, which boosts the crop yields by 25% and adds 5B to the GDP.

*Table 3. Agri Tech Impact in India*

| Metric | Value (2023) |
|---|---|
| Farmers Using Digital Tools (Millions) | 15.2 |
| Crop Yield Increase (%) | 25 |
| Contribution to GDP (Billion) | 5.0 |

*Figure 6. Digital Tool Adoption Among Indian Farmers*

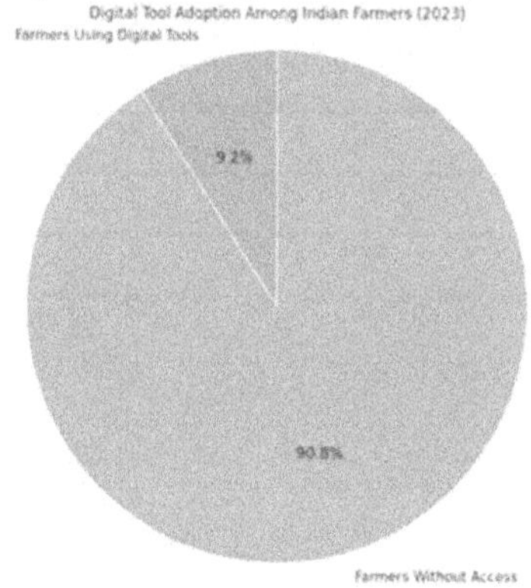


## CONCLUSION

Digital innovation serves as a foundational pillar for progress in developing economies, driving transformative change across sectors and empowering communities. By democratizing access to financial services through platforms like Kenya's M-Pesa, it has unlocked economic participation for millions of unbanked individuals, fostering entrepreneurship and reducing poverty. E-commerce giants such as Jumia and Mercado Libre are reshaping local markets, creating jobs, and connecting informal economies to global opportunities. Meanwhile, agritech tools like India's digital farming solutions have revolutionized agriculture, boosting yields and food security while enhancing rural livelihoods. Digital governance systems, including India's Aadhaar and Rwanda's e-service platforms, have streamlined public service delivery, curbed corruption, and strengthened institutional trust. However, persistent challenges, such as limited internet access in regions like Sub-Saharan Africa and low adoption rates of digital tools among smallholder farmers, highlight the urgent need to bridge the digital divide.

## REFERENCES


Ali, H., & Hajjar, L. (2024). Empowering SMEs with AI and Digital Transformation: A Roadmap to Enhanced Competitiveness and Growth.

Alraja, M. N., Hussein, M. A., & Ahmed, H. M. S. (2021). What affects digitalization process in developing economies? An evidence from SMEs sector in Oman. *Bulletin of Electrical Engineering and Informatics*, *10*(1), 441–448. DOI: 10.11591/eei.v10i1.2033

Aminah, S., & Saksono, H. (2021). Digital transformation of the government: A case study in Indonesia. *Jurnal Komunikasi: Malaysian Journal of Communication*, *37*(2), 272–288.

Bai, T., Qi, Y., Li, Z., & Xu, D. (2023). Digital economy, industrial transformation and upgrading, and spatial transfer of carbon emissions: The paths for low-carbon transformation of Chinese cities. *Journal of Environmental Management*, *344*, 118528. DOI: 10.1016/j.jenvman.2023.118528 PMID: 37392691

Bansal, A., Panchal, T., Jabeen, F., Mangla, S. K., & Singh, G. (2023). A study of human resource digital transformation (HRDT): A phenomenon of innovation capability led by digital and individual factors. *Journal of Business Research*, *157*, 113611. DOI: 10.1016/j.jbusres.2022.113611

Baskar, K., Venkatesan, G. P., & Sangeetha, S. (2020). *A Survey of Workload Management Difficulties in the Public Cloud.* Paper presented at the Intelligent Computing in Engineering: Select Proceedings of RICE 2019. DOI: 10.1007/978-981-15-2780-7_54

Bennett, E. E., & McWhorter, R. R. (2021). Virtual HRD's role in crisis and the post Covid-19 professional lifeworld: Accelerating skills for digital transformation. *Advances in Developing Human Resources*, *23*(1), 5–25. DOI: 10.1177/1523422320973288

Dana, L.-P., Salamzadeh, A., Hadizadeh, M., Heydari, G., & Shamsoddin, S. (2022). Urban entrepreneurship and sustainable businesses in smart cities: Exploring the role of digital technologies. *Sustainable Technology and Entrepreneurship*, *1*(2), 100016. DOI: 10.1016/j.stae.2022.100016

Danladi, S., Prasad, M., Modibbo, U. M., Ahmadi, S. A., & Ghasemi, P. (2023). Attaining sustainable development goals through financial inclusion: Exploring collaborative approaches to Fintech adoption in developing economies. *Sustainability (Basel)*, *15*(17), 13039. DOI: 10.3390/su151713039

Dayıoğlu, M. A., & Turker, U. (2021). Digital transformation for sustainable future-agriculture 4.0: A review. *Journal of Agricultural Sciences*, *27*(4), 373–399.

Du, J., Shen, Z., Song, M., & Zhang, L. (2023). Nexus between digital transformation and energy technology innovation: An empirical test of A-share listed enterprises. *Energy Economics*, *120*, 106572. DOI: 10.1016/j.eneco.2023.106572

Felicetti, A. M., Corvello, V., & Ammirato, S. (2024). Digital innovation in entrepreneurial firms: A systematic literature review. *Review of Managerial Science*, *18*(2), 315–362. DOI: 10.1007/s11846-023-00638-9

Furr, N., Ozcan, P., & Eisenhardt, K. M. (2022). What is digital transformation? Core tensions facing established companies on the global stage. *Global Strategy Journal*, *12*(4), 595–618. DOI: 10.1002/gsj.1442

Ghosh, S., Hughes, M., Hodgkinson, I., & Hughes, P. (2022). Digital transformation of industrial businesses: A dynamic capability approach. *Technovation*, *113*, 102414. DOI: 10.1016/j.technovation.2021.102414

Hendrawan, S. A., Chatra, A., Iman, N., Hidayatullah, S., & Suprayitno, D. (2024). Digital transformation in MSMEs: Challenges and opportunities in technology management. *Jurnal Informasi dan Teknologi*, 141-149.

Jiang, Z., & Xu, C. (2023). Disrupting the technology innovation efficiency of manufacturing enterprises through digital technology promotion: An evidence of 5G technology construction in China. *IEEE Transactions on Engineering Management*.

Kaplinsky, R., & Kraemer-Mbula, E. (2022). Innovation and uneven development: The challenge for low-and middle-income economies. *Research Policy*, *51*(2), 104394. DOI: 10.1016/j.respol.2021.104394

Khan, S. A. R., Piprani, A. Z., & Yu, Z. (2022). Digital technology and circular economy practices: Future of supply chains. *Operations Management Research : Advancing Practice Through Research*, *15*(3), 676–688. DOI: 10.1007/s12063-021-00247-3

Kolade, O., Odumuyiwa, V., Abolfathi, S., Schröder, P., Wakunuma, K., Akanmu, I., Whitehead, T., Tijani, B., & Oyinlola, M. (2022). Technology acceptance and readiness of stakeholders for transitioning to a circular plastic economy in Africa. *Technological Forecasting and Social Change*, *183*, 121954. DOI: 10.1016/j.techfore.2022.121954

Kraus, S., Jones, P., Kailer, N., Weinmann, A., Chaparro-Banegas, N., & Roig-Tierno, N. (2021). Digital transformation: An overview of the current state of the art of research. *SAGE Open*, *11*(3), 21582440211047576. DOI: 10.1177/21582440211047576

Kulandaivel, S., & Kulandaivel, M. (2025). Intelligent and Optical Security Monitoring for 6G Communications. In *RFID, Microwave Circuit, and Wireless Power Transfer Enabling 5/6G Communication* (pp. 417-444): IGI Global Scientific Publishing.

Kurniawan, T. A., Othman, M. H. D., Hwang, G. H., & Gikas, P. (2022). Unlocking digital technologies for waste recycling in Industry 4.0 era: A transformation towards a digitalization-based circular economy in Indonesia. *Journal of Cleaner Production*, *357*, 131911. DOI: 10.1016/j.jclepro.2022.131911

Maiurova, A., Kurniawan, T. A., Kustikova, M., Bykovskaia, E., Othman, M. H. D., Singh, D., & Goh, H. H. (2022). Promoting digital transformation in waste collection service and waste recycling in Moscow (Russia): Applying a circular economy paradigm to mitigate climate change impacts on the environment. *Journal of Cleaner Production*, *354*, 131604. DOI: 10.1016/j.jclepro.2022.131604

Nugraha, D. P., Setiawan, B., Nathan, R. J., & Fekete-Farkas, M. (2022). FinTech adoption drivers for innovation for SMEs in Indonesia. *Journal of Open Innovation*, *8*(4), 208. DOI: 10.3390/joitmc8040208

Oloyede, A. A., Faruk, N., Noma, N., Tebepah, E., & Nwaulune, A. K. (2023). Measuring the impact of the digital economy in developing countries: A systematic review and meta-analysis. *Heliyon*, *9*(7), e17654. DOI: 10.1016/j.heliyon.2023.e17654 PMID: 37501966

Pan, W., Xie, T., Wang, Z., & Ma, L. (2022). Digital economy: An innovation driver for total factor productivity. *Journal of Business Research*, *139*, 303–311. DOI: 10.1016/j.jbusres.2021.09.061

Ragazou, K., Passas, I., & Sklavos, G. (2022). Investigating the strategic role of digital transformation path of SMEs in the era of COVID-19: A bibliometric analysis using R. *Sustainability (Basel)*, *14*(18), 11295. DOI: 10.3390/su141811295

Schneider, S., & Kokshagina, O. (2021). Digital transformation: What we have learned (thus far) and what is next. *Creativity and Innovation Management*, *30*(2), 384–411. DOI: 10.1111/caim.12414

Soluk, J., Kammerlander, N., & Darwin, S. (2021). Digital entrepreneurship in developing countries: The role of institutional voids. *Technological Forecasting and Social Change*, *170*, 120876. DOI: 10.1016/j.techfore.2021.120876

Steiber, A., Alänge, S., Ghosh, S., & Goncalves, D. (2021). Digital transformation of industrial firms: An innovation diffusion perspective. *European Journal of Innovation Management*, *24*(3), 799–819. DOI: 10.1108/EJIM-01-2020-0018

Strielkowski, W., Rausser, G., & Kuzmin, E. (2022). Digital revolution in the energy sector: effects of using digital twin technology. In *Digital transformation in industry: Digital twins and new business models* (pp. 43–55). Springer. DOI: 10.1007/978-3-030-94617-3_4

Thuan, P. Q., Khuong, N. V., Anh, N. D. C., Hanh, N. T. X., Thi, V. H. A., Tram, T. N. B., & Han, C. G. (2022). The determinants of the usage of accounting information systems toward operational efficiency in industrial revolution 4.0: Evidence from an emerging economy. *Economies*, *10*(4), 83. DOI: 10.3390/economies10040083

Uwaoma, P. U., Eboigbe, E. O., Eyo-Udo, N. L., Ijiga, A. C., Kaggwa, S., & Darao-jimba, D. O. (2023). The fourth industrial revolution and its impact on agricultural economics: Preparing for the future in developing countries. *International Journal of Advanced Economics*, *5*(9), 258–270. DOI: 10.51594/ijae.v5i9.646

## KEY TERMS AND DEFINITIONS

**Cyber Security:** It refers to the practice of protecting computer systems, networks, data, and devices from cyber attacks, theft, damage, or unauthorized access.

**Digitization:** The first step or foundation in a larger transformation process — converting analog or physical information into a digital format.

**Gross Domestic Product:** It refers to the total monetary value of all finished goods and services produced within a country's borders in a specific period.

**IoT:** A network of physical devices (like sensors, appliances, vehicles, or machinery) that are attached to the internet, allowing them to collect, exchange, and act upon data.

**Multi-Factor Authentication:** It refers to a security process that requires users to provide two or more forms of identification to gain access to a system, application, or account.

**Smart Systems:** It refers to devices or platforms that can perform tasks intelligently by collecting data, analyzing it, and reacting or adapting to their environment.

# Chapter 13
# Connecting Reality and Virtuality The Power of Digital Twins

**Praveen Kumar Malik**
https://orcid.org/0000-0003-3433-8248
*Lovely Professional University, India*

**Anita Gehlot**
*Division of Research and Innovation, Uttaranchal Institute of Technology, Uttaranchal University, India*

**Fraiz Parveen**
*Division of Research and Innovation, Uttaranchal Institute of Technology, Uttaranchal University, India*

**Mahim Raj Gupta**
*Division of Research and Innovation, Uttaranchal Institute of Technology, Uttaranchal University, India*

**Rajesh Singh**
*Division of Research and Innovation, Uttaranchal Institute of Technology, Uttaranchal University, India*

## ABSTRACT

*Digital twin (DT) users gain access to an enhanced system for testing and assessing complex systems better than standard simulations. DT development requires overcoming several hurdles which consist of complex communication systems alongside inaccessible data storage and processing limitations and incomplete standardization. This study investigates DT enablers followed by sector-based challenges alongside*

DOI: 10.4018/979-8-3373-1534-8.ch013

*opportunities while focusing on Blockchain implementations for trust enhancement and data protection. The article discusses the challenges of evaluating digital twins in urban management because of organizations pursuing different objectives and having unclear priorities. The study supports critical realism because it proves more suitable than postmodernism and ontological flexibility for handling these challenges. It evaluates how DTs function in smart manufacturing through an introduction of deep learning technology that enables monitoring human-robotically interactions.*

## INTRODUCTION

Digitalization is transforming learning, business, and leisure, and the metaverse is evolving as a post-reality realm. This virtual existence replicates reality, which benefits businesses including gaming and education. In the very near future, the metaverse will have a very great influence on the working spaces in the digital world while providing digital teachers that will offer real environments for virtual training. Connecting between virtual worlds and real processes in manufacturing enables such productive employment of these types of environments and workplaces (Kaarlela et al., 2023).

The Metaverse creates a world through the Internet and the Internet of Everything, which brings immersion to individuals where they can essentially carry out activities in the real world, such as socialization, work, and trade in virtual economies. It is really linking the physical and virtual worlds through applications such as IoE, extended reality (XR), artificial intelligence (A)il, machine learning (ML), 6G connectivity, and blockchain technology. Blockchain-enabled IoE (BIoE) is critical for safe service delivery in Metaverse applications (Rafique & Qadir, 2024).

The Metaverse's microorganism analysis and characterization infrastructure has the potential to change biology, medicine, and drug development. Digital twins (DTs) of diverse microbes, including as viruses, fungus, algae, bacteria, protozoa, archaea, and multicellular animal parasites, might help to accelerate the adoption of the Metaverse and upcoming technologies such as Cyber-Physical Healthcare Systems (CPHS) (Jamshidi et al., 2023). Metaverse platforms produce immersive experiences through the integration of DTs and digital artifacts. However, there are still problems to ensuring smooth teleportation and cross-metaverse compatibility for these DTs and artifacts. Furthermore, there is a scarcity of reliable monetization techniques for selling and leasing these DTs and digital artifacts, emphasizing the need for enhanced mechanisms for smooth interoperation (Gebreab et al., 2024).

DTs have the potential for enhancing urban management, however their formulation varies depending on the urban management objectives. They can be smart and inventive in exhibiting technical capability, beneficial interventionist tactics for social engineering, exploratory and collaborative processes for uncovering attitudes, and an emphasis on citizen involvement and social benefits for society transformation. The formulation of DTs differs according on the unique urban management objective (Al-Sehrawy et al., 2023). Artificial Intelligence has brought about revolutions in numerous industries, such as healthcare, education, and services in smart cities. With these advancements comes the need for a fast, secure, and fault-tolerant communication medium. Scalable, immersive, user-centric services have yet to be defined by conventional technologies. As an extension of its predecessor, the metaverse creates an environment, that is an artificial and fictive reality in three dimensions, using cutting-edge technologies such as Virtual Reality and Augmented Reality. However, this ecosystem is still in development and will require significant study to be fully realized (Aloqaily et al., 2022).

The Metaverse is a cutting-edge technical idea that seeks to build immersive virtual worlds in which users interact with mixed-reality surroundings and items. This provides a challenge in terms of improving virtual experiences and developing new use cases and services. Metaverse technologies require a linked, trustworthy, and comprehensive digital representation of individuals and things, based on new information from the real world, in order to create a successful virtual environment (Picone et al., 2023).

New-generation information technology has turned the real world into a virtual digital domain, allowing for the establishment of the Metaverse. This complicated collection of unknown states comprises solid, liquid, gaseous, plasma, and other phases. It also combines tangibles with social interactions, ethics, morality, and law, changing the real world into a virtual one (Lv et al., 2022). VR technology is an effective tool for tackling global concerns because it provides immersive and interactive solutions. It enables users to deal with real-life circumstances in real time, closing the gap between the actual and virtual worlds. This novel concept allows users to recreate real-world circumstances in a virtual environment, so increasing the entire human experience (Dooley et al., 2024).

## The Digital Twin Technologies in Different Industries

Digital Twins (DTs), which were first developed over a decade ago, provide real-time monitoring, modeling, optimization, and precise predictions. However, their theoretical framework and actual implementations do not yet completely realize their ambition on a larger scale. Despite successful deployments in research and industry settings, public information is restricted, making it impossible to analyze components, effectiveness, draw comparisons, discover solutions, and share experiences, therefore progressing and benefitting from DT technique (Sharma et al., 2022).

DTs is an emerging approach that extends beyond standard computer-based simulations and analysis by reproducing components, processes, dynamics, and firmware from a physical system into a digital copy. In table 1 it shows the digital twin technologies used in various sectors. Class input-distribution and action-distribution that works through the real-time connections of data being transferred and activity done. Internet of Things (IoT), Artificial Intelligence technology (AI), 3D models, next-generation mobile communications (5G/6G), Augmented Reality (AR) and Virtual Reality (VR), distributed computing and Transfer Learning, electronic sensing that act as components will help make real-time monitoring, analysis, evaluation, and prediction of a digital counterpart of a real-world system possible using the amalgamated available resources.

That is how input-action distribution works: through real-time data connection and information transfer. Real-time monitoring, analysis, evaluation, and prediction of a digital counterpart of a real-world scenario can be achieved using Internet of Things (IoT), Artificial Intelligence technology (AI), 3D models, next-generation mobile communications (5G/6G), Augmented Reality (AR), Virtual Reality (VR), and distributed computing and Transfer Learning with electronic sensing as core components (Mihai et al., 2022).

Along with IoT adoption, DT and VR technologies mark great advancement in urban development with better service delivery to citizens and communities. This connection therefore provides real-time digital representations of cities that can facilitate enhanced decision making, resource management, and personalized citizen services. This transition attempts to make the urban environment more sustainable, livable, and responsive (del Campo et al., 2024). DTs are transforming supply chain operations by digitally replicating physical assets, processes, and systems. By combining AI and DTs, they may evaluate real-time data, forecast future occurrences, and optimize decision-making processes, increasing operational efficiency, lowering costs, and boosting reaction to disruptions. Data privacy concerns, integration difficulty, and the requirement for defined protocols are among the challenges (Espinosa-Jaramillo et al., 2024).

Although creating digital twins has a lot of advantages, it is also extremely complicated. One must look into the availability and the sufficiency of data with respect to the physical items and appropriate shaping of using such data to create a perfect model for the digital twin. They become efficient in prediction and optimization because they simulate an abstract digital model of a given system-the-model always being a digital-simulation model must continuously adapt and develop itself to keep the properties of DTs. That is how digital twins must evolve. The provision of creating digital twins itself is extremely beneficial as an effort; it has its complications, though. Optical communication is being transformed significantly by heterogeneous hardware resources, flexible and open transmission systems, and virtualized network functions.

These are the computer-simulated virtual replicas of the real-world physical entities acting as abstract digital models of data and behavior. Prediction and optimization are to be done by making an abstract digital model of a given system. Digital simulation models have to be continually updated and evolved to keep the properties of DTs. Making digital twins is complicated despite the benefits that come with it. The availability and sufficiency of information with respect to the physical items must also be investigated, and an appropriate model must be built for applying them (Kang et al., 2021). Optical communication continues to experience rapid change due to the diversity of hardware resources, flexibility of transmission systems, and virtualization of network functions.

*Table 1. The Digital Twin technologies used across various fields and their significance*

| Field | Digital Twin Technologies | Significance |
|---|---|---|
| Smart Manufacturing | Virtual Prototyping<br>Real-Time Monitoring Systems<br>Predictive Maintenance | Improved product design<br>Enhanced operational efficiency<br>Reduced downtime through proactive maintenance |
| Healthcare | Virtual Patient Models<br>Remote Health Monitoring<br>Surgical Simulation | Personalized treatment<br>Accurate medical diagnostics<br>Improved patient outcomes |
| Automotive | Vehicle Simulation Models<br>Autonomous Vehicle Testing<br>Real-Time Performance Monitoring | Faster design iteration<br>Improved safety and performance<br>Cost-effective testing and validation |
| Smart Cities | Infrastructure Monitoring<br>Traffic Management Systems<br>Energy Optimization Models | Optimized urban planning<br>Reduced congestion and energy consumption<br>Enhanced public safety |
| Construction & Architecture | Building Information Modeling (BIM)<br>Virtual Site Simulations<br>Sustainability Modeling | Reduced construction costs<br>Enhanced collaboration<br>Improved sustainability and energy efficiency |
| Agriculture | Precision Agriculture Models<br>Crop Growth Simulation Livestock<br>Health Monitoring | Increased crop yield<br>Efficient use of resources<br>Improved animal welfare and productivity |
| Aerospace & Defense | Flight Simulation Models<br>Mission Planning & Simulation<br>Predictive Maintenance for Aircraft | Improved flight safety<br>Enhanced mission success rate<br>Reduced maintenance costs |

continued on following page

*Table 1. Continued*

| Field | Digital Twin Technologies | Significance |
|---|---|---|
| Energy & Utilities | Smart Grid Management<br>Real-Time Energy Consumption Monitoring<br>Renewable Energy Systems Simulation | Optimized energy distribution<br>Enhanced grid reliability<br>Integration of renewable sources |
| Retail & E-Commerce | Virtual Store Modelling<br>Supply Chain and Inventory Simulation<br>Consumer Behavior Modelling | Improved customer experience<br>Efficient supply chain management<br>Personalized marketing |
| Logistics & Supply Chain | Real-Time Fleet Tracking Warehouse Operations Modelling<br>Demand Forecasting Models | Increased operational efficiency<br>Reduced delivery time<br>Optimized inventory management |
| Telecommunications | Network Optimization Models<br>Infrastructure Planning & Simulation<br>5G Deployment Simulation | Enhanced network reliability<br>Cost-effective infrastructure planning<br>Better service quality |

Traditional management methods face new problems as a result. DTs are a type of technology that uses data, models, and algorithms that imitate and describe physical objects. DTs have been created and widely used in industrial and military settings to improve connection and overcome issues with traditional communication management (Wang et al., 2021).

## Realistic Virtual World with Digital Twins

The DTs is an effective approach that connects the actual and virtual worlds, allowing for substantial improvement in issue solving and challenge resolution. It is based on four infrastructures: networking, processing, digital storage, and data sources. DTs give insights, innovation routes, efficient manufacturing, enhanced service delivery, better experiences, and new business models by bridging the gap between experimentation and learning (Crespi et al., 2023). The digital revolution is blurring the border between the real and virtual worlds, with the notion of DTs, electronic representations of real-world systems, assets, and processes that are updated in real time with data. This technology has the ability to alter different industries by improving knowledge and efficiency.

It can aid in comprehending complicated systems, forecasting maintenance requirements, and optimizing operations (Dashore & Sharma, 2024). Industry 5.0 use DTs to improve manufacturing flexibility and efficiency. Deep learning and other AI approaches have gained appeal in human-robot collaboration as a means of tackling safety and reliability concerns. A deep learning-enhanced DT framework recognizes and classifies human operators and robots, enabling the robot control system to make autonomous decisions and enhance the safety and reliability of collaborative industrial activities (Wang et al., 2024).

The industrial business is adopting digitization, resulting in the growth of DTs as a way of integrating value generation via smart goods. DTs link the real and virtual worlds, but their design is time-consuming, costly, and lacking prescriptive expertise. Design principles (DP) serve to codify design knowledge into prescriptive information, although they are frequently intangible and challenging to apply in software development projects (Wache et al., 2022). The emergence of 5G and IoT has increased the use of big data in product design, with AI emerging as a major tool. The digital twin, which links the actual and virtual worlds, analyzes sensor data employing AI algorithms. Existing sensor technologies, like cameras and microphones, confront issues including excessive power consumption and battery replacement, despite their extensive usage in a variety of applications (Zhang et al., 2022). I oT and data-collection devices have considerably expanded the quantity of data accessible for real-time, data-driven applications. DTs are virtual representations of actual things that can automatically communicate data in both directions. Modern DTs have a five-component design, including a Data Management component that connects other systems, although its role and functionality are uncertain (Correia et al., 2023).

The high-tech sector is quickly creating new technologies, including DTs, are interactive technologies that successfully handle the interaction between the physical and virtual worlds. Since their centrality, integrity, and dynamics, these technologies have grown in favor among academics worldwide. They've been integrated with other technologies and used in a variety of applications, including industrial production, medical modelling, smart cities, aerospace security, and commercial shopping. Due to their importance, integrity, and dynamics, DTs have become a hot topic in a variety of disciplines (Guo & Lv, 2022).

DTs are useful for a variety of applications, including virtual commissioning, defect prediction, and reconfiguration planning. Equipped with artificial intelligence, they can improve on existing benefits or offer new ones, especially cross-phase industrial transfer learning. Improved transfer learning was used in the new task by just considering the previously learned information instead of the data collected through time and space, which is considerably less (Maschler et al., 2021).

Transforming mechanical design processes to those that integrate the physical and virtual worlds is what DT techniques are doing. In addition, this technique goes across the design cycle, simulation, manufacturing, and assessment. The digital twins act as living blueprints for predictive maintenance, rapid prototyping, and risk mitigation. Real-time condition monitoring, performance prediction, and large-scale data analytics give mechanical design strategic advantages, revealing potential for holistic design improvement, as shown in table 2 (Liu, 2023). It is expected that within the decade to come, about 66% of large industrial enterprises will have adopted DTs, but their usage in AEC is yet at a very early stage. Establishing a common process and standards for a specific area of operations and use cases for the sector will allow DTs to be established as practice. (Pregnolato et al., 2022).

*Table 2. Digital Twins (DTs) and Their Applications*

| Topic | Description |
|---|---|
| Digital Twins (DTs) and IoT Integration | DTs act as virtual replicas of physical entities, enabling real-time data exchange. |
| Data Management in DTs | DTs require a structured Data Management component, but its role remains uncertain. |
| DTs in Various Applications | DTs are widely used in smart cities, industrial production, medical modeling, and aerospace. |
| AI-Enhanced DTs and Transfer Learning | AI-powered DTs improve virtual commissioning, defect prediction, and industrial transfer learning. |
| DTs in Mechanical Design | DTs revolutionize mechanical design by integrating real-time monitoring and predictive analytics. |
| Predictive Maintenance with DTs | DTs enable proactive maintenance strategies through real-time condition monitoring. |
| DTs in the Architecture, Engineering, and Construction (AEC) Industry | The AEC sector is in the early stages of DT adoption, requiring standardized practices. |
| Future Adoption of DTs in Industry | Two-thirds of major industrial firms are expected to implement DTs within the next decade. |

## CASE STUDIES

### Rolls-Royce Improves Jet Engine Efficiency

To keep an eye on the engines it manufactures, the multinational aerospace and defence corporation Rolls-Royce has implemented digital twin technology. The business may keep an eye on the pilot's usage of each engine, the flying circumstances, and how each engine performs. For years, the business has provided its customers with engine monitoring as a service; however, Rolls-Royce has been able to customise the service for individual engines because to its digital twin capacity. The company has been able to drastically cut down on its spare parts and components inventory by extending the interval between maintenance for some engines by as much as 50%. To date, Rolls-Royce has saved 22 million tonnes of carbon thanks to the technology's assistance in increasing the efficiency of its engines.

## Mars Uses a Digital Twin to Optimise Its Supply Chain

Mars, a food, pet care, and confectionery firm, has developed a digital model of its manufacturing supply chain to help with its operations. The business uses artificial intelligence (AI) and the Microsoft Azure cloud to process and analyse data produced by production machines in its manufacturing sites. Accenture's digital manufacturing and operations specialists are assisting Mars in enhancing operations across its 160 production locations through the use of Microsoft's Azure Digital Twins IoT service. In order to enhance capacity and process controls, the organisation is developing software simulations. These simulations include increasing machine uptime through predictive maintenance and decreasing waste related to equipment packaging irregular product volumes. A virtual "app store of use cases" that can be applied to all of Mars' business lines can likewise be created using the digital twin construct.

## TIAA Simplifies Client Service

Teachers are assisted in managing their retirement assets by the College Retirement Equities Fund (TIAA), which is administered by the Teachers Insurance and Annuity Association of America. Using a digital twin driven by a graph database, the non-profit financial services company is simplifying the process of onboarding new institutional clients. There are more than 600 features in TIAA's Outsourced Services, which can result in over one trillion customer setups. Technical configurations were manually developed and evaluated against a client's preferred operating model by specialised TIAA teams prior to the deployment of digital twin technology. Because TIAA's associates were highly "functionalised" based on their areas of expertise, they were only able to handle specific kinds of offers. This makes scaling operations challenging as well.

## Using Virtual Factories, Bayer Crop Science Restructures Their Approach

Digital twins have been used by Bayer Crop Science to build "virtual factories" for each of its nine North American maize seed production facilities. The nine locations prepare and bag the seeds once they are gathered from Bayer's fields and then delivered to the farmer. For each of the nine locations, Bayer has produced a dynamic digital representation of the machinery, bill of materials, operating procedures, and process and product flow characteristics. This allows the business to conduct "what-if" assessments for everyone. The company can utilise the virtual factories to determine whether the site is prepared to modify its operations to implement new pricing strategies or seed treatment services as the commercial team offers them.

## DISCUSSION

According to a few authorities on future developments, it should include multi-domain knowledge, real-time data transfer, model creation, and adaptability. Digital Twins provide virtual representation of the physical phenomena-taking place. These will create an internet of things by data collection from sensor devices. Emerging technologies in the form of digital twins, IoT, virtual reality, blockchain, AI, and big data are now combined to become an established R&D system.

The future research shall continue to develop digital twin capabilities for managing smart city services in conjunction with IoT, DT, and VR tools in creating innovations, improving sustainability, and providing resilience. All of these would be achieved through the collaboration of stakeholders, standardized initiatives, and investments made towards technological infrastructure toward realizing the full potential of smart cities in improving the quality of life and sustainable development of cities.

In fact, the major challenges in the deployment of Digital Twins in a Virtual Reality framework are not only associated with the inherent high technological complexity and integration of data but also with associated aspects such as the need for a complex infrastructure to handle real-time data, absence of well-defined models and frameworks that make the interoperability across different systems and IoT devices difficult, thus limiting the efficiency and effectiveness of DT solutions in urban environments

The Digital Transformation is fast progressing towards Industry 4.0, fueled by its enormous potential. Its ever-changing technologies bring us closer to obtaining real DTs, with each step forward.

In the AI era, data is the primary focus, and the DT serves as a vital information loop. It requires carefully selected data to power its complicated Machine Learning algorithms and improve its knowledge through interactive and predictive achievements. As a result, the DT is pushing toward automation of industrial operations.

The advancement of AI technology will lead to the proliferation of digital twins, resulting in more advanced and autonomous solutions. This will transform supply chain management, improving its efficiency, resilience, and sustainability. To remain competitive and survive in today's global businesses must embrace DTs.

Digital transformation is critical for the future development of the global economy as well as academic application of DT. Accurate and robust DT applications are required to acquire user confidence and enable their everyday usage in our lives.

Building Information Modeling (BIM) technology is critical in tackling difficulties in building design, construction, and management since it provides a standardized way to creating comprehensive, consistent, and easily shared digital representations across stakeholders.

BIM technology, when combined with DTs and virtual reality, may dramatically increase modeling accuracy, data management, and interactive capabilities. This is especially important in urban planning, where BIM models speed up the integration of architectural and engineering data, improve real-world simulation, and enable accurate decision-making.

Deep learning algorithms should be successfully trained using synthetic data, proving the efficiency of a fully supervised detection method in real-world settings. A semi-supervised detector is developed to maintain system dependability under varying illumination situations, bridging the detection gap between humans and robots. This method provides system stability under varying illumination conditions.

Legally binding restrictions on data used in DT are critical for effective operation and compliance across businesses, particularly for sites that are spread internationally, because they must follow applicable regulations.

## CONCLUSION

Widespread adoption of this technology depends on existing regulations and standardized security standards and framework guidelines. IoT and AI and data streaming provides more effective systems for DT technology by advancing implementation processes. Scientists have established four fundamental approaches to DT which comprise technologies, disruptors, cognition and humanity. Critical realism theory functions as the most effective combination for uniting these concepts. The technology from IoT MAD Lab based on Virtual Reality enables real-time data collection through IoT integration of urban infrastructure elements including lighting and waste and parking functions. Supply chain digital twins that use AI enhance the capabilities of monitoring functions along with predictive maintenance operations while decision processes achieve both improved efficiency and reduced costs and increased system resilience. Predictive analytics delivered through DT system simulations lets organizations develop intelligent operative functions which actively address potential problems before they materialize.

## REFERENCES


Al-Sehrawy, R., Kumar, B., & Watson, R. (2023). The pluralism of digital twins for urban management: Bridging theory and practice. *Journal of Urban Management, 12*(1), 16–32. DOI: 10.1016/j.jum.2023.01.002

Aloqaily, M., Bouachir, O., Karray, F., Al Ridhawi, I., & El Saddik, A. (2022). Integrating digital twin and advanced intelligent technologies to realize the metaverse. *IEEE Consumer Electronics Magazine, 12*(6), 47–55. DOI: 10.1109/MCE.2022.3212570

Correia, J. B., Abel, M., & Becker, K. (2023). Data management in digital twins: A systematic literature review. *Knowledge and Information Systems, 65*(8), 3165–3196. DOI: 10.1007/s10115-023-01870-1

Crespi, N., Drobot, A. T., & Minerva, R. (2023). The Digital twin in action and directions for the future. In *The Digital Twin* (pp. 1201–1217). Springer International Publishing. DOI: 10.1007/978-3-031-21343-4_38

Dashore, R., & Sharma, N. (2024). Digital Twins in the Modern Context Bridging Realities With Blockchain. In *Ensuring Security and End-to-End Visibility Through Blockchain and Digital Twins* (pp. 1–20). IGI Global. DOI: 10.4018/979-8-3693-3494-2.ch001

del Campo, G., Saavedra, E., Piovano, L., Luque, F., & Santamaria, A. (2024). Virtual Reality and Internet of Things Based Digital Twin for Smart City Cross-Domain Interoperability. *Applied Sciences (Basel, Switzerland), 14*(7), 2747. DOI: 10.3390/app14072747

Dooley, A., Penica, M., McGrath, S., O'Brien, W., Bhattacharya, M., & O'Connell, E. (2024, June). Digital Twin Bridging: Enabling Virtual Worlds to Manifest in the Physical. In *2024 35th Irish Signals and Systems Conference (ISSC)* (pp. 1–6). IEEE.

Espinosa-Jaramillo, M. T., Zuta, M. E. C., Koneti, C., Jayasundar, S., Zegarra, S. D. R. O., & Carvajal-Ordoñez, V. F. M. (2024). Digital Twins in Supply Chain Operations Bridging the Physical and Digital Worlds using AI. *Journal of Electrical Systems, 20*(10s), 1764–1774. DOI: 10.52783/jes.5434

Gebreab, S. A., Musamih, A., Hasan, H. R., Salah, K., Jayaraman, R., Al Hammadi, Y., & Omar, M. (2024). NFTs for accessing, monetizing, and teleporting digital twins and digital artifacts in the metaverse. *Computer Communications, 228*, 107965. DOI: 10.1016/j.comcom.2024.107965

Guo, J., & Lv, Z. (2022). Application of Digital Twins in multiple fields. *Multimedia Tools and Applications*, *81*(19), 26941–26967. DOI: 10.1007/s11042-022-12536-5 PMID: 35194381

Jamshidi, M. B., Sargolzaei, S., Foorginezhad, S., & Moztarzadeh, O. (2023). Metaverse and microorganism digital twins: A deep transfer learning approach. *Applied Soft Computing*, *147*, 110798. DOI: 10.1016/j.asoc.2023.110798

Kaarlela, T., Pitkäaho, T., Pieskä, S., Padrão, P., Bobadilla, L., Tikanmäki, M., Haavisto, T., Blanco Bataller, V., Laivuori, N., & Luimula, M. (2023, May). Towards metaverse: Utilizing extended reality and digital twins to control robotic systems. [). MDPI.]. *Actuators*, *12*(6), 219. DOI: 10.3390/act12060219

Kang, J. S., Chung, K., & Hong, E. J. (2021). Multimedia knowledge-based bridge health monitoring using digital twin. *Multimedia Tools and Applications*, *80*(26), 34609–34624. DOI: 10.1007/s11042-021-10649-x

Liu, X. (2023). *A New Perspective on Digital Twin-Based Mechanical Design in Industrial Engineering*. Innovations in Applied Engineering and Technology. DOI: 10.58195/iaet.v2i1.134

Lv, Z., Xie, S., Li, Y., Hossain, M. S., & El Saddik, A. (2022). Building the metaverse using digital twins at all scales, states, and relations. *Virtual Reality & Intelligent Hardware*, *4*(6), 459–470. DOI: 10.1016/j.vrih.2022.06.005

Maschler, B., Braun, D., Jazdi, N., & Weyrich, M. (2021). Transfer learning as an enabler of the intelligent digital twin. *Procedia CIRP*, *100*, 127–132. DOI: 10.1016/j.procir.2021.05.020

Mihai, S., Yaqoob, M., Hung, D. V., Davis, W., Towakel, P., Raza, M., Karamanoglu, M., Barn, B., Shetve, D., Prasad, R. V., Venkataraman, H., Trestian, R., & Nguyen, H. X. (2022). Digital twins: A survey on enabling technologies, challenges, trends and future prospects. *IEEE Communications Surveys and Tutorials*, *24*(4), 2255–2291. DOI: 10.1109/COMST.2022.3208773

Picone, M., Mariani, S., Virdis, A., & Castagnetti, P. (2023, June). Digital Twin & Blockchain: Technology Enablers for Metaverse Computing. In *2023 IEEE International Conference on Metaverse Computing, Networking and Applications (MetaCom)* (pp. 1–8). IEEE. DOI: 10.1109/MetaCom57706.2023.00017

Pregnolato, M., Gunner, S., Voyagaki, E., De Risi, R., Carhart, N., Gavriel, G., Tully, P., Tryfonas, T., Macdonald, J., & Taylor, C. (2022). Towards Civil Engineering 4.0: Concept, workflow and application of Digital Twins for existing infrastructure. *Automation in Construction*, *141*, 104421. DOI: 10.1016/j.autcon.2022.104421

Rafique, W., & Qadir, J. (2024). Internet of everything meets the metaverse: Bridging physical and virtual worlds with blockchain. *Computer Science Review*, *54*, 100678. DOI: 10.1016/j.cosrev.2024.100678

Sharma, A., Kosasih, E., Zhang, J., Brintrup, A., & Calinescu, A. (2022). Digital twins: State of the art theory and practice, challenges, and open research questions. *Journal of Industrial Information Integration*, *30*, 100383. DOI: 10.1016/j.jii.2022.100383

Wache, H., Hübner, E. J., Hönigsberg, S., & Dinter, B. (2022). Closing the Implementation Gap of Digital Twins. In *AMCIS*.

Wang, D., Zhang, Z., Zhang, M., Fu, M., Li, J., Cai, S., Zhang, C., & Chen, X. (2021). The role of digital twin in optical communication: Fault management, hardware configuration, and transmission simulation. *IEEE Communications Magazine*, *59*(1), 133–139. DOI: 10.1109/MCOM.001.2000727

Wang, S., Zhang, J., Wang, P., Law, J., Calinescu, R., & Mihaylova, L. (2024). A deep learning-enhanced Digital Twin framework for improving safety and reliability in human–robot collaborative manufacturing. *Robotics and Computer-integrated Manufacturing*, *85*, 102608. DOI: 10.1016/j.rcim.2023.102608

Zhang, Z., Wen, F., Sun, Z., Guo, X., He, T., & Lee, C. (2022). Artificial intelligence-enabled sensing technologies in the 5G/internet of things era: From virtual reality/augmented reality to the digital twin. *Advanced Intelligent Systems*, *4*(7), 2100228. DOI: 10.1002/aisy.202100228

# Chapter 14
# Artificial Intelligence in Public Cultural Institutions:
## Shaping the Value Creation Process With Chatbots

**Carmine Sergianni**
https://orcid.org/0009-0004-5881-1436
*University of Naples Federico II, Italy*

**Gesualda Iodice**
https://orcid.org/0000-0001-7084-9091
*University of Naples Federico II, Italy*

## ABSTRACT

*This study investigates the integration of chatbot technology within public cultural institutions, to evaluate its impact on the value creation process. Theoretically based on the Public Service Logic, the research employs a qualitative methodology on a single case study, analyzing data collected from over 3,400 users during a 25-day pilot phase. The results indicate a high level of user acceptance, with 97.4% of visitors opting not to seek human assistance, and an average engagement time of 3 minutes and 9 seconds, reflecting significant interaction and interest. These findings underscore the potential of AI-driven solutions to improve service delivery and value creation process in public cultural settings. The implications suggest that the adoption of digital technologies can play a crucial role in shaping the value creation process and the way public institutions engage with their audiences, ultimately fostering greater accessibility and satisfaction across various sectors.*

DOI: 10.4018/979-8-3373-1534-8.ch014

## INTRODUCTION

Artificial Intelligence (AI) is one of the most discussed and studied topics in the academic community today (Aoki, 2020), thanks to its potential to radically transform the way society, business, and institutions operate. In recent years, interest in this discipline has grown exponentially, accompanied by an acceleration in the development of its practical applications in a wide range of sectors.

The ability of AI to perform complex tasks traditionally performed by humans, such as pattern recognition, decision making, and adaptive learning, has attracted the attention of both academics and practitioners. In particular, AI is increasingly recognized not only as an innovative technology, but also as a revolutionary force destined to redefine the boundaries of technological progress. Indeed, in the coming decades, AI is expected to play a central role in transforming economic and social systems, driving unprecedented change that will have a profound impact on all aspects of daily life (Raj & Seamans, 2019).

Currently, AI seems to have the capacity to radically revolutionise several industries (Raisch & Krakowski, 2021; Bifulco et al., 2016). These developments have led to the emergence of a wide range of highly beneficial opportunities, including effects on many areas of the public sector (Wirtz et al., 2019). This has triggered a transformation process highlighted by the proliferation and increasing adoption of AI-based applications in many public contexts (Misuraca et al., 2020).

Starting from Public Service Logic, which identifies and analyses the logic of value creation in the public sector and assumes a dynamic view of the interactive process between public service organisations (PSOs) and users, Osborne (2018) states the need to also consider technology when describing the value creation process of the many public sectors.

In this context, cultural industry is experiencing a paradigm shift in most processes, affecting value co-creation dynamics (Sacco et al., 2018) for on-site visitors as well as virtual ones (Minkiewicz et al., 2014; Romanelli, 2020; Taormina & Baraldi, 2022). Indeed, to achieve long-term success, cultural organizations must embrace innovation, not only by fostering a forward-thinking internal culture (Ou et al., 2024) but also by finding creative ways to leverage their resources and build stronger, more meaningful connections with their audiences (Sfodera et al., 2020). Governmental agencies are addressing such a change in their policies, to address the critical role such technologies have. In some cases scholars stressed the effects of novel technologies on visitors' experience (Liu, 2020), the support to preservation (Boniotti, 2023), and the viability of such digital transformation, highlighting the availability of low-cost solutions (Ocón, 2023). Value creation in cultural heritage can be driven by digital technologies (Tom Dieck and Jung, 2017); the debate is gathering increasing attention from scholars in different disciplinary fields.

Managerial studies are not exempt from such attention (Del Vacchio et al., 2020), focused on the relationships undertaken between cultural organisations and their stakeholders (Boiano et al., 2018; Orea-Giner et al., 2022; Russo Spena & Bifulco, 2021; Sepe & Marzullo, 2022). In this direction, AI - through intelligent computer systems that establish conversations with users and answer routine questions (Mehr, 2017) - improves efficiency in the value proposition in cultural heritage mostly in terms of reduced response time that is particularly valued by visitors in need to fast access information supported by available devices. At the same time, cultural value driven by digital technologies is proven to encourage user engagement (Huang & Rust, 2021) according to interactive logics by setting enhanced visitor experience (Calvi & Vermeeren, 2023). AI is also infused in chatbots, namely conversational agents capable of interacting with the user using written natural language, recently proliferating at a high pace in various industries (Rapp et al., 2021). The chatbot is an AI solution that has soon developed in the public sector (Maragno et al., 2023), due to its ability to perform several key actions for PSOs, such as answering routine questions and handling a large amount of information (Mehr, 2017). In the cultural sector, Public Organisations use chatbots for a wide and varied range of outputs from the use of human-technology interaction for artwork recognition as well as image analysis (Schaffer et al., 2022) to machine learning, robotics, computer vision, and natural language processing.

However, while different studies have explored the potential benefits of chatbots in the private sector (Rapp et al., 2021), Trischler and Trischler (2022) call for operationalisation of the new technologies in the public sphere also to contribute to a deeper understanding of their impact on the value creation process. In the cultural domain, several studies have paid attention to chatbot applications (e.g., Vidu et al., 2021; Iodice et al., 2024), although research on the latter is still at an early stage if compared with the practical advances, as shown by the success of the AskMona chatbot at Centre Pompidou (Paris). Therefore, this article aims to investigate the impact of chatbots on the value creation process in public-sector cultural institutions.

The article is structured as follows. First, a literature review on value creation in PSL and the role of chatbot AI embedded in public services is presented. Then, the methodology adopted is illustrated, focusing on the case study of the chatbot Nerone, developed for the Archaeological Park of the Colosseum. Finally, the results obtained are presented, followed by a detailed final discussion.

## Value Creation in Public Service Logic

Public service logic (PSL) is an emerging theoretical framework within the field of public administration management (PAM) that emphasises value creation as the core element of public service provision (Osborne, 2018). This perspective departs significantly from traditional paradigms by positioning service as a dynamic process, rather than a static production mechanism. Drawing on fundamental insights from service studies (Grönroos, 2006; Vargo & Lusch, 2008), PSL redefines public services as processes in which value is created through active collaboration between service users, PSOs and third parties. Within this framework, PSOs take on the role of facilitators of the value-creation process (Osborne, 2018).

The conceptual underpinnings of PSL are rooted in Service Logic (SL) and Service-Dominant Logic (SDL) which challenges the traditional goods-dominant view of public services. Indeed, PSL adopts a process-oriented approach that emphasises interactions and resource integration between different actors (Cui & Aulton, 2023), rather than viewing services as outputs to be delivered by PSOs. According to Osborne (2018), PSOs do not create value independently but propose value propositions that users subsequently integrate and elaborate within their contexts. This reconceptualisation places considerable emphasis on the active participation of users in the co-creation process, marking a departure from the passive consumption model traditionally associated with service provision.

Central to the PSL framework is the notion of value as a relational and contextual phenomenon. This approach is closely related to the distinction between "value-in-use" and "value-in-exchange" articulated in the SDL. While value-in-exchange denotes an intrinsic, market-determined value embedded in goods or services, value-in-use emphasises the idea that value is realised during the use of a service and is "uniquely and phenomenologically determined by the beneficiary" (Vargo & Lusch, 2016). This paradigm shift highlights the importance of user-centered processes and collaborative engagement in public service contexts.

Initially, PSL research centred on the concept of co-production, in which service users played a limited role in contributing to service delivery. In this model, user involvement was largely dependent on the time and effort required to participate in service processes initiated and controlled by PSOs (Alford, 2016; Osborne et al., 2013). However, this perspective has evolved to recognise the more complex and dynamic notion of co-creation. As Osborne (2018) notes, co-creation positions users as active agents in the value-creation process, highlighting their ability to integrate the value propositions offered by PSOs into their lived experiences. This shift reflects a growing recognition of the centrality of user agency and the importance of contextualised resource integration in defining value outcomes.

Co-creation also draws on the basic principles of SL (Grönroos, 2006, 2013). According to Grönroos, value creation in service settings occurs in two distinct domains: the joint sphere of interaction between PSOs and users, where co-creation occurs, and the independent sphere of the user, where individual value creation occurs. Osborne et al. (2021) elaborate on this distinction, claiming that the co-creation of value depends on the interaction between users and PSOs, whereas value creation involves broader processes of user-driven resource integration aligned with individual needs. This dual perspective reinforces the notion that PSOs can facilitate, but not unilaterally create value.

PSL marks a departure from the principles of New Public Management (NPM), a dominant paradigm that prioritised internal efficiency within PSOs and considered the market as the optimal mechanism for service delivery. Strongly influenced by private sector principles, NPM viewed users primarily as consumers, neglecting their potential for active engagement and collaboration (Hirschman, 1970). In contrast, PSL highlights the relational and participatory dimensions of public services, emphasising the shared processes of value co-creation. This fundamental difference highlights the limitations of NPM's transactional approach and underlines the transformative potential of PSL in promoting inclusive and user-centred service ecosystems.

PSL's emphasis on value-in-use also contrasts with NPM's focus on value-in-exchange. While NPM prioritised measurable outcomes and cost-effectiveness, PSL advocates a deeper understanding of the experiential and contextual factors that shape users' perceptions of value. This reorientation reflects a broader shift in public administration research towards holistic and user-centred approaches to service design and delivery.

A significant advance in the PSL literature is the recognition that value creation takes place within broader service ecosystems. These ecosystems encompass a wide range of actors, whose interactions collectively contribute to value creation (Eriksson, 2019). This ecosystem perspective challenges traditional linear models of service delivery, highlighting the interconnected and dynamic nature of value creation processes.

Osborne et al. (2013) and Trischler et al. (2023) emphasise the importance of relational dynamics within service ecosystems, arguing that effective value creation requires the collaborative engagement of all relevant actors. This approach is consistent with the broader conceptualisation of ecosystems as interactive networks in which the contributions of individual actors are integrated to achieve common goals (Normann & Ramírez, 1993). Within this framework, users are not only central participants, but also active agents who shape the value creation process through their interactions with other ecosystem members.

The role of PSOs within these ecosystems is multifaceted and includes facilitating the integration of resources, providing enabling infrastructure and fostering collaborative relationships. By fostering these conditions, PSOs can support the value co-creation in ways that extend beyond their immediate organisational boundaries. This ecosystem orientation underlines the importance of systemic thinking and adaptive strategies in public service management, and highlights the need for PSOs to navigate complex networks of stakeholders and align their efforts with broader societal goals.

However, within studies adopting the PSL perspective, numerous scholars have highlighted the urgency of investigating the role and impact of digital technologies in value creation processes in public services through both theoretical and empirical approaches (Leite, 2023; Jiang and Fan, 2024; Zyzak and Martinussen, 2024). The application of digital technologies in public services not only affects traditional operational processes, but also redefines the concept of public value itself, highlighting the need for a more nuanced and evidence-based understanding. In this context, there is an opportunity for researchers to explore the challenges, opportunities and ethical implications arising from the adoption of digital tools in the public sector, thereby enriching the theoretical corpus of PSL and providing practical guidance for policy makers and practitioners.

## AI in Public Services: the role of chatbot

AI is emerging as one of the most transformative and disruptive forces in the ongoing digital revolution. In recent years, a growing number of scientific publications and academic studies have highlighted the potential of AI to fundamentally redefine multiple sectors, including the public sector (Wirtz et al., 2019). This emerging technology offers tools and solutions that can make public services more dynamic, interactive and connected, thus contributing significantly to the modernisation of public institutions (Lee et al., 2011).

One of the most important aspects of the use of AI in the public sector is its ability to improve citizen engagement. With advanced AI-based systems, institutions can provide personalised information presented in a clear, timely and efficient manner, facilitating more direct and accessible communication (Neumann et al., 2022).

Among the many applications of AI, chatbots are one of the technologies that have received the most attention in the academic literature. These innovative tools not only simplify communication between institutions and citizens, but also contribute to reducing administrative burdens while improving the accessibility and quality of the services offered (Maragno et al., 2023; Misuraca et al., 2020).

In the cultural sector, public institutions are increasingly adopting AI to transform the visitor experience. Recent studies have highlighted the growing role of chatbots in providing useful information, guiding visitors along exhibition routes and creating highly interactive and engaging experiences (Schaffer et al., 2022). These tools are proving to be particularly effective not only in enriching the audience experience, but also in collecting meaningful data about visitors. Such data enables institutions to better understand the needs and behaviour of their audiences in order to design increasingly targeted and higher quality services (Barth et al., 2020).

However, despite the progress made, research into the use of AI in public cultural institutions is still at an early stage. This leaves room for numerous opportunities to explore and develop new ways of integrating the technology, thereby expanding the potential of these innovations. Recent literature foresees a future in which tools such as chatbots and other AI-based solutions are increasingly geared towards enhancing the interaction between humans and digital technologies, redefining the way culture is enjoyed and appreciated (Iodice et al., 2024).

This evolving context presents a unique opportunity to redefine the relationship between citizens and institutions, promoting more inclusive, interactive and personalised access to culture.

## RESEARCH METHODOLOGY

To investigate the utilization of chatbot technology within Cultural Institutions associated with the Public Sector and its impact on management practices through innovative methodologies, this study adopts a qualitative research methodology as outlined by Yin (2013). Specifically, a single case study approach is employed (Yin, 2009; Onghena et al., 2019), focusing on the Archaeological Park of the Colosseum as the subject of investigation. This cultural institution was selected due to its recognition as one of the first state-run museums in Italy to incorporate chatbot technology, thereby demonstrating a proactive stance in adapting to modern communication styles and visitor expectations.

The motivation behind this analysis is further underscored by the evidence indicating that, in 2022, a significant 72% of the Colosseum's visitors were individuals under the age of 40 (colosseo.it). This demographic shift highlights the necessity for cultural institutions to evolve and meet the burgeoning needs of younger audiences, who increasingly rely on digital tools for interaction and information.

Central to this investigation is the Technology Acceptance Model (TAM), a theoretical framework developed by Davis (1989), which serves to elucidate the factors that influence technology acceptance within the context of PSOs (Wirtz & Piehler, 2016). According to Piehler et al. (2016), the application of this model provides a nuanced understanding of the specific dynamics present in public administration settings. The TAM integrates two pivotal dimensions: (i) the perceived feasibility of utilising a chatbot as an effective service interface to automate and enhance user communication processes, and (ii) the attitudes and perceptions of users regarding the adoption of chatbots for the automation of communication service delivery.

## Research Framework

The methodological framework used in this study is the TAM, illustrated in Figure 1. This conceptual model serves as a robust lens through which the authors investigate the intricate causal relationships among three key constructs: users' perceptions—specifically their perceived usefulness and perceived ease of use—attitudes towards technology, and behavioral intentions concerning engagement with photographic content. These constructs are pivotal in understanding the degree of acceptance of emerging technologies by end users, as highlighted in Fan (2020). The model's applicability has been reinforced by an array of empirical studies across diverse disciplinary domains, including significant insights into the cultural sector, where added focus has been noted on augmented reality (Cheng et al., 2023), the integration of digital technologies (Nizar et al., 2018; Aziz et al., 2019), and the evolution of digital museum experiences (Shi et al., 2023; Hung et al., 2013).

*Figure 1. Technology acceptance model (Davis; 1989)*

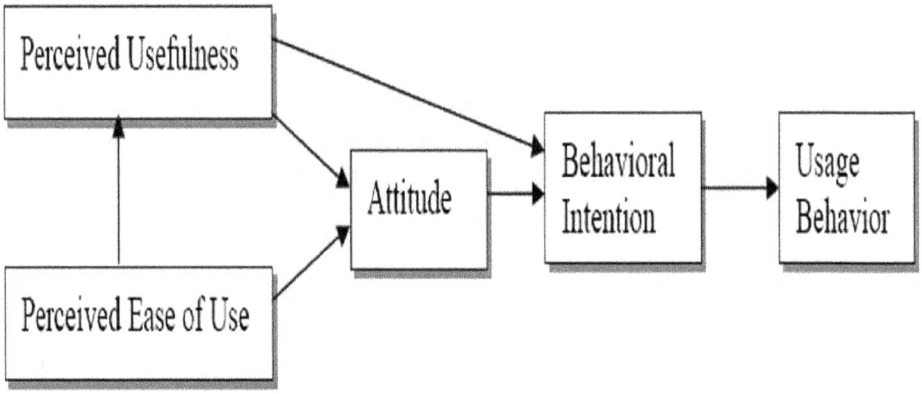


This research draws upon data gleaned during the pilot phase of an interactive chatbot named NERONE, which served as a medium for engagement between the Archaeological Park of the Colosseum and its visitor audience. The chatbot's development process lasted over a year, during which researchers meticulously analysed user interactions and inquiries recorded over five years. This analysis led to the formulation of approximately 120 distinct conversational scenarios tailored to mitigate frequent user queries and enhance communication pathways.

## Data Collection

As outlined by Creswell (2013), the data collection approach of this research is multifaceted and heterogeneous in order to provide a robust understanding of technology adoption in this specific context. The research adopts a mixed methods strategy, integrating different sources of qualitative and quantitative data. In particular, qualitative inputs are drawn from critical analyses of academic articles, curated content from relevant websites, reflective essays, and official reports (Creswell & Creswell, 2017). Collectively, these sources provide a nuanced perspective on the contextual and user-specific variables that influence the adoption of chatbot technology.

As a pilot study, the research focuses on the experimental implementation phase of chatbot technology within the selected case study, a cultural heritage setting. The quantitative dimension of the data collection includes a comprehensive analysis of interactions with a sample of 3,400 users over a period of 25 days. This resulted in an average of over 160 active users per day, with each interaction lasting an average of 3 minutes and 9 seconds. The dataset captures a variety of user behaviors, engagement patterns, and preferences, allowing for granular analysis of the factors that influence technology adoption.

The study's systematic and rigorous data collection process offers a dual benefit: it not only facilitates a deeper understanding of user behavior, but also provides critical insights into the broader implications of using chatbot technology in cultural heritage settings. This rich and diverse dataset serves as a foundational resource for exploring the interplay between technology usability, user engagement, and cultural context, ultimately advancing the broader discourse on technology acceptance in similar fields.

## FINDINGS

The implementation of chatbot technology in the Colosseum Archaeological Park has provided a series of important insights into the way users interact and their perceptions. This research has highlighted how the adoption of advanced technological tools is not only a means to facilitate the use of information, but also a key element in redefining the entire value creation process in a public cultural domain. By analysing data collected from over 3,400 users during 25 days, it's possible to draw meaningful conclusions about the perceived usefulness and ease of use of this technology. The results not only highlight the chatbot's role in facilitating communication between the park and its audience but also underscore the importance of adapting to the evolving needs of visitors, particularly younger generations who are increasingly reliant on digital solutions. The following sections detail the key findings related to user engagement, satisfaction, and the overall impact of the chatbot on the value creation.

## Perceived Usefulness

The construct of perceived usefulness is largely shaped by a multitude of external factors, including but not limited to social dimensions such as individual skills and language proficiencies, cultural nuances, and political influences manifesting through governmental policies and directives (Surendran, 2012). Within the cultural and creative sectors, existing literature underscores that visitors' perceptions of usefulness are intrinsically linked to the outcomes related to the dissemination of culture-centric content and the enhancement of performance-oriented learning experiences (Nizar et al., 2018; Taber, 2014). In the present study, pertinent data gathered from an experimental phase involving the implementation of chatbot technology at the Archaeological Park of the Colosseum reveal exceptionally high levels of perceived usefulness among users. Notably, an overwhelming 97.4% of participants demonstrated a preference for the chatbot as their primary source of support, opting not to seek assistance from human staff. This trend illustrates the urgent necessity for the Archaeological Park of the Colosseum to adapt to contemporary visitor expectations by employing innovative technological solutions, thereby enriching the visitor experience and aligning with evolving linguistic and cultural paradigms, especially among younger demographics.

## Perceived Ease of Use

The notion of perceived ease of use pertains to the ability to achieve optimal performance with minimal effort when employing a given technology (Nizar et al., 2018; Li et al., 2017; Isaais et al., 2017). Empirical evidence indicates that users are more likely to find a technology beneficial when it can be utilised with relative ease (Bruner & Kumar, 2005). Within the cultural and creative landscape, this is particularly salient for younger generations, who are more attuned to leveraging technological interfaces for the effective communication of cultural content. Such integration necessitates a seamless connection between on-site offerings and online resources. In the context of this pilot study, the perceived ease of use is intricately linked to the preparatory phase that involves extensive analysis of potential user inquiries and scenarios. This analysis was informed by systematic monitoring by the Park's Communication Office of user engagements via social media, email correspondence, and telephonic inquiries, thereby allowing for the development of a comprehensive user profile. This preparatory groundwork aids in anticipating frequently recurring questions and logistical concerns. Additionally, the design of the chatbot, embodied by the character of Nerone, functions to bridge the divide between the Archaeological Park and its visitors. This persona enhances user engagement by fostering a more relatable interaction style, thereby enriching the overall user experience. The chatbot's multilingual capabilities—offering support in English, French, and Italian—significantly augment its accessibility. Moreover, strategic placement of the chatbot on the homepage ensures immediate visibility and ease of access for users, who can engage in less formal interactions. In instances where the chatbot encounters inquiries beyond its capacity to address, a robust fallback mechanism is in place, wherein user messages are automatically relayed to the communications department. This ensures continued service provision while simultaneously facilitating the identification of gaps in the chatbot's knowledge base, thereby supporting iterative improvements to the service.

## Attitude

User attitudes are profoundly influenced by both perceived utility and perceived ease of use (Surendran, 2012). In the realm of cultural and creative sectors, numerous studies affirm a predominantly positive disposition towards the incorporation of technology among all stakeholders involved in value creation. Specifically, managerial receptiveness to integrating technology within value propositions is often driven by the prospect of providing cost-effective and easily implementable solutions (Doran et al., 2015). Concurrently, audience engagement reflects favorable attitudes, characterised by heightened interest and interaction levels due to the technology's capability to address fundamental needs, and its expansive integration into daily life (Nizar et al., 2018). This study aims to elucidate how museum visitors' attitudes significantly influence their utilisation of mobile applications. Findings indicate that museum patrons exhibit considerable enthusiasm and positive responses towards mobile technology offerings, as evidenced by both the substantial volume and duration of interactions with the chatbot. This enthusiasm can be attributed to the pervasive integration of mobile technology into visitors' lives, enabling a smooth transition into its application within the museum setting. Moreover, the technological solutions provided by the museum align with visitor needs and aspirations commensurate with 21st-century expectations. These insights collectively affirm that positive visitor attitudes toward mobile technology catalyse the adoption of the museum's application, ultimately enriching the visitor experience.

## Behavioral Intention

Comprehensive literature indicates that behavioral intention is shaped not only by user attitude but also by prevailing subjective norms and perceived behavioral control (Ajzen, 1991). The interplay among these factors is critical to understanding how individuals make decisions regarding the adoption and use of technology in cultural contexts. These decisions influence engagement levels and the overall visitor experience.

The overall evidence collected yields in-depth insights into the enhancement of visitor engagement and satisfaction through the application of digital tools in public cultural institutions. The NERONE chatbot has exhibited exceptionally high perceived usefulness, as evidenced by the finding that 97.4% of users did not feel the need to seek assistance from human operators during their visit. This statistic strongly suggests that visitors view this technology as not only effective but also essential for obtaining timely and accurate information. Furthermore, the positive reception of the chatbot indicates its potential to significantly improve the value creation process by providing instant support and information access, thereby fostering a deeper connection with the cultural heritage presented at the park. This observation aligns with existing literature highlighting that perceived usefulness is shaped by various external, social, and cultural factors, contributing to a seamless visitor experience (Surendran, 2012; Nizar et al., 2018). Furthermore, the perceived ease of use of the chatbot emerges as a critical determinant of its acceptance among users. The thoughtful design of the chatbot, which includes a well-defined user profile based on prior interactions and commonly asked questions, exemplifies a user-centered approach to enhancing the visitor experience. The selection of the character Nerone as a relatable interface further enriches the interaction, rendering it more accessible. Additionally, the chatbot's availability in multiple languages—specifically English, French, and Italian—broadens its reach and caters to a diverse audience, ultimately improving overall usability. Data collected during the testing phase reveal an average engagement time of 3 minutes and 9 seconds, indicative of high levels of user interaction and interest. Such engagement is paramount within the cultural and creative sector, as the effective communication of cultural content significantly influences visitor satisfaction and learning outcomes (Iodice et al., 2025). Notably, the finding that 69.7% of inquiries pertained to ticket information underscores the chatbot's pivotal role in streamlining access to essential services, thereby enhancing the overall visitor experience. Moreover, this study underscores the necessity for continuous improvement in chatbot technology. The mechanism allowing for the redirection of unanswered queries to the communications department not only ensures users receive the requisite information but also generates valuable feedback for the refinement of the chatbot's capabilities. This iterative process of learning and adaptation is essential for maintaining relevance within a rapidly evolving digital landscape.

## DISCUSSION

The results of the study highlight the important role that AI, represented by chatbot technology, can play in the value creation process for the public cultural sector. Specifically, the implementation of the chatbot at the Archaeological Park of the Colosseum provided an opportunity to explore how digital tools can improve the visitor experience, not only by making information more accessible and immediate, but also by customizing it according to users' preferences and needs.

Thanks to the chatbot's ability to interact in real time, visitors can get quick answers to practical questions such as opening hours, access methods and suggested routes, thus improving the organization of their visit. In addition, the chatbot has proven to be able to play an educational role, providing historical and cultural insights in an engaging and adaptable way, increasing the emotional and cognitive engagement of the public.

The analysis highlighted how the use of advanced digital technologies can meet the needs of an increasingly connected and technologically advanced public, making cultural heritage more accessible and inclusive. This approach not only enhances the quality of the individual visitor experience, but also helps to build a stronger and more lasting relationship between cultural institutions and their audiences, opening up new opportunities for the promotion and enhancement of artistic and archaeological heritage.

One of the key points that emerged is the high perceived usefulness of the chatbot, with 97.4% of participants preferring to rely on technology rather than human staff. This data highlights not only the perceived efficiency of the tool, but also the growing expectation of visitors to be able to interact with immediate, personalized digital solutions in contemporary languages. The introduction of the chatbot is not simply a logistical support, but a lever to create a more fluid and engaging cultural experience, capable of integrating the informative and narrative value of the park in a digital context familiar to visitors, especially younger ones.

At the same time, the perceived ease of use proved crucial to the tool's acceptance and success. The care taken in the preparation phase - which included analyzing interactions on social media, emails and phone calls - and the design of a chatbot capable of anticipating visitors' main questions were key elements in ensuring an intuitive experience. The user-centred design, represented by the figure of Nerone, not only made the interaction more immediate and accessible, but also strengthened the bond between visitors and the cultural heritage, proposing an innovative and engaging communication method. The multilingual offer (Italian, English and French) further increased accessibility, confirming the importance of an inclusive approach when designing technologies for the cultural sector.

In terms of attitudes towards digital technologies, the study highlighted that visitors are well disposed towards the use of digital tools in the museum context. The high level of interaction and enthusiasm for the chatbot reflects the effectiveness of this solution in meeting expectations that have already been consolidated through the daily use of digital technologies. This data not only confirms the relevance of AI in the public cultural context, but also suggests that the adoption of digital solutions can be a catalyst to increase visitors' engagement and sense of belonging towards cultural heritage.

The behavioral intention derived from the data shows that the use of digital tools such as the NERONE chatbot is not only well received, but actively contributes to the value creation process. The chatbot's ability to provide timely and accurate information, as well as its role in simplifying key processes such as purchasing tickets, demonstrates how AI can be strategically integrated to improve both the visitor experience and the organizational efficiency of the institution. At the same time, the ability to refer unresolved questions back to the communications department underscores the importance of an iterative approach to continuously improve the system and ensure an increasing ability to respond to user needs.

The analysis shows that the chatbot is a powerful tool for transforming and improving the value creation process, making it more efficient and dynamic. The chatbot's ability to respond immediately and in a personalized manner to users' needs allows for greater accessibility to their needs, favoring an interaction that goes beyond the traditional model.

Rather than passively responding to requests, the chatbot actively transforms the traditional value-added process, adapting it in a flexible and proactive way. Thanks to digitization, cultural institutions can collect useful data on user preferences and behaviors and use it to optimize the services offered and create a richer and more engaging experience. This approach not only improves the quality of the service, but also makes the value creation process more integrated with digital technologies, paving the way for innovative and personalized management of relations with the public.

In this sense, the chatbot becomes a strategic tool for cultural institutions, facilitating the transition to a model of cultural heritage exploitation that is increasingly focused on users' needs and supported by advanced digital technologies. This integration not only meets the expectations of an increasingly connected and tech-savvy public, but also lays the groundwork for a value creation process that is constantly evolving, inclusive and forward-looking.

## CONCLUSIONS

The present study explored the NERONE chatbot's implementation within the Archaeological Park of the Colosseum context, highlighting the transformative potential of AI technologies in enhancing value creation process in public cultural institutions. The Technology Acceptance Model (TAM) is particularly relevant in this examination, as it elucidates the interaction dynamics between cultural institutions and their audience.

The primary evidence that emerged through this study indicates that the integration of chatbot technology at cultural sites such as the Colosseum represents a pivotal advancement in modernising visitor engagement strategies and practices. The favorable reception of the NERONE chatbot implies that analogous initiatives in various cultural settings may yield significant benefits, fostering a more interactive, user-centered approach to the dissemination of information. The high level of acceptance among users signifies the chatbot's effectiveness in addressing its audience's diverse needs, with an especially noteworthy appeal to younger visitors who have grown increasingly accustomed to digital interactions in their everyday lives.

From a theoretical perspective, the successful integration of chatbot technology within the Archaeological Park of the Colosseum offers considerable implications for the field of Public Service Logic, particularly for public institutions tasked with conveying cultural value. Chatbot technologies contribute significantly to the discourse initiated by scholars such as Maragno et al. (2023) by improving the value creation process in the public domain. Additionally, the deployment of chatbots within the cultural heritage sector empowers cultural managers to gather insights regarding visitor preferences, enabling the refinement of strategies to enhance visitor satisfaction (Canonico et al., 2021).

From a managerial perspective, the findings underscore the necessity for cultural institutions to prioritise the continuous development and refinement of AI technologies. As discussed, in instances where the NERONE chatbot cannot comprehensively address a user's inquiry, a mechanism is instituted wherein the user is not left without assistance. Instead, the inquiry is automatically redirected to the communications department, ensuring that a knowledgeable response can be provided, thereby maintaining continuity of service. This feedback mechanism for unanswered queries serves a dual purpose: it enhances the chatbot's performance over time and cultivates a culture characterized by responsiveness and adaptability within the institution. Such an approach is instrumental in enabling cultural institutions to remain attuned to the evolving needs and preferences of their visitors, ultimately enriching their service offerings.

This research is based on the analysis of a single case study. In Italy, among public cultural institutions, only the Archaeological Park of the Colosseum and the Galleria dell'Accademia in Florence currently use a chatbot as a tool for interacting with the public. This limitation is due to the lack of similar cases in the country, which reduces the possibility of internal comparisons and generalisations. Therefore, further research could benefit from studying international contexts where the use of chatbots in public cultural institutions is more widespread, to broaden the understanding of the impact of such technologies in different scenarios.

Furthermore, while the present study contributes valuable insights into the application of chatbot technology within public cultural institutions, it is essential to recognize that further research is critical to comprehensively explore the long-term impacts of such technologies on visitor behavior and satisfaction. While the extant research establishes a substantive foundation, numerous avenues remain either unexplored or underexplored, thus necessitating further scholarly scrutiny.

Firstly, longitudinal studies assessing the long-term impact of chatbot implementation on user engagement and satisfaction within cultural institutions could provide critical insights. Such research endeavors would facilitate the tracking of changes over time, effectively capturing the evolving dynamics of visitor interactions with AI technologies. Secondly, comparative studies that examine the efficacy of diverse chatbot designs and functionalities across varying cultural settings could yield compelling findings. This research may involve analyzing how user demographics influence the reception and utility of chatbots, thereby enabling the customization of AI implementations to meet the diverse needs of distinct audience segments. For instance, the preferences of technologically adept younger audiences may significantly differ from those of older generations. Consequently, an exploration of these user-specific nuances could provide a more nuanced framework for the strategic deployment of chatbots in museums and galleries. Furthermore, research methodologies could be enhanced through the adoption of mixed-method approaches that integrate quantitative assessments of chatbot effectiveness with qualitative insights derived from user experiences. Moreover, the ethical implications of AI in cultural institutions represent another critical research frontier. Issues surrounding data privacy, algorithmic bias, and the digital divide warrant rigorous examination, particularly as institutions navigate the complexities of enriching visitor experiences. Striking a balance between leveraging AI for engagement and ensuring equitable access, alongside ethical usage, is pivotal for fostering public trust and enhancing institutional credibility.

It is pertinent to acknowledge the limitations inherent in the current literature on AI applications, particularly chatbots, within public cultural institutions, which remains in its nascent stages. Although advancements have been made, the empirical basis supporting the effectiveness and efficiency of these technologies is limited. Much of the current research tends to be anecdotal or focused on isolated case studies, potentially failing to represent broader trends within the sector. Additionally, there exists a tendency to overlook the unique challenges encountered by smaller institutions that may lack the necessary resources for extensive technological integration. In conclusion, as the integration of AI, notably chatbots, into public cultural institutions continues to proliferate, the cultivation of a robust research agenda proves imperative. By addressing these projected avenues for future inquiry and acknowledging existing limitations, scholars can contribute to a comprehensive understanding of how AI may be leveraged to augment the value and impact of cultural institutions in the 21st century.

## REFERENCES

Ajzen, I. (1991). The Theory of planned behavior. *Organizational Behavior and Human Decision Processes*, *50*(2), 179–211. DOI: 10.1016/0749-5978(91)90020-T

Alford, J. (2016). Co-Production, Interdependence and Publicness: Extending public service-dominant logic. *Public Management Review*, *18*(5), 673–691. DOI: 10.1080/14719037.2015.1111659

Aoki, N. (2020). An experimental study of public trust in AI chatbots in the public sector. *Government Information Quarterly*, *37*(4), 101490. DOI: 10.1016/j.giq.2020.101490

Aziz, M. N. A., Harun, S. N., Baharom, M. K., Khairulnizam, M., & Ramlie, A. S. M. S. (2019). The Acceptance Level of Digital Natives towards the Interactive Kiosk in the Museum: Tam-Based Research Instrument. *TEST Engineering & Management*, (Nov-Dec), 1032–1044.

Barth, F., Candello, H., Cavalin, P., & Pinhanez, C. (2020, July). Intentions, meanings, and whys: designing content for voice-based conversational museum guides. In *Proceedings of the 2nd Conference on Conversational User Interfaces* (pp. 1-8). DOI: 10.1145/3405755.3406128

Bifulco, F., Tregua, M., Amitrano, C. C., & D'Auria, A. (2016). International Journal of Public Sector Management. *ICT and sustainability in smart cities management*, (2), 132-147.

Boiano, S., Borda, A., Gaia, G., Rossi, S., & Cuomo, P. (2018). Chatbots and new audience opportunities for museums and heritage organisations. Electronic visualisation and the arts, 164-171

Boniotti, C. (2023). The public–private–people partnership (P4) for cultural heritage management purposes. *Journal of Cultural Heritage Management and Sustainable Development*, *13*(1), 1–14. DOI: 10.1108/JCHMSD-12-2020-0186

Bruner, I. I. G. C.II, & Kumar, A. (2005). Applying TAM to consumer usage of handheld Internet devices. *Journal of Business Research*, *58*(5), 553–558. DOI: 10.1016/j.jbusres.2003.08.002

Calvi, L., & Vermeeren, A. P. (2023). Digitally enriched museum experiences–what technology can do. *Museum Management and Curatorship*, •••, 1–22.

Canonico, R., Cozzolino, G., & Sperlì, G. (2021). Suggesting Cultural Heritage Points of Interest Through a Specialized Chatbot. In Advances in Intelligent Networking and Collaborative Systems: The 12th International Conference on Intelligent Networking and Collaborative Systems (INCoS-2020) (pp. 112-120). Springer International Publishing. DOI: 10.1007/978-3-030-57796-4_11

Cheng, A., Ma, D., Pan, Y., & Qian, H. (2023). Enhancing museum visiting experience: Investigating the relationships between augmented reality quality, immersion, and TAM using PLS-SEM. *International Journal of Human-Computer Interaction*, •••, 1–12.

Creswell, J. W. (2013). Steps in conducting a scholarly mixed methods study. In DBER Group Discussion, available at https://digitalcommons.unl.edu/dberspeakers/48/

Creswell, J. W., & Creswell, J. D. (2017). *Research design: Qualitative, quantitative, and mixed methods approaches*. Sage Publications.

Cui, T., & Aulton, K. (2023). Conceptualizing the elements of value in public services: Insights from practitioners. *Public Management Review*, •••, 1–23. DOI: 10.1080/14719037.2023.2226676

Davis, F. D. (1989). *Technology acceptance model: TAM. Al-Suqri, MN.* Information Seeking Behavior and Technology Adoption.

Del Vacchio, E., Gargiulo, R., & Bifulco, F. (2020). Digital Communication and Museum Experience. A Multichannel Approach. *Micro & Macro Marketing*, *29*(3), 513–533.

Doran, R., Hanss, D., & Larsen, S. (2015). Attitudes, efficacy beliefs, and willingness to pay for environmental protection when travelling. *Tourism and Hospitality Research*, *15*(4), 281–292. DOI: 10.1177/1467358415580360

Eriksson, E. M. (2019). Representative co-production: Broadening the scope of the public service logic. *Public Management Review*, *21*(2), 291–314. DOI: 10.1080/14719037.2018.1487575

Grönroos, C. (2006). Adopting a service logic for marketing. *Marketing Theory*, *6*(3), 317–333. DOI: 10.1177/1470593106066794

Hirschman, A. O. (1970). *Exit, Voice and Loyalty: Responses to Decline in Firms, Organizations and States*. Harvard University Press.

Huang, M. H., & Rust, R. T. (2021). Engaged to a robot? The role of AI in service. *Journal of Service Research*, *24*(1), 30–41. DOI: 10.1177/1094670520902266

Hung, S. Y., Chen, C. C., Hung, H. M., & Ho, W. W. (2013). Critical factors predicting the acceptance of digital museums: User and system perspectives. *Journal of Electronic Commerce Research, 14*(3), 231.

Iodice, G., Clemente, L., & Bifulco, F. (2024). The role of chatbot technology in the cultural value proposition: A managerial perspective in human-like interaction. In *The use of artificial intelligence in digital marketing: Competitive strategies and tactics* (pp. 166–187). IGI Global.

Iodice, G., Sergianni, C., Tregua, M., & Bifulco, F. (2025). Shaping Value Propositions in Cultural Heritage From a PSL Perspective: The Case of Grassroots Museums. In *Impacts of Museums on Global Communication* (pp. 213–240). IGI Global.

Jiang, M., & Fan, B. (2024). Citizens' conscious engagement in digital platforms: advancing public service delivery within the public service logic framework. Public Management Review, 1–19. Fan, C. W. (2020). Applied the Technology Acceptance Model to Survey the mobile-learning adoption behavior in Science Museum. In Cross-Cultural Design. Applications in Health, Learning, Communication, and Creativity: 12th International Conference, CCD 2020, Held as Part of the 22nd HCI International Conference, HCII 2020, Copenhagen, Denmark, July 19–24, 2020 [Springer International Publishing.]. *Proceedings, 22*(Part II), 270–280.

Lee, C. P., Chang, K., & Berry, F. S. (2011). Testing the development and diffusion of e-government and e-democracy: A global perspective. *Public Administration Review, 71*(3), 444–454. DOI: 10.1111/j.1540-6210.2011.02228.x

Leite, H. (2023). The impact of virtual reality on private value dimensions: A healthcare case study. *Public Management Review*, •••, 1–27.

Li, R., Chung, T.-L., & Fiore, A. M. (2017). Factors affecting current users' attitude towards eauctions in China: An extended TAM study. *Journal of Retailing and Consumer Services, 34*, 19–29. DOI: 10.1016/j.jretconser.2016.09.003

Liu, Y. (2020). Evaluating visitor experience of digital interpretation and presentation technologies at cultural heritage sites: A case study of the old town, Zuoying. *Built Heritage, 4*(1), 14. DOI: 10.1186/s43238-020-00016-4

Maragno, G., Tangi, L., Gastaldi, L., & Benedetti, M. (2023). AI as an organizational agent to nurture: Effectively introducing chatbots in public entities. *Public Management Review, 25*(11), 1–31. DOI: 10.1080/14719037.2022.2063935

Mehr, H., Ash, H., & Fellow, D. (2017). Artificial intelligence for citizen services and government. *Ash Cent. Democr. Gov. Innov. Harvard Kennedy Sch*, (August), 1–12.

Minkiewicz, J., Evans, J., & Bridson, K. (2014). How do consumers co-create their experiences? An exploration in the heritage sector. *Journal of Marketing Management, 30*(1-2), 30–59. DOI: 10.1080/0267257X.2013.800899

Misuraca, G., van Noordt, C., & Boukli, A. (2020). The use of AI in public services: Results from a preliminary mapping across the EU. In *Proceedings of the 13th international conference on theory and practice of electronic governance* (pp. 90-99). DOI: 10.1145/3428502.3428513

Neumann, O., Guirguis, K., & Steiner, R. (2022). Exploring artificial intelligence adoption in public organizations: A comparative case study. *Public Management Review, 26*(1), 114–141. DOI: 10.1080/14719037.2022.2048685

Nizar, N. N. M., Rahmat, M. K., Nabihah, N., & Nizar, M. (2018). Examining the museum visitors' use of mobile technology through technology acceptance model (TAM). *Journal of Tourism, Hospitality, and Event Management, 3*(11), 14–24.

Normann, R., & Ramirez, R. (1993). From value chain to value constellation: Designing interactive strategy. *Harvard Business Review, 71*(4), 65–77. PMID: 10127040

Ocón, D. (2023). Low-cost digital tools to preserve cultural heritage 'blind spots': The case of Kubor Kassim in Singapore. *International Journal of Heritage Studies, 29*(9), 961–987. DOI: 10.1080/13527258.2023.2231903

Onghena, P., Maes, B., & Heyvaert, M. (2019). Mixed methods single case research: State of the art and future directions. *Journal of Mixed Methods Research, 13*(4), 461–480. DOI: 10.1177/1558689818789530

Orea-Giner, A., Muñoz-Mazón, A., Villacé-Molinero, T., & Fuentes-Moraleda, L. (2022). Cultural tourist and user experience with artificial intelligence: a holistic perspective from the Industry 5.0 approach. Journal of Tourism Futures, (ahead-of-print).

Osborne, S. P. (2018). From Public Service-dominant Logic to Public Service Logic: Are Public Service Organizations Capable of Co-production and Value Co-creation? *Public Management Review, 20*(2), 225–231. DOI: 10.1080/14719037.2017.1350461

Osborne, S. P., Nasi, G., & Powell, M. (2021). Beyond co-production: Value creation and public services. *Public Administration, 99*(4), 641–657. DOI: 10.1111/padm.12718

Osborne, S. P., & Strokosch, K. (2013). It takes Two to Tango? Understanding the C o-production of Public Services by Integrating the Services Management and Public Administration Perspectives. *British Journal of Management, 24*(S1), S31–S47. DOI: 10.1111/1467-8551.12010

Ou, S., Gan, J. L., & Li, L. (2024). Innovative Culture: A Fifty-year (1972-2023) Analysis and Future Direction. *Journal of Innovation Management, 12*(3), 72–101. DOI: 10.24840/2183-0606_012.003_0004

Piehler, R., Wirtz, B. W., & Daiser, P. (2016). An analysis of continuity intentions of eGovernment portal users. *Public Management Review, 18*(2), 163–198. DOI: 10.1080/14719037.2014.965270

Raisch, S., & Krakowski, S. (2021). Artificial intelligence and management: The automation–augmentation paradox. *Academy of Management Review, 46*(1), 192–210. DOI: 10.5465/amr.2018.0072

Raj, M., & Seamans, R. (2019). Primer on artificial intelligence and robotics. *Journal of Organization Design, 8*(1), 1–14. DOI: 10.1186/s41469-019-0050-0

Rapp, A., Curti, L., & Boldi, A. (2021). The human side of human-chatbot interaction: A systematic literature review of ten years of research on text-based chatbots. *International Journal of Human-Computer Studies, 151*, 151. DOI: 10.1016/j.ijhcs.2021.102630

Romanelli, M. (2020). Museums and technology for value creation. Technology and Creativity: Production, Mediation and Evaluation in the Digital Age, 181-210.

Russo Spena, T., & Bifulco, F. (Eds.). (2021). *Digital Transformation in the Cultural Heritage Sector: Challenges to Marketing in the New Digital Era*. Springer Nature. DOI: 10.1007/978-3-030-63376-9

Sacco, P. L., Ferilli, G., & Tavano Blessi, G. (2018). From culture 1.0 to culture 3.0: Three socio-technical regimes of social and economic value creation through culture, and their impact on European Cohesion Policies. *Sustainability (Basel), 10*(11), 3923. DOI: 10.3390/su10113923

Schaffer, S., Ruß, A., Sasse, M. L., Schubotz, L., & Gustke, O. (2022, February). Questions and Answers: Important Steps to Let AI Chatbots Answer Questions in the Museum. In ArtsIT, Interactivity and Game Creation: Creative Heritage. New Perspectives from Media Arts and Artificial Intelligence. 10th EAI International Conference, ArtsIT 2021, Virtual Event, December 2-3, 2021, Proceedings (pp. 346-358). Cham: Springer International Publishing.

Sepe, F., & Marzullo, M. (2022). Making Smarter Museums Through New Technologies. In Handbook of Research on Museum Management in the Digital Era (pp. 75-98). IGI Global. DOI: 10.4018/978-1-7998-9656-2.ch005

Sfodera, F., Mingo, I., Mattiacci, A., & Colurcio, M. (2020). Night at the museum: Technology enables visitor experiences. *Sinergie Italian Journal of Management, 38*(1), 231–250.

Shi, M., Wang, Q., & Long, Y. (2023). Exploring the Key Drivers of User Continuance Intention to Use Digital Museums: Evidence from China's Sanxingdui Museum. *IEEE Access : Practical Innovations, Open Solutions, 11*, 81511–81526. DOI: 10.1109/ACCESS.2023.3297501

Surendran, P. (2012). Technology acceptance model: A survey of literature. International journal of business and social research, 2(4), 175-178.

Taber, M. M. (2014). Use of Mobile Technology Among Museum Visitors: A Case Study. Use of Mobile Technology Among Museum Visitors, 1-50.

Taormina, F., & Baraldi, S. B. (2022). Museums and digital technology: A literature review on organizational issues. *European Planning Studies, 30*(9), 1676–1694. DOI: 10.1080/09654313.2021.2023110

Tom Dieck, M. C., & Jung, T. H. (2017). Value of augmented reality at cultural heritage sites: A stakeholder approach. *Journal of Destination Marketing & Management, 6*(2), 110–117. DOI: 10.1016/j.jdmm.2017.03.002

Trischler, J., Røhnebæk, M., Edvardsson, B., & Tronvoll, B. (2023). Advancing Public Service Logic: Moving towards an ecosystemic framework for value creation in the public service context. *Public Management Review*, •••, 1–29. DOI: 10.1080/14719037.2023.2229836

Trischler, J., & Trischler, J. W. (2022). Design for experience – a public service design approach in the age of digitalization. *Public Management Review, 24*(8), 1251–1270. DOI: 10.1080/14719037.2021.1899272

Vargo, S. L., & Lusch, R. F. (2008). Service-dominant logic: Continuing the evolution. *Journal of the Academy of Marketing Science, 36*(1), 1–10. DOI: 10.1007/s11747-007-0069-6

Vargo, S. L., & Lusch, R. F. (2016). Institutions and axioms: An extension and update of service-dominant logic. *Journal of the Academy of Marketing Science, 44*(1), 5–23. DOI: 10.1007/s11747-015-0456-3

Vidu, C., Zbuchea, A., & Pinzaru, F. (2021). Old Meets New: Integrating Artificial Intelligence in Museums' Management Practices. Strategica. Shaping the Future of Business and Economy, 830-844.

Wirtz, B. W., & Piehler, R. (2016). eGovernment Applications and public personnel acceptance: An empirical analysis of the public servant perspective. *International Journal of Public Administration, 39*(3), 238–247. DOI: 10.1080/01900692.2014.1003384

Wirtz, B. W., Weyerer, J. C., & Geyer, C. (2019). Artificial intelligence and the public sector—Applications and challenges. *International Journal of Public Administration, 42*(7), 596–615. DOI: 10.1080/01900692.2018.1498103

Yin, R. K. (2009). *Case study research: Design and methods* (Vol. 5). Sage Publications.

Yin, R. K. (2013). Validity and generalization in future case study evaluations. *Evaluation, 19*(3), 321–332. DOI: 10.1177/1356389013497081

Yu, C. S., & Tao, Y. H. (2009). Understanding business-level innovation technology adoption. *Technovation, 29*(2), 92–109. DOI: 10.1016/j.technovation.2008.07.007

Zyzak, B., & Martinussen, P. E. (2024). OK computer: Applying the public service logic on digital health services. *Public Management Review, 26*(11), 3342–3364. DOI: 10.1080/14719037.2024.2337080

# Chapter 15
# Application of Digital Twin Technology for Building Sustainable Strategic Business Models in Healthcare

**Eleni Tsianaka**
https://orcid.org/0000-0003-4293-8315
*Rushford Business School, Switzerland*

**Konstantinos Peramatzis**
*Australian University, Kuwait*

## ABSTRACT

*Digital Twin (DT) technology revolutionises personalised healthcare by creating virtual patient replicas for real-time monitoring, predictive modelling, and tailored treatment planning. This chapter examines the current applications of digital twins in healthcare, emphasising their potential to improve clinical outcomes and reduce costs through precision medicine. Key uses include managing chronic diseases, integrating multi-omics data and medical imaging to customise treatment strategies in personalised cancer care and surgical planning, allowing for the rehearsal of complex procedures and custom implant design. Despite its promise, adopting DTs in healthcare faces challenges, particularly in data privacy, model accuracy, and integrating diverse data sources. Ensuring model reliability and addressing ethical concerns about patient data are essential for widespread adoption. This chapter highlights the need for ongoing research and development to overcome these challenges and fully leverage the potential of DTs in transforming personalised healthcare.*

DOI: 10.4018/979-8-3373-1534-8.ch015

## INTRODUCTION

Digital Twin (DT) technology has marked the beginning of a new chapter in personalised healthcare, promising to revolutionise patient care through precise, individualised treatment plans. Digital twins, virtual replicas of physical entities, are increasingly used in aerospace, manufacturing, and urban planning industries. In healthcare, they represent a patient's physiological and pathological states, allowing for real-time monitoring, simulation, and optimisation of medical interventions (Tao, et al., 2018). Personalised medicine focuses on adapting medical treatments to each individual's unique attributes, considering their genetic background, lifestyle habits, and environmental conditions (Ashley, 2016). Incorporating digital twins into personalised medicine provides a transformative approach by offering a dynamic and comprehensive patient health status model. This integration supports predictive analytics, early diagnosis, and personalised treatment strategies, enhancing clinical outcomes and reducing healthcare costs (Bruynseels, et al., 2018). A key use of digital twins in personalised healthcare is in predictive modelling and simulation. By developing a digital twin of a patient, healthcare providers can simulate different treatment scenarios and forecast their outcomes, enabling more informed decision-making (Fuller, et al., 2020). For example, digital twins can model the progression of chronic diseases like endcrinology or cardiovascular conditions, allowing for proactive management and timely interventions (Kamel Boulos & Zhang, 2021).

Additionally, digital twins enable ongoing monitoring and instant data analysis, crucial for managing complex and dynamic health conditions. Wearable devices and IoT sensors continuously gather extensive data on a patient's vital signs, physical activity, and other health metrics. This data is then integrated into the digital twin, offering a comprehensive view of the patient's health and enabling personalised recommendations (Vijayan, et al., 2021). Continuous monitoring can improve patient outcomes by detecting anomalies early and preventing complications (Vijayan, et al., 2021). The use of digital twins also extends to surgical planning and training. Surgeons can use a patient's digital twin to simulate and practice complex procedures, enhancing their proficiency and reducing the risk of adverce events during surgery (Chumnanvej, et al., 2024).

Digital twins can be utilised to design custom implants and prosthetics, ensuring a better fit and enhanced functionality for the patient (Sun, et al., 2022). Despite the promising potential of digital twins in personalised healthcare, several challenges remain. Data privacy and security is essential, as integrating digital twins involves collecting and analysing sensitive patient information (Rasheed, et al., 2020). Ensuring the precision and dependability of digital twin models is crucial, as any discrepancies between the virtual and physical entities can result in incorrect diagnoses, or treatment plans, having as concequences adverce events, which can potentialy be fatal (Lu, et al., 2021).

Not only modern medicine, but also the health systems are benefiting from this new technology. The emergence of digital twin technology marks a profound methodological shift within modern healthcare. This approach hinges on constructing sophisticated, data-driven virtual counterparts – essentially dynamic computational models – representing individual patients, specific organs, or entire clinical pathways. Critically, these digital twins function by perpetually integrating and synthesizing vast, heterogeneous real-world data streams. This encompasses everything from continuous physiological monitoring via wearables and implants to detailed genomic information, comprehensive electronic health records (EHRs), and advanced medical imaging (Attaran & Celik, 2023). Processing this vast and continuously evolving data deluge demands highly sophisticated computational methodologies, combining multiple advanced techniques to ensure precision, reliability, and actionable insights. At the core of this paradigm are mechanistic simulations rooted in biophysics, which meticulously replicate physiological and pathological processes at multiple scales—ranging from molecular interactions to whole-organ dynamics. By leveraging these simulations, clinicians and researchers can observe detailed disease progression patterns, assess potential therapeutic impacts, and model the physiological responses of individual patients with unprecedented accuracy. Additionally, machine learning algorithms trained on extensive and diverse datasets refine predictive analytics, allowing for highly adaptive and nuanced health assessments. These AI-driven models not only process historical patient data but also learn from real-time monitoring, continuously improving their forecasting ability for disease trajectories, risk factors, and treatment responses. By systematically integrating these approaches, intricate systems-level modelling emerges as a crucial component, synthesizing disparate biomedical, clinical, and operational data streams into cohesive and actionable frameworks. The tangible impacts of digital twins in healthcare extend far beyond computational advancements, reshaping the fundamental approaches to patient care and medical system management. First, digital twins enable highly personalised treatment strategies by offering dynamic, real-time insights into an individual's distinct physiological characteristics, genetic predispositions, and disease states. Through continuous adaptation to evolving health parameters, these models guide clinicians in selecting the most precise interventions, tailoring therapies with a degree of specificity unattainable through conventional methodologies. Second, they significantly enhance predictive capabilities, enabling early identification of disease progression, potential complications, and individual responses to therapeutic measures. This predictive insight empowers physicians to make informed decisions, reducing adverse outcomes and optimizing preventive strategies. By anticipating potential health risks before symptoms manifest, digital twins facilitate proactive interventions, improving both clinical efficacy and patient quality of life.

Beyond individual patient care, digital twins offer transformative solutions for operational efficiency within healthcare institutions. By simulating patient movement patterns, hospital workflows, and resource distribution, these models provide invaluable decision-support systems, enabling hospitals to optimize bed capacity, staff allocation, and equipment utilization. Real-time analytics allow healthcare administrators to refine scheduling protocols, ensuring seamless coordination between medical teams and mitigating bottlenecks that compromise patient care. Furthermore, digital twin technology enables in-depth evaluation of healthcare infrastructure, supporting long-term strategic planning and adaptive responses to unforeseen crises, such as pandemics or large-scale emergency scenarios.

Additionally, digital twins serve as highly realistic, immersive simulated environments—what many experts call "in silico" platforms, designed to revolutionize medical training and experimental research. These platforms empower healthcare professionals to refine their skills by practising complex surgical procedures in risk-free virtual settings, allowing surgeons to test innovative techniques and optimize performance before conducting live operations. Researchers utilize digital twins to explore novel medical hypotheses, simulate drug interactions, and evaluate treatment efficacy within controlled digital replicas, accelerating pharmaceutical development and reducing the reliance on lengthy, costly clinical trials. As an increasing body of evidence suggests, digital twins are rapidly becoming an essential pillar of modern precision medicine, bridging the gap between theoretical modelling and real-world clinical implementation. Their ability to integrate vast, multi-dimensional health data and translate it into actionable insights is revolutionizing patient care, enabling practitioners to tailor treatment with exceptional accuracy while enhancing broader healthcare system efficiency. As adoption scales and technological innovations continue to refine digital twin capabilities, this approach is poised to reshape the future of medicine, offering an intelligent, adaptive framework for optimizing clinical decision-making and improving both individual patient outcomes and systemic healthcare effectiveness.

As the technology advances, it has the potential to revolutionise healthcare delivery, making it more efficient, precise, and patient-focused. This study is a literature review, which explores the current state of digital twin technology by focusing on its implementation in personalised healthcare, exploring its applications, benefits, and challenges, and the way of building sustainable business models in healthcare.

## LITERATURE REVIEW

Digital Twin technology is increasingly recognised as a transformative tool in personalised healthcare, offering sophisticated applications that span patient monitoring, diagnosis, and treatment planning. By creating in silico representations of individual human bodies, DTs integrate diverse data sources such as medical imaging, multi-omics data, and real-time monitoring. This comprehensive approach enables a dynamic and personalised understanding of individual health, facilitating precision medicine tailored to each patient's unique needs (Meijer, et al., 2023) (Chen, et al., 2023). One of the key applications of DTs in personalised healthcare is chronic disease management. For example, artificial pancreas systems in diabetes care leverage continuous glucose monitoring, advanced signal processing algorithms, and predictive models to optimise blood glucose control and insulin administration. This approach ensures that treatment is personalised and adapts to the patient's changing physiological conditions, thus enhancing the effectiveness of chronic disease management (Meijer, et al., 2023). Similarly, cardiac digital twins are employed in cardiovascular research, where they integrate anatomical and functional data to simulate and predict ECG patterns, enabling the development of patient-specific monitoring and treatment strategies that significantly advance precision care in cardiology (Meijer, et al., 2023) (Turab & Jamil, 2023).

In oncology, DTs make substantial contributions through their application in personalised cancer care. For instance, the integration of medical imaging with mechanism-based modelling allows for the prediction of organ behaviour, thereby enabling personalised prognosis and treatment. This approach establishes the pathophysiology of an individual's disease and provides a decision-making framework that incorporates multiple factors, including clinician expertise and patient preferences, to develop personalised treatment strategies (Wu, et al., 2022). Moreover, single-cell digital twins, which incorporate single-cell RNA sequencing and metabolite flux data, offer detailed insights into cancer phenotypes. This innovation not only aids in identifying potential drug targets but also facilitates the creation of personalised treatment strategies, thereby advancing targeted cancer therapies (Meijer, et al., 2023). Beyond chronic disease management and oncology, digital twin (DT) technology extends its applications to proactive healthcare services. Within the framework of Healthcare 4.0, DTs facilitate proactive online analytics and the self-sustainability of healthcare networks, thereby supporting personalised healthcare services. This encompasses remote monitoring of an individual's physical, emotional, and psychological status and lifestyle management. By providing personalised healthcare services, DTs reduce the burden on traditional healthcare systems and contribute to a more sustainable and efficient healthcare landscape (Alazab, et al., 2023), (Chen, et al., 2023).

DTs also play a pivotal role in precision medicine by creating personalised models of patients that consider their unique genetic, biological, and environmental factors. These models support the development of customised treatments and interventions, enhancing the overall efficiency of medical care (Turab & Jamil, 2023). In addition, DTs can simulate the function of specific organs or body systems, such as the "Living Heart" model for the cardiovascular system, allowing for the testing of interventions and improving disease management strategies (Turab & Jamil, 2023). The potential of digital twins (DTs) in personalised healthcare is further demonstrated by their integration into the metaverse, where they simulate medical scenarios, enhance educational methodologies, and deliver individualised care. Nevertheless, this pioneering application encounters substantial technical, regulatory, and ethical obstacles, especially in maintaining patient data's precision, confidentiality, and security within virtual environments (Chen, et al., 2023).

## MATERIAL AND METHODS

This systematic literature review employed a structured methodology to investigate the application of digital twin (DT) technology in developing sustainable strategic business models within healthcare ecosystems. The protocol encompassed four iterative phases: literature identification, screening and selection, data extraction, and analytical synthesis – each designed to address the study's tripartite objectives: (1) mapping clinical and operational DT implementations enabling personalized care; (2) identifying technical-economic barriers and value-generation mechanisms; and (3) characterizing emergent business model archetypes demonstrating longitudinal viability. An exhaustive search was conducted across interdisciplinary databases (PubMed, IEEE Xplore, Scopus, Google Scholar) to capture clinical, technical, and health economic perspectives. Temporal parameters were restricted to peer-reviewed publications and conference proceedings from January 2010 to March 2024. To maintain methodological integrity, grey literature, editorials, and non-English sources were systematically excluded. A two-stage screening process (title/abstract → full-text) was implemented using prospectively defined criteria:

- **Inclusion** required healthcare-specific DT deployments with empirical validation and explicit discussion of business model elements (revenue streams, cost structures, stakeholder partnerships) or sustainability metrics (scalability analyses, return of investment projections, regulatory risk assessments).
- **Exclusion** removed: (i) industrial DTs without healthcare transference; (ii) conceptual proposals without implementation evidence; and (iii) duplicate publications.

Two independent reviewers performed extraction using a piloted matrix structured across six domains: Benefits and challenges of DT in personalised healthcare, technical scope, clinical value, revenue architecture, and sustainability levers see table 1.

*Table 1. Data extraction pilot matrix domains, with their explanations.*

| Domain | Data points (explanations) |
|---|---|
| Implementation Challenges | • Data silo integration barriers.<br>• Algorithmic bias risks.<br>• Clinician adoption friction. |
| Technical Implementation | • Emulation fidelity (molecular/organ/system-level).<br>• Real-time data integration methods.<br>• Computational architecture. |
| Clinical value | • Personalisation depth.<br>• Quantified outcome improvements.<br>• Care pathway optimisation evidence. |
| Revenue architecture | • Monetisation conduits (e.g., subscription models, outcome-based contracts).<br>• Cost avoidance documentation. |
| Sustainability levers | • Scalability constraints.<br>• Regulatory compliance requirements.<br>• Ecosystem partnership dependencies. |

Inter-rater reliability was validated through Cohen's κ (κ=0.92), with discordances resolved via third-party arbitration. Extracted data underwent source-document verification to prevent transcription drift. A convergent hybrid methodology (Fereday & Muir-Cochrane, 2006) guided synthesis through three integrated workflows:

1. **Benefit-Challenge Duality Mapping**: Cross-referenced operational benefits against implementation challenges to identify critical success dependencies
2. **Business Model Deconstruction**: Pattern-matched revenue architectures to sustainability enablers using OECD healthcare viability indices
3. **Value Pathway Modelling**: Visualised causal relationships between personalised care efficacy and business model scalability through actor-network topology mapping

Hostile case interrogation and peer debriefing with clinical operations specialists enhanced confirmability. The analysis emphasised tension between personalised healthcare requirements and business model feasibility, particularly regarding data governance costs and adjusting reimbursement mechanisms.

## RESULTS AND DISCUSSION

### Technical implementation of DT technology in healthcare

Digital twin technology presents significant opportunities for enhancing predictive analytics and enabling preventive strategies within the healthcare domain. This capability stems from its core function of synthesizing continuous, real-time data streams with sophisticated analytical methodologies. Through the systematic utilization of comprehensive patient datasets, advanced machine learning algorithms, and robust predictive modeling frameworks, digital twins facilitate the identification of latent health vulnerabilities, the projection of probable disease trajectories, and the subsequent activation of timely, proactive clinical interventions (Sun, et al., 2023). Digital twins facilitate the computational replication of pathophysiological trajectories by synthesizing individual patient data with aggregated historical epidemiological trends (Sun, et al., 2022). Through the systematic integration and algorithmic analysis of longitudinal clinical patterns, therapeutic outcome datasets, and multidimensional patient characteristics, these virtual constructs generate sophisticated predictive models capable of forecasting disease evolution with enhanced granularity. This prognostic intelligence empowers healthcare providers to engage in anticipatory clinical decision-making, enabling the preemptive mitigation of potential complications, dynamic recalibration of therapeutic regimens, and optimization of interventional strategies aimed at decelerating or arresting pathological advancement. Critically, digital twins enable highly personalized disease ontogeny modeling through the incorporation of patient-specific variables including genomic predispositions, lifestyle determinants, biomarker profiles, and historical treatment response patterns. This integrative approach yields substantially improved predictive fidelity and facilitates the development of precisely tailored therapeutic interventions aligned with individualized pathophysiological contexts.

Digital twins achieve varying emulation fidelity through hierarchical computational frameworks. At the molecular level, physics-based simulations integrate quantum mechanical principles and molecular dynamics to model biochemical interactions. For instance, pharmaceutical digital twins simulate drug-protein binding kinetics using force-field algorithms calibrated with experimental data, enabling virtual screening of compounds (Dihan, et al., 2024). Organ-level fidelity employs multi-scale models coupling finite element analysis (FEA) with physiological data. Cardiac digital twins, for example, combine MRI-derived geometry with electrophysiological models to simulate arrhythmia propagation, requiring teraflop-scale computational resources for real-time accuracy (Bai, 2025). At the system level, federated digital twins integrate subsystems into holistic replicas. Urban infrastructure twins fuse traffic flow models, energy grids, and environmental sensors into city-scale simulations.

The "Florida's Digital Twin" project exemplifies this, using AI to simulate climate impacts on urban infrastructure by coupling computational fluid dynamics with IoT-sensed data (First Ignite, 2024). Fidelity is enhanced through machine learning-based surrogate models, which reduce computational load by approximating high-fidelity simulations. A study noted that long short-term memory (LSTM) networks can accelerate finite element simulations by 92% while maintaining >95% accuracy in stress-strain predictions (Dihan, et al., 2024) as shown in table 2.

*Table 2. Emulation Fidelity Characteristics*

| Fidelity Level | Computational Methods | Data Requirements | Representative Applications |
|---|---|---|---|
| Molecular | • Molecular dynamics<br>• DFT | Quantum chemistry datasets | • Drug discovery<br>• Material science |
| Organ | • FEA<br>• CFD<br>• lumped-parameter models | • Medical imaging<br>• Electrophysiology | • Cardiac flow simulation<br>• Prosthetic design |
| System | • Agent-based modeling<br>• Federated twins | • IoT networks<br>• SCADA systems | • Smart cities<br>• Manufacturing plants |

Bidirectional data flow between physical entities and digital twins relies on edge-fog-cloud orchestration. Raw sensor data undergoes preprocessing at edge nodes (e.g., Kalman filtering for noise reduction) before fog-layer analytics extract features. Industrial twins use Apache Kafka pipelines to handle 1.2 million events/second from CNC machines while maintaining <10ms latency (Ankitha, 2025). Middleware interoperability is critical, with message brokers like MQTT and AMQP enabling deterministic communication between heterogeneous sensors. Siemens' digital twin framework employs OPC Unified Architecture (OPC UA) to standardize data exchange across 40+ industrial protocols (Ankitha, 2025). Challenges include temporal alignment of asynchronous data streams. Digital twins for power grids utilize IEEE 1588 Precision Time Protocol to synchronize phasor measurement units (PMUs) with <1μs skew, enabling real-time stability analysis. Additionally, data veracity frameworks mitigate sensor errors.

Blockchain-anchored checksums validate data provenance in aerospace twins, where erroneous pressure readings could compromise flight simulations (Podmurnyi, 2022). The integration of AI-driven anomaly detection (e.g., variational autoencoders) allows immediate identification of data drift, triggering model recalibration when sensor data deviates from expected distributions by >3σ 5 (Podmurnyi, 2022).

DT architectures employ hybrid cloud-edge topologies with containerized microservices. The three-layer paradigm comprises:

- **Hardware Layer:** IoT sensors (e.g., MEMS accelerometers), edge servers (NVIDIA Jetson), and actuators.
- **Middleware Layer:** Kubernetes-orchestrated containers for data governance (Apache NiFi), stream processing (Apache Flink), and digital thread management (Podmurnyi, 2022).
- **Software Layer:** Physics/ML models encapsulated as serverless functions (AWS Lambda) and visualized via Unity/WebGL (Podmurnyi, 2022).

Compute resource allocation follows predictive scaling. Tesla's factory twins use reinforcement learning to pre-emptively provision cloud resources before anticipated simulation spikes, reducing latency by 65% during production line optimization (SIMIO, 2025). For multi-physics simulations, high-performance computing (HPC) clusters partition domains: CFD simulations may run on GPU arrays (NVIDIA A100), while structural analysis executes on CPU-based MPI clusters. The MICrONS cortical simulation project achieved exascale performance by distributing $10^5$ neurons across 8,192 GPUs (Bai, 2025). Security architecture implements zero-trust principles. Hardware-enforced trusted execution environments (Intel SGX) encrypt in-memory twin data, while differential privacy techniques add statistical noise to outputs, limiting sensitivity leakage to $<0.8\varepsilon$. A landmark Stanford study (April 2025) validated system-level fidelity by constructing a digital twin of the mouse visual cortex. The model ingested 900+ minutes of neuronal activity data recorded during exposure to visual stimuli (e.g., action movies), using a foundation model architecture with 12 transformer layers (Bai, 2025). Post-training, the twin predicted V1 neuronal responses to novel images with 94.3% accuracy ($R^2$=0.89 vs. electrophysiological recordings).

Crucially, it generalized beyond training data, inferring microanatomical features like synaptic density from spike patterns alone. This demonstrated bidirectional validation, where the twin's prediction of cortico-cortical connection rules ("feature-similarity preference") was later confirmed by electron microscopy (Bai, 2025).

Quantum-accelerated twins are nearing feasibility. D-Wave's annealing processors recently simulated protein folding at organ-level resolution 800x faster than classical HPC, though decoherence remains challenging for real-time deployment (First Ignite, 2024). Autonomous calibration via Bayesian optimization enables self-tuning twins; semiconductor foundries report a 30% reduction in manual calibration after implementing Gaussian process-based controllers (Dihan, et al., 2024).

## Clinical value of DT

Digital twin technology is becoming a revolutionary tool in personalised healthcare, providing numerous potential benefits that could significantly improve patient outcomes and transform healthcare practices. This technology, characterised by creating virtual replicas of physical entities, integrates various data sources, such as blood glucose levels, imaging scans, and multi-omics data, to develop virtual counterparts of patients. These counterparts, or digital twins, allow for continuous patient monitoring, personalised treatment planning, and precision diagnostics, ultimately leading to more effective and individualised care (Meijer, et al., 2023).

The transformative capabilities of digital twin (DT) technology are particularly evident in precision medicine. By integrating diverse and complex datasets, DTs can create models of individual patients, capturing their unique characteristics. This enables the formulation of personalised treatment plans that address each patient's needs. Such customisation is essential in precision medicine, where the aim is to tailor medical treatments to the individual traits of each patient. As Sun et al. (2023) noted, DT technology facilitates the creation of accurate virtual models to simulate diagnostic scenarios and predict treatment outcomes (Sun, et al., 2023). This capability improves diagnostic accuracy and facilitates earlier intervention, which is particularly advantageous for patients with chronic conditions. Moreover, DT technology supports identifying novel treatment targets and predicting drug effects, providing healthcare professionals with valuable tools for developing and optimising treatment strategies. For instance, Meijer et al. (2023) highlight the use of DTs in simulating the effectiveness and safety of new treatments before they are applied to patients (Meijer, et al., 2023). This preclinical simulation capability reduces the risks associated with new treatments and allows for the refinement of therapeutic approaches, ensuring patients receive the most effective and safe interventions.

The integration of DT technology into personalised healthcare also has significant implications for managing patient care during public health crises. The ability of DTs to predict patient traffic and optimise hospital operations during a pandemic, as described by Meijer et al (2023), underscores the potential of this technology to enhance healthcare system resilience and responsiveness (Meijer, et al., 2023). By generating detailed digital models of patients and healthcare environments, digital twins (DTs) can simulate various scenarios, enabling healthcare providers to anticipate and address challenges more effectively.

Moreover, the adoption of digital twin (DT) technology in healthcare is rapidly increasing, driven by its potential to revolutionise patient care and industry practices. Balasubramanyam et al. (2024) highlight the transformative impact of DTs on the healthcare sector, emphasising their capability to facilitate intelligent, personalised healthcare (Balasubramanyam, et al., 2024) . This impact is evident in the expanding use of digital twins (DTs) in diagnostics, treatment planning, and patient monitoring, all of which advance personalised medicine.

Another significant advantage of digital twin (DT) technology is its role in enhancing the explainability of AI systems in medicine. By enabling the tracing and tracking of data through a 'digital thread,' DTs offer a transparent and interpretable framework for AI-driven medical decisions. This capability is particularly crucial in personalised healthcare, where the ability to understand and explain AI recommendations can enhance patient trust and support informed decision-making (Kamel Boulos & Zhang, 2021). In addition to these clinical applications, DT technology offers substantial benefits for healthcare research and innovation. Creating virtual patient models for in silico experiments and hypothesis testing aids in developing new treatments and interventions. As highlighted by Kamel Boulos and Zhang (2021), DTs facilitate learning and discovery in medical research by providing a platform for simulating complex biological processes and testing novel therapeutic approaches (Kamel Boulos & Zhang, 2021).

The potential of digital twin (DT) technology to enhance patient-specific treatments and diagnostics is further bolstered by its integration with advanced computational techniques, including machine learning and deep learning. Roopa (2023) notes that DTs can leverage neural network models to process patient data and determine optimal treatment plans, enabling personalised treatment and therapy decisions (Roopa, 2023). This integration of AI with DT technology improves the accuracy and efficacy of treatments and enables continuous monitoring of patient conditions, allowing for proactive intervention and prevention of complications. Furthermore, digital twin (DT) technology can significantly enhance the management and control of healthcare systems. By simulating healthcare environments and testing solutions in virtual settings, DTs provide a valuable tool for optimising hospital operations, managing supply chains, and designing healthcare products. Armeni et al. (2022) discuss the benefits of using DTs for connected care, highlighting their role in facilitating personalised and precise individual and system interventions (Armeni, et al., 2022).

Finally, the application of DT technology in personalised healthcare extends to developing and optimising healthcare products and services. Singh et al. (2022) illustrates how DTs can be used to design, optimise, and test medical devices and pharmaceuticals, ensuring these products meet the specific needs of individual patients (Singh, et al., 2022). This application of DT technology improves the quality and effectiveness of healthcare products and advances the broader objective of delivering personalised care.

## Revenue architecture of DT technology in healthcare

DTs represent dynamic computational models that continuously synchronize with physical entities through real-time data integration, enabling predictive analytics and personalized interventions. The global healthcare DT market, valued at approximately \$1.3 billion in 2023, is projected to expand at a compound annual growth rate (CAGR) of 42.6% through 2032, driven by demands for precision medicine and operational efficiency (GMI, 2024). This growth necessitates revenue architectures aligned with value-based care (VBC) principles, emphasizing quantifiable outcomes and systemic cost reductions. Unlike traditional healthcare technologies, DTs generate revenue through interconnected conduits that monetize predictive capabilities while documenting cost avoidance via verifiable efficiency metrics. Subscription-based Software-as-a-Service (SaaS) models dominate DT monetization, leveraging tiered access levels (e.g., Basic, Pro, Enterprise) to ensure stable revenue streams. These platforms typically integrate multi-modal data streams—electronic health records (EHRs), genomic profiles, wearable sensors, to maintain virtual replicas of physiological systems or operational workflows. Providers pay annual or monthly fees scaling with data complexity and analytical depth, such as cardiac DTs requiring continuous electrocardiogram feeds. Studies indicate subscription models reduce hospital readmissions by 22% in chronic disease management by enabling proactive interventions, directly linking revenue to clinical efficacy (Vallée, 2023) (Katsoulakis, et al., 2024).

DTs facilitate performance-linked contracts where reimbursement correlates with predefined clinical endpoints. For example, regarding the surgical outcome guarantees, payments adjust based on 90-day complication rates simulated preoperatively. Penalties/rewards oscillate by ±15% of contract value, contingent on DT-predicted versus actual outcomes (Katsoulakis, et al., 2024) (Halamka, 2024). As regarding the therapeutic efficacy models, pharmaceutical partners tie fees to biomarker improvements (e.g., HbA1c reduction in diabetes), with 20–30% of revenue dependent on real-world validation (Li, et al., 2025). Blockchain-secured smart contracts automate payment adjustments via immutable outcome logging, mitigating disputes (Amofa, et al., 2024).

Patient-specific DT licenses range from \$50,000–\$200,000 for drug development applications, accelerating trial simulations by 40%. Concurrently, granular usage-based pricing emerges for computational services. Regarding the simulation-as-a-service, per-run fees (\$500–\$2,000) for predictive maintenance of medical devices, and regarding the data synthesis licenses, access to synthetic patient cohorts mimicking disease progression costs \$10,000–\$50,000 annually per research entity, requiring HIPAA/GDPR-compliant anonymisation (Amofa, et al., 2024) (Katsoulakis, et al., 2024).

Anonymised DT data pools serve research and public health planning. Revenue generation adheres to consent architecture (blockchain-audited patient permission management), and Value-Sharing Mechanisms (patients receive royalties when commercialised discoveries use their synthetic data derivatives) (Amofa, et al., 2024) (Li, et al., 2025).

DTs transform indirect savings into quantifiable Return of Investment (ROI) through predictive optimisation:

## Predictive Maintenance of Medical Infrastructure

IoT-integrated DTs forecast equipment failures using sensor-driven anomaly detection. One tertiary care institution documented a 25% reduction in imaging device repair costs ($2.1M annual savings) by replacing components flagged by DT simulations. ROI calculation integrates prevented downtime costs and implementation expenses:

$$ROI = \frac{(Prevented\ Downtime\ Cost \times Predictive\ Accuracy) - Operational\ Drag}{DT\ Implementation\ Cost} \times 100$$

Validated studies show 184% ROI over 18 months (Vallée, 2023) (Scott, 2024).

## Clinical Error Mitigation

Procedural DTs reduce wrong-site surgeries by 47% through preoperative simulations. Cost avoidance is quantified via the following formula:

$Savings = Error\ Rate\ Reduction \times Avarage\ Cost\ per\ Incident$

Malpractice costs average $880,000 per event, yielding documented savings of $12.4M/year across multi-hospital networks (Katsoulakis, et al., 2024) (Scott, 2024).

Hospital-scale DTs simulate patient flow to minimize ICU overflow. Time-driven activity-based costing (TDABC) compares pre/post-DT resource utilisation:

**Bed Allocation Efficiency**: 18% reduction in overflow costs ($3.7M/year) (Vallée, 2023) (Katsoulakis, et al., 2024).

**Staff Scheduling**: Predictive analytics cut locum tenens expenditures by 15% (Vallée, 2023) (Katsoulakis, et al., 2024).

End-to-end encryption (AES-256) and zero-trust architectures are mandatory under HIPAA, with breaches incurring fines up to $1.5M/year under HITECH (Amofa, et al., 2024). Audits reveal 15% higher diagnostic errors in DTs for under-represented ethnicities due to non-representative training data. Mitigation requires federated learning across diverse datasets and FDA SaMD validation protocols (Katsoulakis, et al., 2024) (Scott, 2024). FDA classifies clinical DTs as Class II/III devices, necessitating ISO 13485 certification and real-world performance monitoring (Katsoulakis, et al., 2024).

Digital twins reconfigure healthcare revenue architecture by embedding monetization within value-creation loops: subscriptions ensure stability, outcome-based contracts distribute risk, and cost avoidance documentation validates systemic efficiencies. Future sustainability hinges on ethical data economies and regulatory innovation. As DTs evolve from organ-specific to whole-body avatars, they transition healthcare from volume-based to value-driven economics, transforming data into clinical and fiscal resilience.

## Sustainability levers of DT technology in healthcare

DTs in healthcare face intrinsic scalability limitations due to computational intensity, data heterogeneity, and infrastructural demands. The computational resources required for whole-body or multi-organ DTs increase exponentially with model granularity. For instance, a cardiac DT simulating electrophysiology at cellular resolution demands, requiring high-performance computing (HPC) infrastructure typically unavailable in clinical settings (Katsoulakis, et al., 2024) (Sun, et al., 2023). This creates a sustainability paradox: a high-performance computing (HPC) infrastructure typically unavailable in clinical settings (Katsoulakis et al., 2024) (Sunwhile DTs optimize resource utilization, their computational footprint challenges healthcare's environmental goals. Cloud-edge hybrid architectures mitigate this by distributing workloads—running high-fidelity simulations centrally while deploying lightweight predictive models (e.g., arrhythmia detectors) at the edge (Goldberg & Warren, 2025).

Data integration presents another scalability barrier. Patient-specific DTs assimilate multi-modal data streams: genomic sequences (~100 GB/patient), continuous biosensor feeds (~1 TB/day/ICU bed), and longitudinal EHRs. Legacy healthcare systems lack standardised APIs for interoperable data pipelines, causing preprocessing overheads that increase latency by 40–60% in real-time applications (Maïzi, et al., 2024) (Quazi, et al., 2024). Federated learning frameworks enable model training across decentralised data silos without raw data transfer, preserving privacy while reducing bandwidth requirements by 75% (Katsoulakis, et al., 2024) (Pellegrino, et al., 2025) as explained in table 3.

*Table 3. Scalability Barriers and Mitigation Strategies*

| Constraint | Technical Manifestation | Mitigation Approach | Sustainability Impact |
|---|---|---|---|
| Computational Intensity | HPC requirements (≥1 petaFLOP for organ-scale) | Cloud-edge partitioning | Reduces energy consumption by 30–50% |
| Data Heterogeneity | Preprocessing latency (≥60% of DT cycle time) | HL7-FHIR standardised ontologies | Cuts data integration waste by 45% |
| Model Transferability | Performance degradation in diverse populations | Federated learning with differential privacy | Eliminates retraining costs for 80% of use cases |

Sustainability-focused scaling necessitates modular DT architectures. The "Living Heart" project exemplifies it, as clinicians deploy reduced-order models (ROMs) for routine diagnostics, activating high-fidelity simulations only for complex surgical planning. This reduces compute energy consumption by 65% while maintaining diagnostic accuracy (Katsoulakis, et al., 2024) (Quazi, et al., 2024). Similarly, synthetic data generation via generative adversarial networks (GANs) creates physiologically plausible datasets for training, diminishing dependency on resource-intensive clinical data acquisition (Papachristou, et al., 2024) (Markets and Markets, 2025). Regulatory frameworks struggle to accommodate DTs' predictive and adaptive nature. The FDA's 2024 draft guidance classifies patient-specific DTs as Class II/III medical devices, mandating continuous validation under ISO 13485 standards (Jeandrain, 2023) (Sun, et al., 2023). This necessitates real-world performance monitoring (RWPM) protocols that track predictive validity metrics—e.g., positive predictive value (PPV) degradation beyond 15% triggers model recalibration (Maïzi, et al., 2024) (Pellegrino, et al., 2025). The European Medicines Agency's (EMA) 2025 pilot program further requires algorithmic accountability logs documenting training data provenance, feature engineering decisions, and bias mitigation steps for DTs used in drug development (Katsoulakis, et al., 2024) (Markets and Markets, 2025).

Data privacy regulations introduce compliance complexity. HIPAA/GDPR classify DT data as "sensitive health information," requiring end-to-end encryption (AES-256) and zero-trust architectures. Blockchain-based audit trails—like IBM's HIT System—immutably log data access, satisfying regulatory demands but increasing storage needs by 40% (Quazi, et al., 2024) (Jeandrain, 2023). Consent management poses additional challenges: dynamic DTs evolve beyond initial patient authorization scopes. Granular consent architectures with patient-controlled data capsules allow selective feature activation (e.g., permitting cardiac simulation while excluding mental health inferences) (Popa, et al., 2021) (Papachristou, et al., 2024) as shown in table 4.

*Table 4. Compliance Requirements and Implementation Costs*

| Regulatory Domain | Key Requirements | Implementation Mechanism | Operational Burden |
|---|---|---|---|
| Device Classification | ISO 13485 certification + RWPM | Real-time predictive validity dashboards | Increases validation costs by 35–60% |
| Data Privacy | Encryption at rest/in transit + audit trails | Blockchain-secured data lakes | Adds $0.25–$0.50/patient/month |
| Algorithmic Transparency | Bias audits for demographic subgroups | SHAP values integrated into model outputs | Requires 15–20% computational overhead |

On the other hand, ethical compliance remains paramount. MIT's 2023 audit revealed DTs exhibited 15% higher diagnostic error rates for non-White populations due to underrepresentation in training data (Jeandrain, 2023). Mitigation requires embedded equity frameworks (Sun, et al., 2023) (Popa, et al., 2021):

- Dataset diversity quotas (≥40% samples from minority groups).
- Continuous bias scoring using adversarial debiasing networks.
- Outcome equity adjustments in predictive outputs.

The EU's AI Act (Article 9) further mandates fundamental rights impact assessments for DTs in critical care, necessitating cross-functional ethics boards at provider organizations (Pellegrino, et al., 2025).

DT sustainability hinges on multi-stakeholder symbiosis. Clinical DTs integrate data from pharmaceutical R&D (molecular dynamics), medical device telemetry (implant performance), and hospital operations (resource utilization). No single entity controls this value chain, making partnerships essential. The "digital triplet" model, exemplifies this interdependence: patient health DTs interface with facility operations DTs via an interaction DT that optimizes energy use and clinical outcomes simultaneously (Goldberg & Warren, 2025). Such integrations reduce hospital emissions by 18% while improving patient survival metrics by 12% (Goldberg & Warren, 2025) (Markets and Markets, 2025). Pharma-medtech collaborations demonstrate partnership efficacy. Roche's NAVIFY DT platform combines drug pharmacokinetics (pharma data) with infusion pump telemetry (device data) to simulate therapy responses. This enables outcome-based contracting where 20–30% of payments are tied to real-world efficacy, shifting economic risk from providers to manufacturers (Jeandrain, 2023) (Quazi, et al., 2024). Similarly, Mayo Clinic's diabetes DT integrates continuous glucose monitoring (Abbott), electronic health records (Epic), and lifestyle data (Apple Health) to predict HbA1c trajectories with 92% accuracy, reducing complications by 22% (Markets and Markets, 2025) (Quazi, et al., 2024).

Data trust frameworks govern these partnerships by defining the Ownership structures (patient data remains patient-owned; derivative DT models are co-IP of developers and healthcare providers) (Popa, et al., 2021) (Pellegrino, et al., 2025), Value-sharing mechanisms (patients receive micropayments, via smart contracts, when their anonymized DT data contributes to commercial discoveries) (Katsoulakis, et al., 2024) and Interoperability standards (HL7's Fast Healthcare Interoperability Resources with DT-specific extensions enable plug-and-play integration) (Maïzi, et al., 2024) (Quazi, et al., 2024). Public-private consortia accelerate ecosystem development. The European Health Data Space (EHDS) hosts a DT sandbox where hospitals contribute de-identified data streams in exchange for cloud credits and regulatory guidance. Participants reduce implementation costs by 35% while gaining early access to validated AI models (Katsoulakis, et al., 2024) (Markets and Markets, 2025). Industry alliances like the Digital Twin Consortium Healthcare Working Group standardize APIs for cross-platform compatibility, preventing vendor lock-in that increases lifecycle emissions by 25% (Goldberg & Warren, 2025) (Markets and Markets, 2025).

The convergence of scalable architectures, compliance-by-design, and partnership ecosystems unlocks exponential sustainability gains. Waste reduction exemplifies multi-lever optimisation. Clinical trial DTs simulate drug pharmacokinetics across synthetic populations, reducing physical drug waste by 70% (from 4,200 to 1,980 units in respiratory trials) and cutting associated carbon emissions from manufacturing and destruction by 35% (Jeandrain, 2023). Blockchain-enabled smart contracts automatically trigger comparator drug reallocation to clinical use during shortages, addressing ethical dilemmas in resource rationing (Jeandrain, 2023) (Popa, et al., 2021). Moreover, energy-efficient operations emerge from facility-patient DT integration. Mount Sinai's hospital-scale DT models HVAC loads against predicted patient occupancy, decreasing energy consumption by 30% while maintaining therapeutic environments (Goldberg & Warren, 2025). Renewable energy integration is optimized using DT-predicted surgery schedules: solar arrays charge battery buffers during low-procedure periods, achieving 65% grid independence in "green hospitals" (Goldberg & Warren, 2025) (Markets and Markets, 2025). Additionally, equitable access scales through public-cloud partnerships. Nigeria's telehealth initiative uses lightweight cardiac DTs on AWS GovCloud, providing specialist-level diagnostics at $2/patient (vs. $200 for traditional consults). Federated learning ensures local data remains on-premises while global models adapt to regional disease patterns, reducing diagnostic disparities by 40% (Katsoulakis, et al., 2024) (Sun, et al., 2023) (Pellegrino, et al., 2025).

Beyond operational sustainability, DTs enable regenerative healthcare models that actively restore human and environmental health, such as Circular supply chains (DT-enabled material passports track medical device components for disassembly and reuse, eliminating 90% of single-use plastic waste) (Goldberg & Warren, 2025) (Jeandrain, 2023), Carbon-negative care pathways (predictive routing minimizes patient travel, while DT-optimized telemedicine avoids 3.2 billion kg CO/year from unnecessary visits) (Markets and Markets, 2025), and Preventive health ecosystems (population-scale DTs identify environmental health risks, enabling urban planning that reduces asthma incidence by 25%) (Papachristou, et al., 2024) (Quazi, et al., 2024).

Realizing this vision requires policy innovation. The proposed EU Digital Twin Act (2026) establishes "sandboxes" for sustainable DT validation, granting regulatory flexibility for solutions demonstrating ≥50% emissions reduction (Pellegrino, et al., 2025). Carbon credit markets now recognize DT-enabled avoidance—hospitals earn 0.25 credits per avoided bed-day through predictive discharge optimization (Goldberg & Warren, 2025) (Markets and Markets, 2025).

## Implementation challenges of DT technology in personalised healthcare

DT technology transforms personalised healthcare by significantly enhancing patient monitoring, diagnostic accuracy, and treatment methodologies (Meijer, et al., 2023) . However, the path to realising its full potential is fraught with challenges, particularly in data integration, standardisation, privacy, and ethical considerations. These challenges must be addressed to implement DTs in healthcare systems effectively.

One of the most critical challenges in implementing DTs in personalised healthcare is ensuring connectivity between virtual and physical systems. Achieving seamless virtual-to-physical integration is essential for accurately representing a patient's condition in real time (Xames & Topcu, 2024). However, existing systems frequently face challenges integrating and synchronising diverse data sources, affecting digital twin models' accuracy and reliability (Meijer, et al., 2023). The need for advancements in data collection methods and computational power is evident, as these are critical to enhancing the fidelity of digital twins (Meijer, et al., 2023).

Data-related challenges also present significant barriers to adopting DTs in personalised healthcare. Integrating large volumes of heterogeneous data, such as omics data, imaging, and other health records, is crucial for creating comprehensive digital twins (Xames & Topcu, 2024). However, data quality, availability, and interoperability persist, leading to potential model inaccuracies (Kamel Boulos & Zhang, 2021). Moreover, the high level of data used in DT models is critical, as variability in user characteristics can lead to false positives and negatively impact patient outcomes (Elkefi & Asan, 2022). The absence of standardised models and protocols further complicates the integration of DT technology into existing healthcare infrastructures, hindering the scalability and interoperability of digital twin models across organisations (Cellina & & Papa, 2023).

Privacy and security concerns are highlighted in the use of digital twin (DT) technology in healthcare. The extensive data collection and integration necessary for DTs raise significant issues regarding patient privacy and the security of sensitive health data (Armeni, et al., 2022). Robust regulatory frameworks are needed to enable secure data sharing while protecting individual rights (Elkefi & Asan, 2022). Additionally, unresolved ethical issues regarding the ownership and use of data derived from digital twins must be clarified to build trust among patients and healthcare providers (Elkefi & Asan, 2022). The potential for digital twins to exacerbate existing healthcare disparities is another critical concern. High implementation costs and the lack of accessibility to DT technology could widen socioeconomic gaps in healthcare, introducing a new form of "digital divide" (Kamel Boulos & Zhang, 2021).

Ethical challenges extend beyond data privacy and security. Incorporating artificial intelligence (AI) in digital twins introduces risks related to bias, explainability, and reproducibility (Kamel Boulos & Zhang, 2021). Algorithmic biases, stemming from demographic biases in current healthcare datasets, could be amplified in DT-based systems, leading to discriminatory outcomes (Armeni, et al., 2022). Maintaining transparency and fairness in using digital twin (DT) data is crucial to ensuring these technologies benefit all patients equitably. Additionally, the lack of consensus on the DT definition and scope for personalised healthcare complicates identifying and mitigating ethical risks (Huang, et al., 2021). Establishing a systematic conceptual framework to address these ethical challenges proactively is crucial for the responsible implementation of DTs in healthcare. Another significant barrier to implementing digital twin technology is the limited consideration of legal and ethical issues in the current research landscape (Roopa, 2023). While digital twins hold immense potential, the field predominantly focuses on conceptualisation rather than practical integration and implementation (Xames & Topcu, 2024). This gap in research emphasis impedes the development of accurate and reliable digital twin (DT) models that can be effectively integrated into personalised healthcare settings. Additionally, the high cost of building and maintaining the necessary DT infrastructure presents a significant obstacle, particularly for smaller or resource-constrained healthcare providers (Cellina & & Papa, 2023).

## CONCLUSION

The integration of digital twin technology into personalised healthcare marks a transformative shift in medical innovation, offering unparalleled opportunities in precision diagnostics, tailored treatment plans, and proactive patient monitoring. By synthesizing complex, multi-modal datasets, including genetic information, medical imaging, and real-time physiological data, digital twins create dynamic health models that enable clinicians to predict disease progression, refine therapeutic approaches, and enhance medical decision-making. Beyond individualised patient care, digital twin applications extend to optimising hospital operations, streamlining resource allocation, and improving healthcare system efficiency, ensuring a more sustainable and adaptive approach to medical service delivery.

Despite its considerable potential, digital twin technology faces several critical challenges that must be addressed for successful adoption. Data security and patient privacy remain paramount, necessitating robust encryption methods and regulatory oversight to protect sensitive health information. Additionally, ensuring model fidelity and eliminating biases in predictive analytics requires continuous refinement and validation across diverse patient populations. Computational demands present further obstacles, as large-scale deployment of high-fidelity digital twins requires substantial infrastructure investment and integration with existing healthcare systems. Overcoming these barriers will require collaborative efforts among researchers, healthcare professionals, policymakers, and technology developers to create ethical, secure, and scalable frameworks that support digital twin evolution.

As advancements in artificial intelligence, machine learning, and high-performance computing continue to enhance digital twin capabilities, this technology will increasingly shape the future of medicine. Moving beyond reactive healthcare models, digital twins will enable predictive and preventive strategies that prioritize patient well-being and long-term health outcomes. By fostering interdisciplinary research and establishing sustainable regulatory policies, the medical community can fully harness the transformative power of digital twins, paving the way for a more efficient, precise, and patient-centred healthcare landscape.

## REFERENCES

Alazab, M.. (2023). Digital Twins for Healthcare 4.0—Recent Advances, Architecture, and Open Challenges. *IEEE Consumer Electronic Magazine. Volum, 12*, 29–37.

Amofa, S. et al., (2024). Blockchain-secure patient Digital Twin in healthcare using smart contracts. *PLoS One*, 29 February, 19(2), p. e0286120.

Ankitha, V. (2025). *Toobler.* [Electronic] Available at: https://www.toobler.com/blog/real-time-integration-in-digital-twins [Access 6 June 2025].

Armeni, P. et al., (2022). Digital Twins in Healthcare: Is It the Beginning of a New Era of Evidence-Based Medicine? A Critical Review. *Journal of Personalized Medicine,* Volum 12.

Ashley, E. (2016). The precision medicine initiative: A new national effort. *Journal of the American Medical Association, 315*(8), 793–794. PMID: 25928209

Attaran, M., & Celik, B. (2023). Digital twin: Benefits, use cases, challenges, and opportunities. *Decis Anal J, 6*(6), 100165. DOI: 10.1016/j.dajour.2023.100165

Bai, N. (2025). *Stanford Medicine.* [Electronic] Available at: https://med.stanford.edu/news/all-news/2025/04/digital-twin.html [Access 8 Jube 2025].

Balasubramanyam, A., Ramesh, R., Sudheer, R., & Honnavalli, P. (2024). Revolutionizing Healthcare: A Review Unveiling the Transformative Power of Digital Twins. *IEEE Access. IEEE Access : Practical Innovations, Open Solutions, 12*, 69652–69676. DOI: 10.1109/ACCESS.2024.3399744

Bruynseels, K., Santoni de Sio, F., & van den Hoven, J. (2018). Digital twins in health care: Ethical implications of an emerging engineering paradigm. *Frontiers in Genetics. Volum, 9*, 31. PMID: 29487613

Cellina, M. C. M. A. M. I. G. I. S. C. E. F. D. G., & Papa, S. (2023). Digital Twins: The New Frontier for Personalized Medicine? *Applied Sciences (Basel, Switzerland), 13*(13), 7940. DOI: 10.3390/app13137940

Chen, J.. (2023). Networking Architecture and Key Supporting Technologies for Human Digital Twin in Personalized Healthcare: A Comprehensive Survey. *IEEE Communications Surveys & Tutorials. Volum, 26*, 706–746.

Chumnanvej, S., Chumnanvej, S. & Tripathi, S., (2024). Assessing the benefits of digital twins in neurosurgery: a systematic review. *Neurosurg Rev,* 18 January, 47(1), p. 52.

Dihan, M. S., Akash, A. I., Tasneem, Z., Das, P., Das, S. K., Islam, M. R., & Hasan, M. M. (2024). Digital twin: Data exploration, architecture, implementation and future. *Heliyon*, *10*(5).

Elkefi, S. & Asan, O., (2022). Digital Twins for Managing Health Care Systems: Rapid Literature Review. *Journal of Medical Internet Research,* Τόμος 24.

Fereday, J., & Muir-Cochrane, E. (2006). Demonstrating Rigor Using Thematic Analysis: A Hybrid Approach of Inductive and Deductive Coding and Theme Development. *International Journal of Qualitative Methods*, *5*(1), 80–92. DOI: 10.1177/160940690600500107

First Ignite. (2024). *First Ignite.* [Electronic] Available at: https://www.firstignite.com/exploring-the-latest-advancements-in-digital-twins-for-2025/ [Access 5 June 2025].

Fuller, A., Fan, Z., Day, C., & Barlow, C. (2020). Digital twin: Enabling technologies, challenges and open research. *IEEE Access*. *Volum*, *8*, 108952–108971.

GMI. (2024). *Global Market Insights.* [Electronic] Available at: https://www.gminsights.com/industry-analysis/healthcare-digital-twins-market [Access 5 June 2025].

Goldberg, B., & Warren, P. (2025). *CGI.* [Electronic] Available at: https://www.cgi.com/en/podcast/energy-utilities/building-resilient-sustainable-healthcare-digital-twins-and-innovation [Access 2 June 2025].

Halamka, J. (2024). *Mayo Clinic Platform.* [Electronic] Available at: https://www.mayoclinicplatform.org/2024/12/23/digital-twin-technology-has-potential-to-redefine-patient-care/

Huang, P., Kim, K. & Schermer, M., (2021). Ethical Issues of Digital Twins for Personalized Health Care Service: Preliminary Mapping Study. *Journal of Medical Internet Research,* Volum 24.

Huang, P., Kim, K. & Schermer, M., (2021). Mapping the Ethical Issues of Digital Twins for Personalised Healthcare Service. *jmir.*

Jeandrain, A. (2023). *Applied Clinical Trials.* [Electronic] Available at: https://www.appliedclinicaltrialsonline.com/view/how-digital-twins-are-driving-sustainability-in-clinical-supply-planning [Access 2 June 2025].

Kamel Boulos, M., & Zhang, P. (2021). Digital twins: From personalized medicine to precision public health. *Journal of Personalized Medicine*, *11*(8), 745. DOI: 10.3390/jpm11080745 PMID: 34442389

Katsoulakis, E., Wang, Q., & Wu, H. (2024). Digital twins for health: A scoping review. *npj. Digital Medicine*, *7*(77). PMID: 38519626

Li, X., Loscalzo, J., Mahmud, A., Aly, D. M., Rzhetsky, A., Zitnik, M., & Benson, M. (2025). Digital twins as global learning health and disease models for preventive and personalized medicine. *Genome Medicine*, *17*(11), 11. DOI: 10.1186/s13073-025-01435-7 PMID: 39920778

Lu, Y.. (2021). Digital twin-driven smart manufacturing: Connotation, reference model, applications and research issues. *Robotics and Computer-Integrated Manufacturing. Τόμος*, *63*, 101897.

Maïzi, Y., Arcand, A., & Bendavid, Y. (2024). Digital twin in healthcare: Classification and typology of models based on hierarchy, application, and maturity. *Internet of Things : Engineering Cyber Physical Human Systems*, *28*, 101379. DOI: 10.1016/j.iot.2024.101379

Markets and Markets. (2025). *Markets and Markets.* [Electronic] Available at: https://www.marketsandmarkets.com/blog/HC/digital-twins-future-of-precision-healthcare [Access 8 June 2025].

Meijer, C., Uh, H. & S. el Bouhaddani, S., (2023). Digital Twins in Healthcare: Methodological Challenges and Opportunities. *Journal of Personalized Medicine,* Volum 13.

Papachristou, K. et al., (2024). Digital Twins' Advancements and Applications in Healthcare, Towards Precision Medicine. *J Pers Med*, 11 November, 14(11), p. 1101.

Pellegrino, G., Gervasi, M., Angelelli, M., & Corallo, A. (2025). A Conceptual Framework for Digital Twin in Healthcare: Evidence from a Systematic Meta-Review. *Information Systems Frontiers*, *27*(1), 7–32. DOI: 10.1007/s10796-024-10536-4

Podmurnyi, S. (2022). *Hackernoon.* [Electronic] Available at: https://hackernoon.com/how-to-implement-digital-twin-architecture [Access 5 June 2025].

Popa, E., van Hilten, M., Oosterkamp, E., & Bogaardt, M.-J. (2021). The use of digital twins in healthcare: Socio-ethical benefits and socio-ethical risks. *Life Sciences, Society and Policy*, *17*(6), 6. DOI: 10.1186/s40504-021-00113-x PMID: 34218818

Quazi, F., Raju, N., Viradia, V. & Aparna, B., (2024). Digital Twin Technology in Healthcare: Applications, Challenges, and Future Insights. *International Journal of Global Innovations and Solutions (IJGIS).*

Rasheed, A., San, O., & Kvamsdal, T. (2020). Digital twin: Values, challenges and enablers from a modeling perspective. *IEEE Access. Volum*, *8*, 21980–22012.

Roopa, M. (2023). A Systematic Review: Digital Twin in Healthcare. *International Journal for Research in Applied Science and Engineering Technology*.

Scott, J. (2024). *HealthTeck*. [Electronic] Available at: https://healthtechmagazine.net/article/2024/01/what-are-digital-twins-and-how-can-they-be-used-healthcare [Access 23 May 2025].

SIMIO. (2025). *Simio*. [Electronic] Available at: https://www.simio.com/how-will-digital-twins-software-transform-your-business-in-2025/ [Access 20 May 2025].

Singh, M., Srivastava, R., Fuenmayor, E., Kuts, V., Qiao, Y., Murray, N., & Devine, D. (2022). Applications of Digital Twin across Industries: A Review. *Applied Sciences (Basel, Switzerland)*, *12*(11), 5727. DOI: 10.3390/app12115727

Sun, T. et al., (2022). The Digital Twin in Medicine: A Key to the Future of Healthcare?. *Frontiers in Medicine,* Volum 9.

Sun, T., He, X. & Li, Z., (2023). Digital twin in healthcare: Recent updates and challenges.. *Digital Health,* Volum 9.

Tao, F.. (2018). Digital twin-driven product design, manufacturing and service with big data. *Int J Adv Manuf Technol. Volum*, *94*, 3563–3576.

Turab, M., & Jamil, S. (2023). A Comprehensive Survey of Digital Twins in Healthcare in the Era of Metaverse. *BioMedInformatics*, *3*(3), 563–584. DOI: 10.3390/biomedinformatics3030039

Vallée, A., (2023). Digital twin for healthcare systems. *Frontiers in Digital Health,* Volum 5.

Vijayan, V., Connolly, J. P., Condell, J., McKelvey, N., & Gardiner, P. (2021). Review of Wearable Devices and Data Collection Considerations for Connected Health. *Sensors (Basel)*, *21*(16), 5589. DOI: 10.3390/s21165589 PMID: 34451032

Wu, C. et al., (2022). Integrating mechanism-based modeling with biomedical imaging to build practical digital twins for clinical oncology.. 3(2), p. Biophysics Reviews.

Xames, M., & Topcu, T. (2024). A Systematic Literature Review of Digital Twin Research for Healthcare Systems: Research Trends, Gaps, and Realization Challenges. *IEEE Access. IEEE Access : Practical Innovations, Open Solutions*, *12*, 4099–4126. DOI: 10.1109/ACCESS.2023.3349379

# Chapter 16
# Navigating GDPR Compliance:
## Transforming Digital Marketing in a Privacy-First World

**Mohammad Irfan**
https://orcid.org/0000-0002-4956-1170
*Chandigarh University, Lucknow, India*

**Saanvi Singhal**
https://orcid.org/0009-0002-2676-6816
*Christ University, Bengaluru, India*

**Ajay Kumar Yadav**
*Chandigarh University, Lucknow, India*

**Fadillah Mansor**
https://orcid.org/0000-0003-0423-953X
*Universiti Malaya, Malaysia*

## ABSTRACT

*The General Data Protection Regulation (GDPR), implemented by the European Union (EU), safeguards the personal data and privacy of individuals, granting them greater control over how their information is used and stored. This legislation mandates businesses to maintain transparency and responsibility in handling personal data, ensuring enhanced privacy and security for users. GDPR has significantly transformed the landscape of digital marketing by redefining data collection and usage practices. It necessitates that marketers adopt clear and transparent methods to respect user rights, fostering trust and compliance. Moreover, GDPR aims to prevent digital fraud and manipulation, particularly on E-commerce platforms, by*

DOI: 10.4018/979-8-3373-1534-8.ch016

*reinforcing stringent data protection measures. As a result, GDPR not only strengthens individual privacy but also compels businesses to adopt ethical data practices, shaping a privacy-first approach to digital marketing.*

## INTRODUCTION

The data-driven marketing strategy transformed digital markets by launching data-based markets in that period. Big enterprises use controlled private data to operate their modified advertising programs and customized content delivery services and personalized customer engagement systems (Li, 2024). Data-centric business operations performed by companies have led to dangerous security risks (Singh & Bishnoi, 2024). The European Union established GDPR to control business operations that involve personal information management through new regulations.

GDPR achieves its objectives through dual mechanisms which grant users control over their data while businesses must establish their data practice responsibilities (Smith et al., 2024). The businesses under this regulation need clear user permission for handling data before showing operational data practices while giving users complete control over processing their data. The new digital marketing ethical standards acquire mandatory status for businesses due to these requirements (Granata et al., 2024). Implementing GDPR framework strategies enables businesses to create customer loyalty by developing privacy protection measures. Organizations that follow customer privacy rights will gain better competitive advantages versus other companies managing digital business operations. People prefer brands which protect their data rights since such organizations maintain privacy standards as their selection criterion (Li, 2024).

An analysis by the author demonstrates how GDPR affects digital marketing through a combination of review-based challenges and benefits (Miller et al., 2024). The document provides both GDPR compliance methods and effective promotional procedures (Singh & Bishnoi, 2024). This paper conducts a detailed study which aims to present vital information regarding implementing GDPR requirements while taking advantage of data privacy as a foundation for developing business advantages.

### Global Prospect of General Data Protection Regulation (GDPR)

When GDPR took effect in 2018 it transformed into an active expansion of European Union (EU) data protection legislation (Negri-Ribalta et al., 2024). Organizations performing responsible transparent data management under the GDPR gain access to a generalized protection system. The universal nature of GDPR supports privacy legislations worldwide thus it leads digital marketing evolution and encourages other authorities to adopt matching privacy laws (Smith et al., 2024).

GDPR activates outside suggestions that affect areas beyond European Union territories. The personal data protection laws of Brazil through Lei Geral de Proteção de Dados and the United States via California's Consumer Privacy Act resultant in their framework from GDPR to start privacy regulations in their respective territories (Miller et al., 2024). The standards that all data privacy jurisdictions follow include basic concepts of user authorization and they work toward creating digital systems built around privacy (Granata et al., 2024).

Multinational corporations through their global data processing reforms have established privacy-friendly systems which protect privacy rights while building trust with customers (Yadav et al., 2024). Companies have deployed advanced encryption technologies together with consent collection systems and cookie management solutions which satisfy these requirements per Granata et al. (2024). The introduction of GDPR mandates organizations to create both consent monitoring technology solutions as well as data audit systems and secure storage technology platforms.

GDPR stands as the European regulation which revolutionizes every digital marketing practice (Miller et al., 2024). The necessity for digital marketing approach modifications has emerged because data collection must happen ethically and privacy limits for users need protection (Li, 2024). The evolution of personalization marketing requires marketers to develop privacy-respecting consumer engagement approaches due to increased difficulties in the field (Singh & Bishnoi, 2024).

However, challenges continue. Companies face multiple obstacles because they must manage conflicting rules for data protection between different regions and solve discrepancies between data-based marketing and privacy regulations at the same time while paying for maintaining agreement standards (Negri-Ribalta et al., 2024). Digital marketing development demands the protection of consumer information because GDPR deploy a uniform data protection standard beyond regional boundaries. Organizations that leverage GDPR for strategic gain maintain market leadership through developing trust with sensitive privacy consumers in digital markets.

## The General Data Protection Regulation (GDPR) on Economic Contribution

The General Data Protection Regulation (GDPR) reformed economic situations by reforming digital marketing rules while defining new procedures. Through GDPR organizations faced new rules about client data transparency which transformed their business activities for acquiring and handling information storage. The economic value activated by GDPR spreads through multiple domains where businesses put protective measures into place to secure data while simultaneously building trust with consumers and using technological strategies for enabling present-day business operations (Yadav et al., 2024). The economic benefits brought by GDPR represent an obstacle for business operations. GDPR imposes two mandatory requirements through continuous monitoring and financial support systems to handle complex regulatory needs (Granata et al., 2024). Customer trust demands businesses to tackle the ethical concerns about their biased algorithms while providing clear data usage explanations to their clients. According to Smith et al the primary economic highlights of GDPR surpass the costs of compliance because they make the overall agreement process more efficient.

(2024). Companies using the initiative increase technological development and market performance by investing in security measures which generates heightened trust among consumers. Organizations that adopt GDPR strategically as a growth opportunity will obtain substantial business success while strengthening their market position in digital markets (Yadav et al., 2024).

## LITERATURE REVIEW

During 2018 the EU introduced an innovative data protection standard in the General Data Protection Regulation (GDPR) that other jurisdictions worldwide needed to align with. Under the General Data Protection Regulation (GDPR) companies needed to fulfil strict monitoring requirements for maintaining data security of personal information. Organizations handling personal data of EU citizens must uphold GDPR data privacy standards regardless of their physical location because GDPR demands businesses to prove clear data procedure practices accompanied by robust data security measures as well as obtaining authorized consent. Research reveals the GDPR controls all parts of digital marketing sales operations through its impact on consumer trust functions as well as operational and technological developments and evolving data protection patterns (Miller et al., 2024).

GDPR used previous EU data protection rules to create its system through the 1995 Data Protection Directive yet this directive only established guidelines without appropriate performance requirements. A fundamental requirement of GDPR makes all EU nations follow strict regulations by law. The key data protection elements receive their official definition within this regulation by requiring precise responsible data management and limited scope data handling. GDPR has achieved its main goal of establishing better international data protection norms according to Voigt and Bussche (2017). The GDPR created a standard for privacy policies which multiple non-EU countries adopted in their privacy regulations as demonstrated by Brazil using LGPD and the United States through CCPA (Negri-Ribalta et al., 2024). The regulatory change brought by GDPR demonstrates its ability to establish worldwide privacy principles that force organizations to transform their operational practices toward better regulatory alignment.

Digital marketing practices are influenced by the General Data Protection Regulation which both requires data clearness regulations for collection activities and controls the use of third-party cookies. Traditional marketers who require extensive user data establish new practices which move their advertising towards privacy-oriented methodologies Li (2024). Organizations need updated methods to keep customer ties during GDPR initiatives by utilizing relative advertising combined with first-party data collection along with artificial intelligence analytics. Under GDPR requirements businesses need to find equilibrium between delivering modified content and protecting user privacy throughout their preference of user choices (Yadav et al., 2024).

Most academic research about GDPR investigates trust with customers as its primary matter. The CDPR develops enhanced user data control capabilities that strengthens customer brand trust relationships. After gaining clearer insights into their data rights users now prefer to provide support to companies that reveal proper practices for protecting personal information. Two negative aspects of GDPR appear because people accept cookie policies without understanding their consequences (Buckley et al., 2022). Organizations face financial consequences because of failing to fulfill GDPR requirements based on GDPR enforcement against Google Facebook and Amazon (Dutta & Dhal, 2024).

Under GDPR companies needed to demonstrate their data collection methods while third-party cookies faced access limitations. The former trust in big data by traditional marketers now needs a complete transformation to foster adoption of privacy-first advertising solutions as described by Li (2024). The implementation of GDPR required businesses to create new advertisement solutions which enhanced first-party data gathering ability and expanded their AI analytical programs (Singh & Bishnoi, 2024). Organizations adopting GDPR need to manage the delivery of personalized content against protecting user privacy during their efforts to respect user preferences (Yadav et al., 2024).

Consumer trust remains the core subject matter in research studies about GDPR. Data control activities under CDPR turn brands into trusted providers of data security for their user community. The research reveals users become conscious about their data rights since they will perform business with organizations that promise secure and transparent information handling. GDPR leads to unexpected complications through the acceptance behavior of users regarding cookie policies though they remain unaware about information handling mechanisms (Buckley et al., 2022). The GDPR noncompliance activities of businesses result in substantial financial penalties demonstrated by Google Facebook and Amazon (Dutta & Dhal, 2024).

GDPR's upcoming course of action depends on technological developments considered alongside regulatory developments. Accurate projections from Buckley et al. (2022) show that future AI-driven compliance systems will arrive in 2030 to conduct advanced procedures which cut down human labor while advancing regulatory compliance. Multiple countries have adopted GDPR-style data protection laws to merge international data defense standards as different nations implement similar regulatory approaches (Zaguir et al., 2024). Consumer trust remains the core subject matter in research studies about GDPR. Data control activities under CDPR turn brands into trusted providers of data security for their user community. The research reveals users become conscious about their data rights since they will perform business with organizations that promise secure and transparent information handling. GDPR leads to unexpected complications through the acceptance behavior of users regarding cookie policies though they remain unaware about information handling mechanisms (Buckley et al., 2022). The GDPR noncompliance activities of businesses result in substantial financial penalties demonstrated by Google Facebook and Amazon (Dutta & Dhal, 2024).

GDPR will transform through the combination of advancing technologies together with developing regulatory requirements. The development of AI-driven compliance systems for 2030 aims to execute advanced difficulty levels that decrease manual operations while enhancing regulatory acceptability according to Buckley et al. (2022).

Academic researchers have established that GDPR generates ethical challenges during data management although it operates outside economic frameworks and regulatory parameters. Academic scholars focus on researching moral limitations that stem from methods of personal data management alongside advertising schemes in their approach to marketing authorization (Negri-Ribalta et al., 2024). The standards for managing customer data as per both ethical and legal requirements establish a clear proof of liability that results in sustained effects on brand reputation and corporate accountability (Franke et al., 2024).

Companies now need to operate worldwide under global data protection rules as GDPR has brought numerous changes to digital marketing operations. This regulation delivers two major aspects that help protect user privacy and deliver transparent marketing together with strategic operational changes. Organizations required GDPR to establish proper data system management but these requirements elevated their corporate governance capability. Businesses focusing on privacy first management encounter three main operational hurdles that involve large compliance expenses along with structural operational changes coupled with intricate regulatory frameworks that particularly affect medium and small businesses across multiple jurisdictions.

The difficulties of GDPR implementation let businesses build better consumer trust by ensuring data security that created market distinctiveness advantages. The regulation arouses technical progression through artificial intelligence solutions which help with data governance alongside blockchain solutions for secure data storage and privacy protection technologies for better transparency and security standards. These technological developments serve two purposes by helping organizations maintain GDPR compliance and advancing their mission to perform digital transformation responsibly.

GDPR will keep developing as it establishes modern global data protection policies for the future. Jurisdictions adopting GDPR policies will create potential development of a global privacy standard which would simplify worldwide business data compliance obligations. Investigations should create long-term economic analysis of GDPR together with the study of AI-based application methods and the strength of privacy-by-design methods in protecting data from risks.

Organizations that make GDPR compliance a key operational component will achieve both risk reduction and improve their brand image and customer loyalty. The data-driven economy demands businesses to focus on privacy which leads to enduring success in the market. The future developments of GDPR will involve both simplification and expansion to deal with upcoming challenges as it evolves within the quickly transforming digital environment. Organizations must stay alert about threats and adapt by using innovation together with ethical data procedures which defend consumers and increase market reliability.

## RESEARCH METHODOLOGY

The research approach features secondary data analysis with case study methods to complete its investigation about how GDPR affects digital marketing compliance. The methodology delivers extensive knowledge about GDPR's influence on digital marketing practice standards as well as privacy duties for companies and personal data usage patterns. The study uses several research methods to analyze complex regulatory transformation in a manner that reveals business effects. A structured evaluation process is formed from the use of secondary data analysis together with qualitative methods and case study evaluations to investigate GDPR's legal framework and ethical standards and operational functions. The methodology delivers complete evidence-based research which combines academic theory with practical uses for the developing digital marketing industry.

### Secondary Data

This study draws its information from secondary data obtained through academic literature together with industry reports and regulatory guidelines as well as corporate compliance documents. The evaluation of GDPR's regulatory structure together with its law enforcement dynamics and business impacts on data-driven marketing requires secondary data. Official GDPR guidelines from the European Commission and reports from the European Data Protection Board (EDPB) and research on multinational companies deliver fundamental understanding regarding organizational approaches to GDPR compliance executions. Businesses and researchers can understand GDPR's marketing techniques and trust mechanisms together with innovation through academic studies and market research reports and legal journal publications. The research draws from secondary data sources which allows it to rest on verified industry data including research findings and regulatory guidelines instead of mere speculation. The data-driven structure enables researchers to conduct an essential evaluation of GDPR's influence on digital marketing elements like advertising platforms and customer consent systems and protection procedures.

## Mixed-Methods Approach

The study follows a mixed-methods evaluation structure which combines quantitative research methods together with qualitative research techniques. The assessment of GDPR compliance rates together with consumer behavior changes as well as marketing restrictions and penalties against non-compliant organizations depends on quantitative analysis methods. The data-based technique reveals numerical trends along with the patterns regarding regulatory enforcement procedures. The research evaluates GDPR enforcement statistics together with consumer consent statistics and digital advertising monetary changes to produce measurable evidence about its effects on business operations.

The evaluation of business approaches to GDPR compliance consists of both quantitative methods which measure data rates and qualitative studies that examine compliance implementation practices. The study examines privacy policies as well as ethical factors as well as corporate strategies which preserve trust and maintain legal compliance with consumers. The qualitative findings demonstrate how organizations handle the marketing and privacy conflict induced by GDPR regulation and help establish the advantages and restrictions of these regulations. The mixed-methods evaluation method provides complete GDPR assessment because it shows quantitative effects while documenting specific business strategic actions which arise because of legal requirements.

## Case Study Methods

Additional research includes studies about leading technology and digital marketing corporations that experienced GDPR impacts. Google, Meta (Facebook) and Microsoft and Amazon as well as other large multinational corporations serve as the primary businesses for analysis. The company's existing in data-heavy domains encounter major changes to their advertising models alongside their user consent practices and data management procedures because of GDPR. The compliance measures implemented by these organizations display concrete GDPR effects on the operation of digital marketing practices.

These individual case studies review three main elements of business procedures: data collection adaptation and GDPR-compliant marketing strategy creation along with techniques for sustaining consumer interest because of strengthened privacy requirements. Case studies reveal several difficulties from GDPR implementation which includes lowering dependency on third-party cookies but also includes handling consumer consent fatigue and handling high compliance costs. The investigation evaluates GDPR compliance strategies of leading digital marketing firms to extract key practices which all businesses can apply. The examined case studies teach important principles about data ethics and customer privacy protection as well as strategic marketing advancement which helps organizations succeed with GDPR compliance standards.

The research design which combines secondary data and both mixed-methods and case studies provides a strong basis to explore the varied effects that GDPR produces on digital marketing practice. The study delivers practical solutions to businesses that need to find equilibrium between privacy regulations and marketing performance through an examination of market patterns and organizational case research. Organizations can create enduring privacy-first marketing plans through the research findings which will achieve legal compliance alongside consumer trust as well as long-term digital marketing excellence despite growing data protection laws.

As shown in Figure 1, the following flowchart reveals how GDPR influences digital marketing while showing essential factors of data security investment together with consumer trust development and sustainable business expansion. A new flowchart will show the forecast analysis in the subsequent development stage.

*Figure 1. GDPR's influence on digital marketing activities*

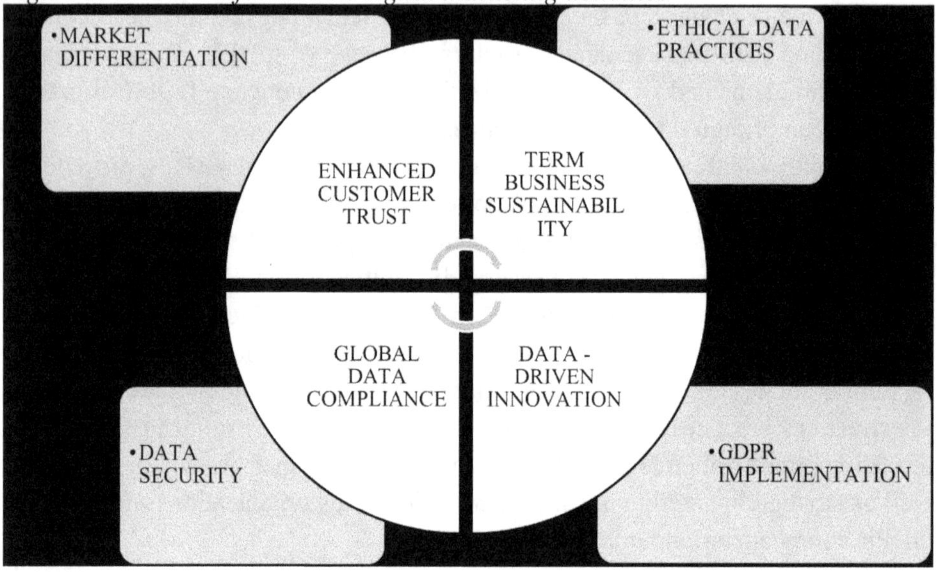


The diagram displays an extensive sequence which demonstrates that GDPR compliance actively contributes to digital marketing efforts by shaping different business elements starting from market unique positioning through consumer confidence to responsible data handling and worldwide standards and secure data management and marketing innovation. The GDPR stands today as a global leader in data protection laws because the EU launched it in 2018 to protect personalized information and privacy. The requirement to be GDPR compliant offers businesses strategic benefits which enhance their reputation together with consumer trust while creating stability that lasts for the long term within digital marketing operations.

Businesses who observe GDPR gain market competitive advantage by showing customers their commitment to handle data with transparency and accountability during the first step of the process. Companies which take steps to defend user data properly along with obtaining clear user permissions and following legal data handling protocols experience increased market success because aware consumers value such businesses. These businesses gain customer loyalty because people see them as trustworthy ethical providers who focus on customers hence achieving better retention rates. The direct connection between market differentiation and better consumer trust leads to successful business operation in the digital market.

A company that successfully builds consumer trust establishes itself for sustained business success in the long run. Today's customers exhibit great reservation about their personal data sharing while breaches of privacy together with improper data usage result in serious reputational harm. GDPR compliance creates reassurance for customers about responsible data management thus attracting more customer involvement along with brand loyalty. By fostering trust with customer's brands can enhance their engagement levels and boost conversion numbers in addition to developing customer support which contributes to sustainable revenue operation and enduring business expansion.

GDPR compliance demands ethical data management that centers on open disclosure and consumer rights and also requires responsibility in data handling. Businesses must only seek data-collection when necessary with straightforward accountability reasons before giving end-users access to their personal information rights for modification or deletion. Data security requires businesses to reduce data storage duration while they must protect personal information from cyber-attacks. Ethical data practices must be prioritized because they protect both necessary legal frameworks and develop positive brand reputation as well as enduring consumer trust.

Digital marketing strategies for businesses must prioritize global data compliance since organizations now conduct operations throughout various international regions. GDPR exists among various global data protection statutes that also include the USA's CCPA and Brazil's LGPD and India's PDPB and China's PIPL. The compliance of these regulations by companies leads to unhindered operation expansion while protecting their customers' trust and blocking financial penalties across all markets. Through GDPR compliance businesses prepare themselves for future changes in privacy laws worldwide and derive advantages in worldwide regulatory conditions.

The primary outcome of ethical data practices and global compliance allows organizations to generate innovative solutions using data-driven approaches. Businesses can study customer data through privacy-protected methods without legal noncompliance after businesses receive proper consent from customers. GDPR-compliant data analysis lets marketers use data responsibly for creating personalized marketing campaigns as well as custom-targeted ads and predictive analytical models. Due to regulatory compliance, organizations can utilize data insights efficiently to drive innovation because they keep to ethical guidelines.

GDPR compliance demands organizations to build robust security measures to defend consumer data from breaches together with cyberattacks and unauthorized access. Companies need to employ encryption together with anonymization as well as multi-factor authentication plus regular cybersecurity audits to protect personal data protection. Any GDPR breach must be notified to the authorities within 72 hours resulting in financial penalties that reach either €20 million or 4% of the organization's global annual turnover. Organization-wide focus on safeguarding data leads to both legal compliances together with customer trust because people feel secure transacting with brand entities that monitor against misuse of their information.

The GDPR implementation stage serves to make businesses embed privacy-first principles throughout their digital marketing activities. Companies must perform marketing willingly through consent methods since customers need to explicitly accept promotional content before receiving it. Privacy policies must include straightforward and concise language while business entities need to implement systems which enable users to control their data choices and cancel data collection along with the ability to remove stored information. Organizations should provide GDPR guidelines training to their staff members while keeping continuous observation on compliance through active monitoring of changing legal standards.

The compliance requirements of GDPR for digital marketing serve to create businesses that can maintain trust with consumers while operating sustainably in a legally correct fashion. The base of responsible marketing rests on four fundamental elements which include ethical data handling and global standards as well as data security and innovation systems. Firms which focus on upholding privacy together with security and transparency achieve consumer trust along with market distinction and prolonged business expansion. Businesses gain status as responsible digital marketing leaders by embracing GDPR and similar regulations which create legal safety while ensuring customer satisfaction.

## 2030 GDPR will substantially influence businesses

By 2030 GDPR will substantially influence businesses with enhanced power because technology development will transform the way organizations handle regulatory compliance and digital advertising. Data privacy regulations continue to evolve so businesses need to implement innovative systems which ensure compliance alongside operational efficiency in data-driven environments. AI-driven compliance emerges as a vital development because industries from different sectors will depend heavily on artificial intelligence tools to monitor regulatory requirements. Businesses can use these AI-based systems to verify consent automations while real-time compliance risk monitoring and process security become automated functions. Organizations that embed AI into their compliance frameworks can detect changing regulations better as they simultaneously minimize their risks of non-compliance. From data protection every business will gain efficiencies through AI solutions that both minimize mistakes in audits and make audits more precise while simplifying security efforts.

The advancements in data security technologies emerge from GDPR's impact on industry throughout the year 2030. Businesses need to put blockchain-based encryption together with zero-trust architecture into practice since these are modern security protocols to protect sensitive consumer information. The decentralized system of blockchain technology ensures unalterable data security with reduced vulnerability to breaches along with unauthorized access The zero-trust architecture principle "never trust always verify" will establish itself as the conventional security framework that protects data through continuous verification and controlled access function. The developed innovations represent essential solutions for security threats protection which strengthens digital platform user trust. AI ethical practice will establish itself as a core marketing strategy component that maintains compliance of AI-driven personalization methods with privacy standards. AI marketing systems will adopt privacy-based design principles through federated learning along with differential privacy methods that protect data exposure while allowing personalization for customers. Business adaptation to ethical AI practices will enable them to achieve regulatory compliance while maintaining optimal marketing performances that respect privacy standards.

Business success will depend heavily on consumer empowerment together with trust because privacy-first customers will favor brands which prove their commitment to data protection. The businesses which maintain transparent data practices together with visible privacy policies and simple consent management features will achieve better market position against their competitors. Improved trust from consumers leads to better brand loyalty together with elevated customer retention because people feel safer giving their private data to organizations that maintain strict privacy protecting measures. The alignment of different protection laws across jurisdictions with GDPR standards will establish uniform global data privacy measures because of which a harmonized regulatory framework appears. Multiple nations worldwide will join efforts to develop common data protection guidelines which enables multinational businesses to expand internationally with ease while maintaining international datum privacy compliance. The alignment enables regulatory standards to simplify data transfer regulations which boosts both legal simplicity and regulatory unity of the digital economy.

## 2047 GDPR will substantially influence businesses

GDPR alongside newer advancements will direct businesses worldwide along with digital systems during the approaching timeframe leading to 2047. Universal data privacy standards represent a major development through which one standardized framework will protect data worldwide. A unified global standard will make businesses follow compliance protocols effortlessly and eliminate diversity between local privacy practices to enable worldwide information exchanges. Organizations can achieve efficiency gains through data privacy unity which leads to secure consumer data protection. Consumer data sovereignty will gain prominence because it equips people with increased autonomy to supervise their private data. The integration of advanced consent management systems alongside decentralized identity solutions provides consumers simple methods for both giving and taking back data access rights to preserve their privacy demands permanently. The evolution of data practices will establish consumer dominance for monitoring business activities that impact their data control and distribution.

The sustainable digital ecosystem evolution emerges as a fundamental long-term effect of GDPR. The business sector will fund cooperative data-sharing structures that protect both user security and encourage ethical technological developments. Through cooperation between businesses and technology providers and regulators they will build agreements about data-sharing which protect consumer rights while stopping improper use of personal information. Through the creation of these sustainable systems businesses will achieve product innovations alongside moral standards to access data insights that protect user privacy. Economic growth will increase when businesses adopt privacy-focused business strategies through data protection commitments which allow them to gain market distinction. A complete adoption of privacy-first practices enables organizations to satisfy regulation standards and welcome consumers interested in both confidentiality and data protection transparency.

Business innovation together with growth receives its fundamental direction from GDPR principles since companies implementing privacy-first methods gain market supremacy by demonstrating adherence to these principles. Sophisticated data handling strategies combined with responsible practice allow businesses to create long-term customer support while improving their reputation along with sustained revenue growth. Elements of compliance combined with security and ethical usage of AI technologies make organizations successful in a market where data privacy concerns govern consumer actions.

GDPR exists beyond its enforcement stage because it will drive a digital economy which merges secure customer-business interactions with transparent data management throughout the next decades. Future predictions indicate that GDPR will become a fundamental pillar of data security which will promote trust alongside economic growth through technological innovation until at least the next three decades.

*Figure 2. GDPR compliance process enabling ethical digital transformation and business growth*

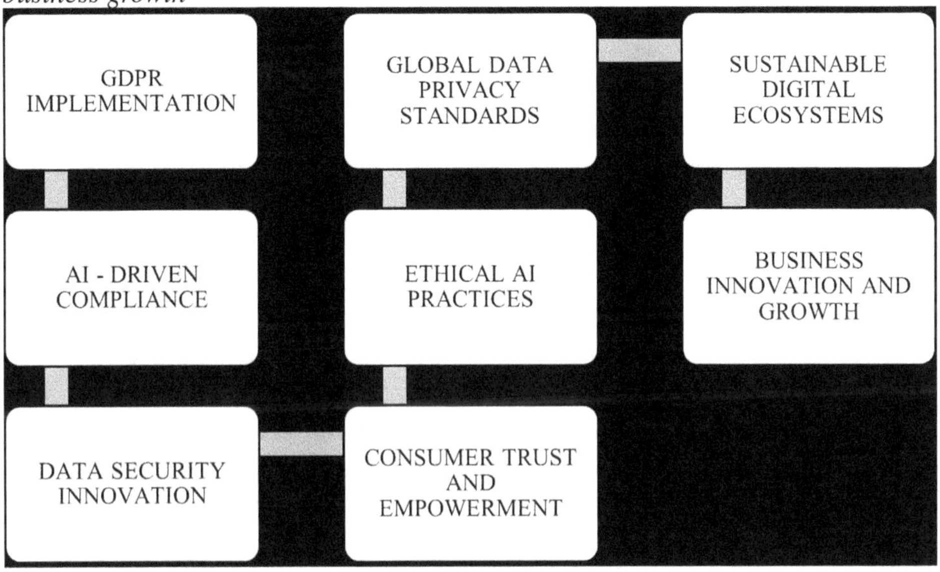


*Sources: Google Scholar, ResearchGate, IEEE Xplore*

As illustrated in Figure 2, the flowchart presents a systematic approach to business innovation and growth which ensures implementation of GDPR requirements in digital marketing. The key function of GDPR implementation creates secure data conditions which promote both ethical AI techniques and user confidence and develops standard worldwide privacy rules leading to a sustainable digital environment. Organizations which follow this method will both observe international privacy regulations and use AI compliance tools for process optimization. Ethical data handling together with consumer trust and innovation-based business development can be achieved using these measures.

GDPR Implementation sets the base for data privacy compliance through its initial procedures. The European Union established GDPR as a legal framework which determines the rules businesses must follow during data collection and storage and processing and data sharing of personal information. The regulations enforce precise rules that guide user acceptance procedures along with minimum data storage practices yet demand full openness and responsibility from organizations. A GDPR implementation demands companies to build their own protection policies and name data protection officers and maintain legal protected data while securing all storage points. Those who fail to comply will face substantial monetary penalties in addition to potential legal consequences as well as negative impact on their reputation. User data collection must happen only with clear consent from users while giving them complete authority to view their information and request changes as well as data deletion. Organizations that comply with GDPR regulations obtain market competitiveness through their transparent data handling practices that improves their digital marketplace credibility.

A successful execution of data security innovations automatically drives Consumer Trust & Empowerment within businesses. The European Union through GDPR established data criteria which grants individuals the authority to direct the collection process and storage and usage practices of their personal information. Brand engagement from consumers becomes higher when companies display data privacy commitment thus creating stronger customer relationships with increased brand loyalty. Consumers achieve empowerment through businesses that maintain open data practices along with specific privacy guidelines and intuitive user management choices. Service providers furnish privacy dashboards in combination with simple opt-in and opt-out systems together with thorough consent documentation to enable data choices for their users. Companies that focus on ethical data practices and transparency build better consumer loyalty which produces increased user engagement during digital marketing initiatives.

Consumer trust in AI technologies continues to be sustained through business adoption of Ethical AI Practices for the current age of artificial intelligence-driven operations. The application of AI and machine learning (ML) in digital marketing serves directly two main purposes: it produces customer profiles and delivers forecasted analytics which result in exactly directed marketing campaigns. AI systems that receive improper management have the potential to create biased choices and infringe privacy protection and misuse data unethically. Ethical AI practices establish guidelines that make AI models thoroughly accountable and transparent with unbiased performance. Businesses need to deploy bias detection software and create explainability tools and follow AI principles to stop outdated or unwanted marketing practices. GDPR requires vendors to make AI decisions easily understandable since users must receive information about how automated processes shape their experiences. The combination of privacy-enhancing technologies such as differential privacy allows AI-driven marketing to achieve personalization through analysis that does not expose individual identities. AI governance frameworks form part of ethical AI practices through their operational purpose of matching AI models to essential legal and ethical standards of data privacy.

The adoption of GDPR by organizations with ethical AI principles leads to the formation of global privacy standards. GDPR serves as an EU regulation that directs international data privacy legislation including the California Consumer Privacy Act (CCPA) and Brazil's Lei Geral de Proteção de Dados (LGPD) and India's Digital Personal Data Protection Act (DPDPA). International businesses need to follow various data protection laws which demand a common approach to personal data protection. The standardized privacy regulations allow organizations to handle their data across national borders efficiently without violating multiple legal requirements. Global data privacy standards create seamless operations between different jurisdictions because they allow businesses to maintain user privacy and function without interruptions.

Businesses can advance towards Sustainable Digital Ecosystems through global data privacy standardization which embeds ethical data practices within business operations. A sustainable digital ecosystem provides responsible personal data utilization which limits massive data acquisition and prohibits improper data monetary profits. The present focus for digital marketing strategies should be on privacy-first strategies that include contextual advertising which functions without tracking methods that violate privacy. The adoption of secure multi-party computation (SMPC) alongside homomorphic encryption represents two privacy-protecting technologies which businesses can use to evaluate data without revealing underlying user information. The transition of organizations toward sustainable digital marketing delivers benefits that guarantee GDPR compliance together with better internet privacy and ethical practices.

The result establishes a single framework for data protection which becomes known as Universal Data Privacy Standards to govern all sectors and geographic areas. A collaborative effort between businesses and regulators together with technology providers results in joint development of universal privacy standards until regulatory complexities disappear. Organizations benefit from operational improvements through universal standards because they get to use one privacy policy which meets worldwide regulations. These standards enable the smooth operation of data exchange processes between multinational companies to accomplish common regulations throughout multiple jurisdictions. The push for worldwide data privacy uniformity drives the development of Privacy by Design methodologies that protect data protection systems throughout business processes.

All these initiatives produce Business Innovation & Growth as their final outcome. Customers recognizing organizations which effectively implement GDPR compliance and use AI compliance tools together with ethical data practices and align with global privacy standards recognize these organizations as leaders in responsible digital marketing. Having a privacy-first approach brings businesses more clients and lessens their legal exposure and enables them to develop data-powered innovative solutions. Businesses which protect their data acquire market superiority because they stand apart from organizations that do not meet privacy standards. Security-focused advancements in innovation help develop stronger customer experiences through AI support and protected analytics which supports permanent business expansion.

The flowchart serves as an essential strategic tool enabling businesses to handle GDPR compliance requirements within digital marketing practice. Business victory in the present and future depends on privacy-first methods and AI-compliant solutions and augmented data security with ethical AI practices which enable lasting business success and consumer confidence. The implementation of worldwide data protection standards gives businesses a balanced digital operating environment which lets them succeed using data but still follow regulatory rules. GDPR compliance stands as both a law and an innovative driver which enables responsible digital transformation with sustainable business expansion.

*Figure 3. Projected impact of GDPR compliance factors from 2025 to 2045*

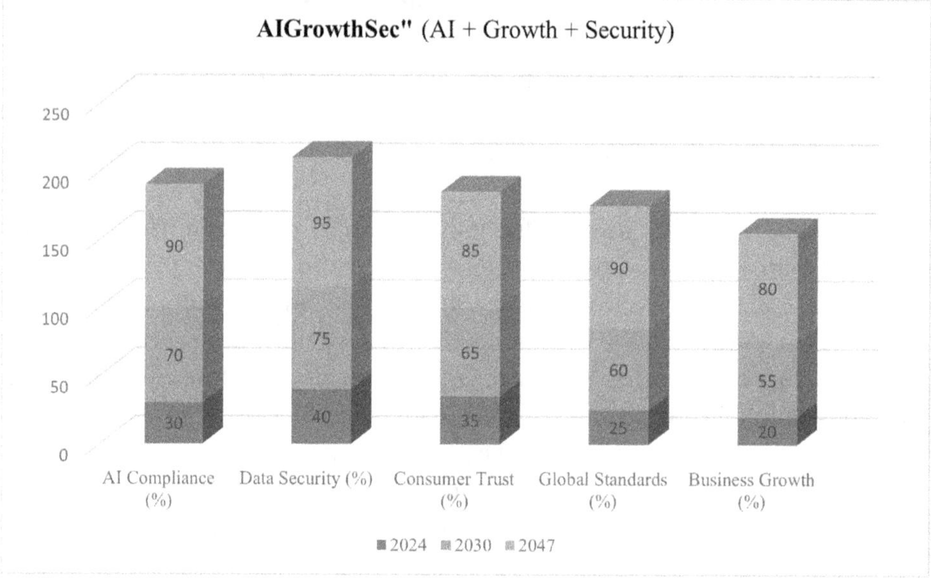


As demonstrated in Figure 3, the chart shows how GDPR compliance will affect various factors through three different timeframes that span from 2024 to 2030 and extending up to 2047. It functions as a visual monitoring system that studies forecasted regulatory transformations during different periods and business responses to developing data privacy standards. The chart arranges GDPR aspects into four distinct categories which include AI-driven compliance along with data security measures and consumer trust levels and global standard development and business expansion aspects. The transformation of digital marketing along with business operations under regulation depends fundamentally on these important factors which predict GDPR compliance changes.

The graph displays Year as a Y-axis factor producing a distorted scale because years belong in the categorical axis. Mathematical representation of the year variable as numbers produces visual distortion compared to other factors thus misrepresenting the underlying data structure. Comparing trends between variables becomes challenging because the presented visualization is off balance. For optimal viewing it would be better to show years through X-axis labels along with a display method that respects the numerical proportions of remaining factors. The modification of this problem would enhance the clarity aspects of the chart together with its interpretation capabilities.

The analysis shows that AI-driven compliance represents an essential variable expected to experience major development through time. New technological advancements allow businesses to expect that they will depend more heavily on AI automated tools for regulatory compliance functions which require consent verification and continuous data monitoring along with cybersecurity enhancements. Organizations benefit from an increasing number of AI-driven solutions which helps them maintain compliance effectively as new privacy laws and GDPR changes occur. The adoption of AI compliance systems will become standard practice in all digital marketing and data governance strategies by 2047 thus eliminating manual oversight work and protecting against data breaches.

A successful execution of data security innovations automatically drives Consumer Trust & Empowerment within businesses. The European Union through GDPR established data criteria which grants individuals the authority to direct the collection process and storage and usage practices of their personal information. Brand engagement from consumers becomes higher when companies display data privacy commitment thus creating stronger customer relationships with increased brand loyalty. Consumer power originates from transparent data procedures and well-defined privacy guidelines combined with easily accessible user management tools. User privacy dashboards together with simple options for consent and denial and specific forms ensure customers possess sufficient knowledge when deciding about their data usage. Companies committed to using data ethically together with transparent practices will earn trustworthy consumer interactions which results in both higher customer retention and stronger engagement during digital advertisement campaigns.

The success of business operations in AI-driven markets depends on their adherence to Ethical AI Practices for retaining customer trust. AI technology together with machine learning functions as standard digital marketing tools for generating profiles of customers while delivering predictions and focusing advertisements. The improper control of AI generates problems with biased choices in addition to infringements of privacy and unacceptable data utilization. Fairness together with transparency and accountability are all fundamental elements which ethical AI practices provide for AI models. Businesses need to deploy bias detection software and create explainability tools and follow AI principles to stop outdated or unwanted marketing practices. GDPR requires vendors to make AI decisions easily understandable since users must receive information about how automated processes shape their experiences. AI-driven marketing systems should achieve data personalization compatibility with user-privacy needs through implementation of differential privacy along with other privacy technologies. AI governance frameworks serve to maintain adherence between AI models and both legal and ethical data privacy rules while operating under ethical AI practices.

A successful execution of data security innovations automatically drives Consumer Trust & Empowerment within businesses. The European Union through GDPR established data criteria which grants individuals the authority to direct the collection process and storage and usage practices of their personal information. Brand engagement from consumers becomes higher when companies display data privacy commitment thus creating stronger customer relationships with increased brand loyalty. Consumers develop greater control because organizations disclose their data procedures while establishing detailed privacy specifications and making their personal settings accessible for users. Users benefit from privacy dashboards alongside obvious opportunities to choose whether they want to participate along with specific forms that explain data handling in detail. Organizations implementing transparent policies about data management along with their ethical standards achieve customer loyalty and increase marketing campaign success rates among users.

The rising use of AI technology demands businesses to follow Ethical AI Practices if they aim to keep consumer trust. The fields of digital marketing heavily depend on AI and machine learning technologies to run customer profiling operations and carry out predictive analytics for targeted advertising campaigns. AI systems produce biased choices as well as privacy breaches when their management processes are inadequately controlled thus leading to unethical data handling. A system of ethical AI standards maintains the fairness along with transparency and accountability of artificial intelligence models. Businesses need to deploy bias detection software and create explainability tools and follow AI principles to stop outdated or unwanted marketing practices. GDPR requires vendors to make AI decisions easily understandable since users must receive information about how automated processes shape their experiences. The admittance of AI marketing systems into customer experiences requires privacy-protecting additions that implement differential privacy solutions to confirm privacy while delivering individual-targeted content. The implementation of AI governance frameworks serves as an ethical practice to guarantee that AI models comply with privacy regulations along with ethical standards. Figure 4 explains AI Compliance, Data Security, Consumer Trust, Global Standards, Business Growth

*Figure 4. AI Compliance, Data Security, Consumer Trust, Global Standards, Business Growth*

This line chart provides multi-year insights about essential factors related to data governance and compliance as well as business growth under growing digitalization and expanding data security stressors. The chart provides data about five core elements from 2025 until 2045 which includes AI compliance and global standards together with data security and consumer trust alongside business growth. These essential elements provide essential guidance about the relationship between businesses and the GDPR and other emerging global standards as well as regulatory bodies and consumers. The projection illustrates what organizational data security and management approaches will look like during the upcoming twenty years.

The different factors have their own lines drawn in this chart with distinct color distinctions. According to the presented analysis the blue line represents AI compliance while projecting substantial expansion during the time horizon. The utilization of artificial intelligence to maintain legal data protection standards and policies constitutes AI compliance. The technology regularly performs automated compliance evaluations and monitors user data handling through AI systems and uses analytical prediction to discover potential regulatory violations. The implementation of AI compliance began at a low level in 2025 and later accelerated its growth to reach 70% by 2030. The data demonstrates that organizations increasingly use AI-based solutions for simplifying compliance work and strengthening their regulatory reporting and reducing manual tasks. AI compliance will remain consistently on the rise during the upcoming ten-year period as evidence indicates that artificial intelligence will be essential for data governance by performing automated complex regulatory processes to improve operational efficiency.

Global data standards enhance from their initial low base in 2025 to achieve approximately 60% standards by 2030 according to the dark blue line. Global standards denote the standardization process of universal data protection regulations which follow GDPR principles and multiple regional privacy measures. Data exchanges between different international entities will encourage governments and businesses to normalize privacy regulations thus creating standard and comparable worldwide enforcement. Millions of countries and industries will acknowledge the need for standardized data protection policies for safe international business operations by 2045 because of continuous global standard advancement. The rising global data privacy standards direct the world toward more structured and unified digital security frameworks that harmonize legal inconsistencies among different states.

According to the research data security begins at a higher starting point in 2025 and demonstrates consistent growth until it reaches 75% in 2030. Many security technologies forming data security comprise encryption together with access control and threat detection systems and mitigation methods. The increasing data security numbers indicate that businesses are developing their protection systems progressively to combat security threats and data breaches and unauthorized access attempts. Organizations will need to allocate substantial financial resources for security protocols because cyberattacks combined with digital fraud incidents continue to escalate. The security of data will keep rising over the course of the next thirty-five years since businesses understand data integrity threats better and value safe customer information within digital interconnection.

Consumer trust tracks a progressive growth pattern to reach 65% during 2030 according to the purple line. Business success depends heavily on consumer trust during periods of increasing data security concerns which threaten to harm corporate reputation. Customer trust continues to rise because organizations enhance protection of personal data and strengthen their security systems and maintain clear data operations thus customers develop increased comfort sharing individual information. The retention of customers depends on trust and similarly trust drives brand reputation and sustainable business success over many years. Organizations which maintain ethical data practices besides providing transparent user communication will develop stronger customer relationships together with obtaining market advantages because consumer trust continues rising. Analyze of expected consumer trust at 2045 shows further growth which supports companies implementing privacy-first strategies because they will achieve better long-term business success.

The pink line reveals business expansion starting at a low rate compared to other elements and gradually growing to 55% during the 2030 period. Business growth depends on regulatory compliance according to this portion of the data chart. Businesses that see data privacy regulations as burdensome at first discover later on that compliance leads to enduring business expansion. The integration of modern data protection requirements by enterprises establishes consumer trust and reduces both legal difficulties and strengthens their industry leadership. Business growth shows a continuous upward trend because organizations which spend on data protection alongside AI compliance and clear customer relationship management achieve better expansion and financial success. The projected business growth up to 2045 demonstrates that following data privacy regulations creates opportunities for both innovation and market distinction as well as enhanced profit potential.

A central element of the chart displays future 2030 data values through its tooltip mechanism. The tooltip in the chart functions as a vital reference so that viewers can examine projected changes in factors for 2030. The tooltip reveals that 2030 will bring AI compliance to 70% while global standards fall to 60% and data security reaches 75% and consumer trust hits 65% and business growth attains 55%. The presented data shows how data governance and compliance together with security and business success operate as interconnected elements. AI compliance together with data security rates are raising because companies employ advanced technologies to fulfil regulatory needs while safeguarding their sensitive information. Modern data protection standards continue to ascend worldwide because countries actively cooperate to protect data which results in a better structured regulatory structure. The trust of consumers continues to improve because companies have implemented better security along with ethical data procedures that has boosted their faith in handling personal information. The rising business growth demonstrates that compliance requirements provide organizations with beneficial strategic advantages in addition to their legal requirements.

The chart delivers an extensive vision about how businesses alongside regulatory authorities need to handle data security complications during the next two decades. The chart emphasizes the role of AI in regulatory compliance as well as the changing international data standards and secure data management methods and consumer trust factors alongside regulatory adherence effects on business growth. The data indicates that GDPR alongside other data protection legislation provides organizations with both legal compliance requirements and enables the creation of a safe thriving digital structure that people can trust. Businesses which focus on data governance and privacy protection in 2045 will achieve better market competitiveness as they gain devoted customers along with minimized legal risks and enduring business development.

The chart delivers an extensive vision about how businesses alongside regulatory authorities need to handle data security complications during the next two decades. The graphic shows why AI-powered regulatory compliance matters as well as how worldwide data rule standards evolve and why solid cybersecurity measures matter and because well-trusted customers drive business success while data protection compliance directly links to financial expansion. The chart establishes how GDPR and related data protection laws have developed into vital components for establishing a protected digital framework that trusts its users. Businesses which focus on data governance and privacy protection in 2045 will achieve better market competitiveness as they gain devoted customers along with minimized legal risks and enduring business development.

## ANALYSIS

### The Role of GDPR in Microsoft, Google, AWS, IBM, and Meta and Its Impact on the Indian Economy

As a regulation of the European Union the General Data Protection Regulation (GDPR) completely transformed business practices in data privacy management and protection. All major technology firms including Microsoft Google and Amazon Web Services (AWS) IBM and Meta together with numerous worldwide corporations needed to totally revise their data management frameworks to meet GDPR standards. Major corporations have two functions under GDPR: maintaining their own compliance standards and delivering tools and systems that enable other businesses to meet data protection criteria. GDPR as a European regulation affects global businesses including those operating in India because their data management needs to follow its standards when working with multinational firms. Insight into the digital age data privacy implications comes from examining how top companies integrate GDPR into their operations together with their framework on the Indian economy.

### Microsoft: Leader in GDPR-Compliant Cloud Services

The GDPR compliance leader Microsoft implements data protection features that run across its cloud services range from Microsoft Azure to Office 365 and Dynamics 365. Through its broad cloud infrastructure Microsoft provides GDPR-compliant technologies to its platform users that include data encryption together with access management systems and protection against data loss scenarios. Microsoft Compliance Manager enables organizations to test their GDPR compliance level while delivering control implementation solutions. Microsoft provides its customers with strong user privacy guarantees by enabling them to store their data within selected geographical areas to meet GDPR data location requirements.

The European market demands GDPR compliance for Microsoft so the company can operate without incurring substantial penalties. Microsoft achieves customer trust through strong data privacy practices which also creates distinct advantages in its marketplace against competitors. Microsoft serves as a blueprint for companies to follow when they aim to meet rigorous data security regulations although its GDPR readiness provides operational framework for other businesses.

Microsoft provides GDPR-compliant services that serve as essential foundations for IT and corporate organizations within the Indian market. The use of Microsoft Azure and Office 365 by many Indian businesses for their cloud and business needs brings built-in GDPR compliance tools directly to them. Indian businesses that operate with European clients benefit greatly from the use of Microsoft's GDPR-compliant services since these businesses can demonstrate secure and compliant data management capabilities to their clients. The strong focus on privacy by Microsoft motivates both Indian government officials and business leaders to develop robust data protection laws throughout India.

## The Google Corporation Uses GDPR Compliance Standards to Strengthen Data Privacy for All Users

Google included GDPR principles in every section of its cloud services along with its advertising platforms and consumer applications to demonstrate its dedication to data protection. Google Cloud provides its customers access to various data privacy management tools including encryption functions as well as security logs and identity management system. The toolset of Google Cloud enables businesses which depend on its services to meet their GDPR requirements. The Data Protection Officer (DPO) created by Google provides companies with assistance in managing their customer data operations in a responsible manner.

The core reason why GDPR impacts Google consists of their advertising business structure that depends on accumulated user information. The GDPR demands companies such as Google to disclose data collection practices while achieving user consent for processing personal data before starting operations. Through these precautionary steps Google prevents regulatory penalties while upholding its reputation to its enormous user community.

The GDPR compliance requirements that Google adopted directly affect Indian operations on multiple levels. A significant number of Indian organizations particularly start-ups alongside digital enterprises conduct their business operations by using the services of Google Cloud and Google Ads. Businesses using Google services for operations must establish privacy-based policies to stay effective under

GDPR regulations. Indian advertisers must understand strict European advertising rules since they need to maintain GDPR alignment for their European customer-targeting campaigns. The change will make Indian digital marketers implement best data privacy practices which strengthens their nation's developing digital ecosystem.

## AWS as a Leading Cloud Service Provider Maintains a Security Strategy That Protects Data in Cloud Environments

AWS operates as a leading cloud service provider which maintains numerous comprehensive initiatives to fulfil GDPR standards. DW provides a secure cloud infrastructure including features for compliance which include encryption technology for data protection and authentication systems with multi-factor authentication and identity access management solutions. AWS Artifact delivers businesses a useful tool which produces regulatory compliance reports to assist organizations meet their obligations.

The vast number of businesses which utilize AWS's cloud platform makes GDPR compliance essential for the platform. By following GDPR policies AWS maintains its legal operational status for European regional customers. The market position of AWS becomes stronger because customers view its services as both secure and privacy-oriented cloud solutions.

The GDPR compliance standards implemented by AWS create economic benefits for its operations throughout India. Indian information technology firms at both startup and enterprise levels depend on AWS to operate their applications and store their data. A reliance on AWS's secure infrastructure enables these companies to fulfil international data protection standards that become essential for European market entry. AWS promotes Indian businesses to enhance their cybersecurity posture and thus supports the nationwide growth of digital security.

## IBM: Pioneering Data Privacy Solutions

IBM leads the market in data privacy solutions by developing GDPR-oriented technologies which safeguard sensitive business information. Organizations that use IBM InfoSphere Optim Data Privacy receive the ability to shield personal data through security measures which protect against data breaches. IBM Cloud delivers GDPR-compliant infrastructure solutions as part of its built security system to help enterprises handle customer information in a secure way.

The strategic advantage gained from GDPR compliance by IBM has built its status as a reliable technology partner and delivered it an enhanced reputation. Companies looking to IBM for data security guidance leads the market to request more of its protective solutions together with consulting help.

The Indian economy needs IBM's GDPR-compliant technical solutions because they serve IT service providers who deliver services to European clients. The security solutions from IBM provide Indian firms with data protection tools that satisfy international standards for regulatory compliance. As an extension of its operations IBM provide study on GDPR compliance which helps train Indian experts in worldwide data security protocols. India experiences a continuous expansion of its cybersecurity and regulatory compliance talent base as a result.

### Meta (Facebook): Navigating GDPR Challenges

Facebook as Meta has found it difficult to satisfy GDPR requirements because of its business model based on extensive data practices. Meta received several penalties from authorities due to non-compliance so the company needed to update its privacy standards. The GDPR compliance measures adopted by Meta include fresh user consent processes together with increased data control functions and reports about collection activities. Meta has released privacy-oriented tools for business entities using Facebook and Instagram which enable responsible management of user information.

The data protection rules of GDPR directly impact Meta because the company depends on advertising for its main revenue stream. Meta must gain specific consent from users as a condition to process their data for advertising purposes before implementing any new data collection rule. Operating in the European market requires GDPR compliance to protect business operations against additional penalties.

The data collection framework of GDPR directs Meta toward specific adjustments in their digital marketing operations throughout India. Businesses operating in India should make sure their Facebook advertising adheres to GDPR regulations when marketing to European users. The change has occurred so companies can use advertising data ethically. The compliance actions of Meta function as an example which promotes India's developing data protection laws toward stricter privacy requirements that respect global standards.

## RECOMMENDATIONS

Businesses need to implement a privacy-first approach between legal requirements and ethical practices alongside technological advancement for maintaining effective digital marketing practice and GDPR compliance. Businesses must establish open data acquisition practices which demand explicit and aware consumer permission before they process private information. Businesses can secure GDPR compliance and build customer trust by using design measures that reduce data collection extent and control storage times and maintain secure encryption features. Companies need to invest in both artificial intelligence and automation tools which will automate their compliance process tasks. Through artificial-intelligence tools businesses receive automated assistance for managing consent rights while conducting risk assessments and monitoring compliance activities thus they operate with better efficiency for official regulatory requirements.

The compliance of GDPR depends heavily on the extent to which consumers understand its regulations. Organizations need to improve their data visibility through thorough explanations of their data collection and storage procedures as well as processing methods. Privacy policies can win trust from users through friendly interfaces and visual tools combined with easy permission settings which encourage proper data sharing practices. Strong cybersecurity measures need to be adopted because data breaches as well as cyber threats require protection for sensitive information. Every organization needs to implement encryption protocols and conduct security audits as well as practice multi-factor authorization to stop unauthorized personal data access.

The worldwide status of GDPR as a key benchmark for data privacy standards requires businesses to monitor continuously developing privacy regulations throughout the world. A specific data governance team within businesses allows them to meet changing compliance needs and support global business operations across multiple markets effectively. Organizations maintain ethical and consumer-focused digital marketing by making privacy-first principles central to their marketing strategies and thus follow regulatory demands. The success of businesses in the long run depends on their proactive steps to protect customer data which will help them gain market strength and reduce compliance risks while building strong bonds with their customers.

## CONCLUSION

The General Data Protection Regulation (GDPR) strongly transformed digital marketing space through enforced strict data protection rules that provide people with better control of their data. GDPR forces businesses to utilize new data management procedures which focus on privacy by demanding clear user consent and full organizational transparency and accountability. The GDPR regulatory framework imposed various challenges including higher compliance expenses alongside advertising restrictions and system reorganization although it gives businesses chances to construct trust with customers and establish solid brand recognition.

GDPR compatibility has enabled businesses to outpace their competitors through proven ethical data behavior as they attract more loyal customers.

Google along with Meta Microsoft and Amazon demonstrate business profitability through restructuring their advertising systems along with privacy strategies to meet GDPR requirements according to the research. Expert opinion suggests that GDPR motivated regulators worldwide to create data protection policies such as the California Consumer Privacy Act (CCPA) in the United States and both Brazil's Lei Geral de Proteção de Dados (LGPD) and India's Digital Personal Data Protection Bill (DPDPB). The widespread effect of GDPR reveals its position as a worldwide standard for privacy enactment. The process of maintaining GDPR compliance remains a difficult ongoing struggle since small and medium-sized enterprises generally lack resources needed for complex implementation of compliance measures.

Businesses should recognize GDPR compliance beyond legal requirements by using it to craft market strategies which prioritize consumer privacy and center marketing initiatives on consumers. The combination of technological progress including AI compliance tools with secure data processing frameworks enables businesses to stay compliant with regulations and at the same time achieve business expansion and build enduring digital marketplace trust.

## REFERENCES


Buckley, G., Caulfield, T., & Becker, I. (2022). It may be a pain in the backside but... Insights into the resilience of business after GDPR. *arXiv*. [Add link if available]

Dutta, M., & Dhal, S. (2024). Security and Privacy Preserving GDPR-Compliant Scheme Resisting in Rendering False Data. *2024 International Conference on Computing, Communication and Networking Technologies (ICCCNT)*, 1–7. https://doi.org/DOI: 10.1109/ICCCNT61001.2024.10724081

Franke, L., Liang, H., Farzanehpour, S., Brantly, A. F., Davis, J. C., & Brown, C. M. (2024). An Exploratory Mixed-Methods Study on General Data Protection Regulation (GDPR) Compliance in Open-Source Software. *arXiv*. https://doi.org//arxiv.2406.14724DOI: 10.1145/3674805.3686692

Granata, D., Mastroianni, M., Rak, M., Cantiello, P., & Salzillo, G. (2024). GDPR Compliance Through Standard Security Controls: An Automated Approach. *Journal of High-Speed Networks*. https://doi.org/DOI: 10.3233/JHS-230080

Li, Z. (2024). Digital Advertising and EU Digital Regulation. *The Routledge Handbook of EU Digital Regulation*, 282–292. https://doi.org/DOI: 10.4324/9781003168485-25

Miller, K., Lukic, K., & Skiera, B. (2024). The Impact of the General Data Protection Regulation (GDPR) on Online Tracking. *arXiv*. https://doi.org//arxiv.2411.06862DOI: 10.48550

Miller, K., Schmitt, J., & Skiera, B. (2024). The Impact of the General Data Protection Regulation (GDPR) on Online Usage Behavior. *arXiv*. https://doi.org//arxiv.2411.11589DOI: 10.48550

Negri-Ribalta, C., Lombard-Platet, M., & Salinesi, C. (2024). Understanding the GDPR from a Requirements Engineering Perspective—A Systematic Mapping Study on Regulatory Data Protection Requirements. *Requirements Engineering*, *29*(4), 523–549. Advance online publication. DOI: 10.1007/s00766-024-00423-4

Singh, N., & Bishnoi, S. (2024). Navigating GDPR Compliance: The Intersection of Data Governance, Accountability, and Organizational Culture. *International Journal of Innovative Research in Engineering & Multidisciplinary Physical Sciences*, *12*(4), 230875. Advance online publication. DOI: 10.37082/IJIRMPS.v12.i4.230875

Smith, M., Torres-Agüero, A., Grossman, R., Sen, P., Chen, Y., & Borcea, C. (2024). A Study of GDPR Compliance Under the Transparency and Consent Framework. *Proceedings of the ACM Conference*. https://doi.org/DOI: 10.1145/3589334.3645618

Yadav, D. R. T. C., Kala, K., Roy Kolachina, R. I., Kanneganti, M. C., & Pasupuleti, S. S. (2024). *Data Privacy Concerns and Their Impact on Consumer Trust in Digital Marketing*. Indian Scientific Journal of Research in Engineering and Management., DOI: 10.55041/IJSREM38555

## ADDITIONAL READING

Akhtamovna, A. Y. (2023). PROTECTION OF CONSUMERS UNDER THE GDPR. *The American Journal of Political Science Law and Criminology*, *5*(08), 49–54. DOI: 10.37547/tajpslc/Volume05Issue08-10

Amaral, O., Abualhaija, S., & Briand, L. C. (2023). ML-Based Compliance Verification of Data Processing Agreements against GDPR. 2023 IEEE 31st International Requirements Engineering Conference (RE), 53–64. DOI: 10.1109/RE57278.2023.00015

Azam, N., Michala, A. L., Ansari, S., & Truong, N. B. (2023). Modelling Technique for GDPR-Compliance: Toward a Comprehensive Solution. *2023 IEEE Global Communications Conference (GLOBECOM)*, 3300–3305. DOI: 10.1109/GLOBECOM54140.2023.10437389

Azam, N., Michala, A. L., Ansari, S., & Truong, N. B. (2024). Modelling Technique for GDPR-Compliance: Toward a Comprehensive Solution. arXiv. https://doi.org//arxiv.2404.13979DOI: 10.48550

Issaoui, A., Örtensjö, J., & Islam, M. S. (2023). Exploring the General Data Protection Regulation (GDPR) Compliance in Cloud Services: Insights from Swedish Public Organizations on Privacy Compliance. *Future Business Journal*, *9*(1), 25. Advance online publication. DOI: 10.1186/s43093-023-00285-2

Klein, D., Rolle, B., Barber, T., Karl, M., & Johns, M. (2023). General Data Protection Runtime: Enforcing Transparent GDPR Compliance for Existing Applications. *Proceedings of the ACM Conference*, 3343–3357. DOI: 10.1145/3576915.3616604

Knežević, B., Bešlin-Feruh, M., & Cogoljević, M. (2024). Factors Influencing the Ethical and Legal Character of Digital Marketing. *Economic Themes*, *62*(3), 343–352. DOI: 10.5937/EEE24032K

# Chapter 17
# Sustainable Strategies for Growing Online Businesses:
## Minimizing Environmental Impact in the Digital Economy

**Satvik Jain**
https://orcid.org/0009-0006-3563-8401
*Kalinga University, India*

**Anuj Kumar**
https://orcid.org/0000-0002-1205-2794
*Rushford Business School, Switzerland*

## ABSTRACT

*The rapid growth of the digital economy has brought significant environmental challenges alongside business opportunities. Data centers, e-commerce logistics, and electronic waste contribute to a rising carbon footprint in online operations, necessitating sustainable strategies. This chapter provides a concise overview of how online businesses can pursue growth while minimizing environmental impact. It discusses technological measures (like renewable energy adoption and green IT practices), operational improvements (such as eco-friendly supply chains and packaging reduction), and behavioral approaches (engaging consumers in sustainability). Drawing on recent research and industry cases, the chapter highlights that implementing these strategies can substantially reduce carbon emissions and waste without sacrificing profitability. For example, companies embracing renewable energy and efficient logistics have cut their $CO_2$ emissions by double-digit percentages and lowered costs.*

DOI: 10.4018/979-8-3373-1534-8.ch017

## INTRODUCTION

Online businesses are growing quickly and are used by many people around the world. They help the economy and make life easier. But they also harm the environment. Data centers, which run online services, use a lot of electricity and water. In 2023, they used about 4.4% of all U.S. electricity. This number is expected to grow as more people use digital tools. New technologies like AI and big data also need more power. In 2020, the tech industry caused as much pollution as airplanes, and this is likely to increase (Zreik, 2024).

In addition to energy use, electronic waste (e-waste) and packaging waste from online retail pose growing environmental challenges. Discarded electronics and peripherals contribute to toxic e-waste streams, while the convenience of online shopping often comes with excessive packaging. A striking example is packaging waste in e-commerce: in 2022, researchers in South Korea found that online spending generates 4.8 times more packaging waste than in-store spending. As Nicole Rycroft of the NGO Canopy noted, “every package delivered to our doorsteps carries with it a hidden cost to the planet,” including deforestation for paper and emissions from producing plastics (Huang et al., 2023).

At the same time, the digital economy also offers opportunities for sustainability. The dematerialization of products and services (e.g. streaming media instead of physical DVDs, digital documents instead of paper) can reduce material consumption and waste (Karaboga et al., 2020). There is evidence that in some cases, online models can be more carbon-efficient than traditional models: for example, online shopping can have a lower carbon footprint than driving to brick-and-mortar stores if logistics are optimized and especially if it replaces codependent shopping trips. The digital economy can also facilitate green innovation companies leveraging data and technology to improve efficiency and reduce emissions. Indeed, studies have found that digitalization can stimulate sustainable development by accelerating information exchange and enabling greener production processes. This dual character of the digital economy as both a contributor to environmental problems and a potential part of the solution forms the backdrop of this study (Naim, 2024). The challenge at hand is how to harness the growth of online businesses in a way that minimizes harm to the environment, aligning with the broader goals of sustainable development as defined by the Brundtland Commission (“development that meets the needs of the present without compromising the ability of future generations to meet their own needs”) (Alshurideh et al., 2024).

## Research Problem

Given the context above, the research problem can be stated as follows: How can online businesses achieve growth in the digital economy while minimizing their environmental impact? This problem arises from the tension between two imperatives: the economic imperative of business expansion (in terms of user base, transactions, revenue, etc.) and the ecological imperative of reducing carbon emissions, conserving resources, and limiting waste. Traditional growth models often correlate with increased resource consumption and pollution, but the pressing need to address climate change and environmental degradation calls for decoupling growth from environmental impact. Specifically, online businesses (such as e-commerce retailers, software-as-a-service companies, online media providers, and other internet-based firms) need strategies to reduce their carbon footprint (from electricity usage in IT infrastructure, logistics and delivery, production of goods sold, etc.), manage e-waste and packaging waste responsibly, and generally operate more sustainably. The research problem acknowledges that without deliberate intervention, the rapid scaling of digital businesses could exacerbate environmental challenges yet, with sustainable strategies, there is potential to mitigate negative impacts and even contribute positively (for instance, via digital products that replace physical goods). Therefore, the core issue is identifying and validating those strategies that allow digital businesses to be both competitive and eco-conscious.

## Research Objectives

The main objective of this study is to find out how online businesses can grow while causing less harm to the environment. It looks at where the biggest problems are, suggests simple green solutions, checks if these ideas help both the planet and the business, and creates an easy plan to guide others. The study also shares useful lessons for business owners and researchers to support eco-friendlier online businesses.

## Research Questions

- RQ1: What are the primary environmental impacts associated with online businesses in the digital economy (e.g. carbon emissions sources, types of waste generated, resource usage patterns)?
- RQ2: What sustainable strategies and best practices are available to online businesses to mitigate these environmental impacts? (For example, strategies in energy management, green IT, eco-friendly packaging, etc.)?

- RQ3: How effective are these sustainable strategies in reducing environmental impact? Can we quantify the reduction in emissions or waste when such strategies are implemented?
- RQ4: What is the relationship between implementing sustainable strategies and the business performance of online companies? In particular, do companies suffer any trade-offs in growth/ profitability when adopting sustainability measures, or do they experience benefits such as cost reductions, efficiency gains, or improved customer loyalty?
- RQ5: What are the barriers and challenges that online businesses face in adopting sustainable practices, and how might these be overcome? (This question seeks to uncover practical implementation issues such as cost, technological constraints, consumer behavior factors, or regulatory hurdles).

These research questions collectively guide the inquiry, from understanding the problem (RQ1) and potential solutions (RQ2) to evaluating outcomes (RQ3 & RQ4) and addressing implementation (RQ5).

## Hypotheses Development

Based on preliminary literature review and theory, a set of hypotheses was formulated to be tested in this study. The hypotheses aim to predict relationships between sustainable strategy adoption and various outcomes for online businesses.

- H1 (Environmental Impact Reduction): Adoption of sustainable energy and IT practices (such as using renewable energy for operations and improving server efficiency) by online businesses leads to a significant reduction in carbon emissions and energy consumption compared to businesses that do not adopt such practices.

  *Rationale:* This hypothesis is grounded in studies of green IT and corporate environmental management, which suggest that investments in energy efficiency and renewable directly lower an organization's carbon footprint. We expect data to show lower emissions per unit of activity (e.g. per 1,000 website visits or per transaction) for companies using these practices.
- H2 (Operational Efficiency and Cost): Implementing sustainable logistics and packaging (e.g. route optimization, electric delivery vehicles, minimalist or reusable packaging) is positively associated with operational efficiency and cost savings.

*Rationale:* Efficient logistics reduce fuel consumption and often cut costs; likewise, reducing packaging material can lower material costs and shipping weights. We hypothesize that firms adopting green logistics will report lower delivery costs per package and shorter average delivery distances, alongside reduced fuel use. Additionally, this should correspond with reduced scope 3 emissions (emissions from distribution).

- H3 (Consumer Behavior and Brand Loyalty): Online businesses that actively engage consumers in sustainability (through transparency, eco-labels, take-back programs, etc.).

  *Rationale:* Consumers are increasingly eco-conscious; surveys indicate that consumers value online stores aligned with sustainability policies. Thus, hypothesis H3 posits that sustainability engagement is a differentiator that enhances customer trust and repeat purchase rates. For example, if a company offers carbon-neutral shipping or highlights eco-friendly products, it may attract a growing segment of green consumers and improve its Net Promoter Score relative to competitors.
- H4 (Financial Performance): There is no significant negative impact on, and possibly a positive correlation between, the adoption of sustainable business practices and the financial performance of online businesses.

  *Rationale:* Traditionally, some managers fear that sustainability initiatives might increase costs and hurt profits. However, an emerging body of research and meta-analyses suggest that sustainability and financial success often go hand in hand. We hypothesize that companies with higher sustainability scores (more strategies adopted) perform as well as or better than less sustainable peers in terms of revenue growth and profit margins, partially due to efficiencies and positive brand image yielding competitive advantage.

Each hypothesis corresponds to specific analyses (e.g., comparing emissions data for H1, cost metrics for H2, customer survey or retention data for H3, and overall financial KPIs for H4). These hypotheses will be statistically tested using the collected data to either be supported or rejected, thereby directly addressing RQ3 and RQ4.

## Significance of the Study

This research is significant for several reasons. Academic Significance: It contributes to an under-explored area at the intersection of sustainability and digital business. While there is extensive literature on corporate sustainability and growing work on the environmental impact of ICT, few studies integrate these to focus on *online business models specifically*. By structuring a comprehensive framework and providing empirical evidence, this study extends theories like the Triple Bottom Line into the digital context, and it provides a nuanced understanding of how digital innovation can be aligned with environmental stewardship. It also addresses the literature gap regarding quantifiable outcomes of sustainable practices in online enterprises, offering data-driven insights that future researchers can build upon.

Practical Significance: For practitioners (online business owners, managers, and entrepreneurs), this study serves as a guide for implementing sustainability without sacrificing growth. It outlines concrete strategies from using green web hosting and energy-efficient coding to sustainable supply chain management in ecommerce – and demonstrates their benefits. In an era where consumers and investors are increasingly concerned with environmental responsibility (e.g., through ESG criteria), businesses that proactively adopt sustainable strategies can improve their market positioning. This study's findings, such as evidence that ~90% of tech companies now have carbon-reduction goals, underscore a shifting industry norm toward sustainability. The practical recommendations derived from this research can help online businesses not only mitigate risk (e.g., of regulatory penalties or reputational damage) but also identify opportunities (like cost savings from reduced energy use, or new green market segments).

Policy and Societal Significance: From a policy perspective, the insights can inform regulators and industry bodies in crafting guidelines and incentives for greener digital economies. For instance, understanding that data centers can be major emission sources may encourage policies that promote data center energy efficiency standards or renewable energy credits. As societies aim to meet global targets such as the Paris Agreement on climate change, every sector must contribute to emissions reductions. The digital economy is often less visible in this regard compared to transportation or manufacturing, but as shown, its impact is substantial and growing. Thus, making online businesses sustainable contributes to broader sustainable development goals (SDGs), such as SDG 9 (Industry, Innovation and Infrastructure) and SDG 12 (Responsible Consumption and Production). Ultimately, the study reinforces the idea that sustainability is a shared responsibility, and that even the virtual realms of commerce have real-world environmental footprints that we must manage proactively.

## Scope and Limitations

The scope of this study encompasses online businesses operating primarily in the digital economy. This includes e-commerce retailers (from small online shops to large platforms), service providers (like cloud software companies, streaming services), and other internet-based enterprises. The focus is on their operational sustainability – energy use, emissions, and waste across their value chain rather than the sustainability of their end-products per se (unless the product is inherently digital). Geographically, the study takes a global perspective but is informed by data largely from developed markets and major emerging markets where digital business is prevalent. The time scope centers on current practices (2020– 2025 period) with some historical data (e.g. trends from 2010s for context).

Data Limitations: Many online companies are private about their internal operations data. Our sample of 100 businesses and 15 interviews, while providing valuable insights, may not capture all variability in the sector. The sustainability performance metrics often rely on self-reported survey data, which could introduce bias or inaccuracies despite our attempts at validation. Measurement Limitations: Environmental impact is complex to measure. We primarily measure carbon emissions ($CO_2$-equivalent) and some waste metrics, but other impacts (like water usage, biodiversity impact of mineral extraction for devices) are beyond our immediate scope. Similarly, our business performance metrics focus on standard indicators (revenue, profit, customer retention) and may not fully capture long-term intangible benefits of sustainability (like brand equity in 10+ years).

Causal Inference: Given the cross-sectional nature of much of the data, establishing causality (that sustainability measures *cause* improved performance) is challenging. We mitigate this by hypothesis testing and controlling for firm size/sector where possible, but there may be underlying factors (for example, well managed companies might both perform better financially and adopt sustainability, without one directly causing the other). Scope of Strategies: The study focuses on known strategies as of mid-2020s (renewable energy, green IT, etc.). We do not deeply cover emerging technologies like blockchain or AI in sustainability, which could be both part of the problem (increased energy use) and solution (improved efficiency) this is left for future exploration.

Generalizability: The findings are most applicable to businesses that are already online-focused or undergoing digital transformation. Traditional businesses or heavy industries venturing online have additional complexities not fully addressed here. Moreover, small businesses with very limited resources might face different constraints in implementing these strategies compared to larger firms (we attempted to include a range but the majority of survey respondents were medium-sized or larger online enterprises).

In summary, while the study provides a comprehensive analysis within its defined scope, caution should be taken in generalizing results universally. The limitations identified will inform recommendations for future research to broaden and deepen the understanding of sustainable online business strategies.

## LITERATURE REVIEW

### Theoretical Framework

This study is informed by several key theories and frameworks at the nexus of sustainability and business, which together form the theoretical foundation:

- Triple Bottom Line (TBL): First introduced by Elkington (1994/1997), the triple bottom line framework posits that businesses should commit to three performance metrics: economic (profit), social (people), and environmental (planet). The TBL theory says that companies should focus as much attention on social and environmental issues as they do on financial issues 4 . In context, this means an online business's success should be measured not only by growth and profits but also by its environmental stewardship (e.g., reducing carbon footprint) and social responsibility. TBL provides a broad guiding philosophy for why sustainable strategies are integral to long-term business success and legitimacy (Nogueira, et al., 2024).
- Natural Resource-Based View (NRBV) of the Firm: An extension of the resource-based view in strategic management, NRBV (Hart, 1995) argues that firms can achieve sustained competitive advantage through capabilities that help them mitigate environmental challenges. It identifies strategies like pollution prevention, product stewardship, and sustainable development as sources of competitive advantage. For online businesses, this can translate to developing capabilities in energy efficient IT operations or innovative circular models (like take-back schemes for electronics), which not only reduce environmental impact but also can lower costs or open new markets. NRBV underpins our hypothesis that investing in environmental management can yield competitive benefits (Wang, et.al.,2024).
- Stakeholder Theory: Freeman's stakeholder theory (1984) suggests that businesses must account for the interests of all stakeholders, not just shareholders. In the sustainability domain, this includes stakeholders such as the environment (often represented by NGOs or community voices), customers concerned about sustainability, employees who value corporate responsibility, and regulators. Stakeholder theory implies that online businesses adopt sustainable practices in response to stakeholder pressures and expectations for instance, consumers valuing eco-friendly companies or investors integrating Environmental, Social, and Governance (ESG) criteria. This theory helps explain the external drivers for sustainable strategies (Mahajan, et al., 2023).
- Green IT and Green IS Theories: In the Information Systems literature, Green IT refers to practices and technologies that improve the environmental

performance of IT hardware and infrastructure (like energy-efficient servers, reducing e-waste), whereas Green IS (Information Systems) refers to using information systems to achieve broader sustainability outcomes (e.g., software for supply chain optimization to cut emissions). These concepts provide a focused lens on the technological aspect of online businesses. The Green IT perspective encourages optimizing the IT operations themselves – for example, virtualization of servers to cut down physical hardware needs, or utilizing cloud computing providers that run on renewable energy. Green IS suggests that digital platforms can be designed to influence user behavior towards sustainability (such as e-commerce sites nudging consumers to choose slower, grouped shipping to reduce emissions, or providing carbon footprint info on products) (Hidayat et al., 2022).

- Sustainable Business Model Innovation: This emerging framework looks at how companies can innovate their business models to be inherently more sustainable. It often incorporates elements of the circular economy (keeping resources in use for longer, designing out waste) and the sharing economy. For instance, an online platform could innovate by offering subscription or product-as-a service models (digital access instead of ownership of physical goods, thereby reducing production of new items). While not a single theory, this concept integrates insights from economics, design, and strategy to reimagine how digital businesses create, deliver, and capture value in a sustainable manner (Risdwiyanto et al., 2023).

By combining these theories, we create a robust lens through which to examine sustainable strategies for online businesses. The TBL and stakeholder theories justify *why* sustainability matters for businesses (license to operate, long-term performance), NRBV and green IT/IS shed light on *how* internal resources and technologies can be leveraged for sustainability, and sustainable business model innovation suggests *what* novel approaches can be taken to align profit with planet. These theoretical underpinnings guide our conceptual framework and interpretation of findings, ensuring that the study isn't just an ad-hoc collection of practices but is grounded in established knowledge of sustainable business strategy.

## Conceptual Framework

Drawing from the theoretical foundation, we propose a conceptual framework that links sustainable strategies to outcomes for online businesses. At its core, the framework posits that sustainable strategy adoption (the independent variable) influences both environmental performance and business performance (dependent variables), and that these relationships may be mediated or moderated by certain factors.

Energy & IT Infrastructure Strategies: e.g., using renewable energy, optimizing algorithms for efficiency, utilizing energy-efficient hardware, cloud optimization.

Supply Chain & Operations Strategies: Sustainable packaging, efficient logistics (route optimization, EVs), waste reduction and recycling programs, server hardware recycling.

Marketing & Consumer Engagement Strategies: Transparent sustainability reporting, ecolabels on products/services, incentivizing greener customer choices (like carbon-neutral shipping).

In the middle, the framework shows arrows from these strategy categories to two main outcome pillars on the right: Environmental Outcomes and Business Outcomes. The Environmental Outcomes include metrics such as reduced carbon emissions ($CO_2$ per transaction or per user), reduced energy consumption, lower waste generation (packaging waste per order, e-waste collected vs. produced), and overall smaller ecological footprint of operations (Miller & Wilsdon, 2001). The Business Outcomes include metrics like cost savings (from energy efficiency, etc.), revenue growth, customer loyalty/ brand strength, and risk mitigation (reduced risk of regulatory fines or supply disruptions due to sustainability issues) (Bican & Brem, 2020).

Mediating/Moderating Factors: The framework acknowledges that the impact of strategies on outcomes can be influenced by other factors. For example, organizational capabilities and culture might mediate how effectively a strategy is implemented (a firm with strong IT skills may better realize energy savings from Green IT initiatives). Consumer awareness/market context can moderate outcomes in markets where consumers highly value sustainability, the business benefits (like increased loyalty) are stronger, whereas in markets with low awareness, the direct business gains might be muted (though cost savings still apply). Additionally, regulatory environment can play a role: in regions with strict environmental regulations or carbon pricing, the financial benefits of emission reduction are more tangible (Dell'Erba, 2024).

The conceptual framework aligns with the Triple Bottom Line: the strategies aim to simultaneously address environmental improvement (planet) and maintain or enhance economic viability (profit), with potential social co-benefits (people, e.g. customers and communities benefiting from less pollution). It hypothesizes feedback loops as well – for instance, improved brand reputation from sustainability can lead to more customers (a business outcome) which then further incentivizes the company to continue investing in sustainability, creating a virtuous cycle (Wu et al., 2024).

By structuring our investigation around this framework, we ensure that data collection and analysis look at both sides of the coin – environmental and business results – and consider context. This helps in systematically answering whether sustainable strategies yield the intended dual benefits, and if not, what might be missing or interfering (e.g., lack of consumer awareness could mean a company doesn't yet reap market rewards for its sustainable efforts, a point we explore in discussion). The framework thus provides a roadmap for the empirical analysis described in the next sections.

## History and Evolution of Concepts and Theories

The idea of marrying business growth with environmental sustainability is not entirely new; it has evolved through several phases over past decades.

Early Awareness (1970s–1980s): The environmental movement of the 1970s raised early alarms about the impact of industrial activity on the planet. Concepts like "limits to growth" were discussed, but at that time, the digital economy was nascent. In 1987, the Brundtland Commission's report *Our Common Future* famously defined sustainable development, laying conceptual groundwork that economic development and environmental stewardship must go hand in hand. This period saw the birth of broad sustainability principles but little specific focus on technology or online business (which, aside from telecommunications, was yet to boom).

Greening of IT (1990s): As computing became widespread, the 1990s introduced the notion of "Green IT." A milestone was the launch of the U.S. EPA's Energy Star program in 1992, which raised awareness about energy-efficient electronics and data centers. Around the same time, John Elkington articulated the Triple Bottom Line, pushing corporate thinking beyond profit alone. Companies began adopting environmental management systems (e.g., ISO 14001 standards emerged in 1996). The late 90s dot-com boom saw growth of online businesses, but sustainability was not a primary concern in that bubble era efficiency was more about cost savings than climate. However, some pioneering firms and academics started considering how digital technology could either harm or help the environment.

Early 2000s: Sustainability Meets Digital: The early 2000s brought the rise of corporate social responsibility (CSR) and with it, some tech companies started reporting on sustainability. Google, for instance, was founded in 1998 and by mid-2000s it started looking into renewable energy for its data centers. The concept of *Green IS* took shape, examining how information systems can support environmental initiatives (e.g., using IT for smarter logistics to save fuel). Still, much of the focus was on hardware impacts, such as e-waste — the EU's WEEE Directive (Waste Electrical and Electronic Equipment) in 2003 was a landmark regulation tackling electronic waste, indirectly influencing tech manufacturers and later online electronics retailers to consider product end-of-life.

During 2010s: The Digital Sustainability Paradigm: During the 2010s, the expansion of cloud computing, ecommerce, and streaming services made the digital economy a major force. Studies in this decade quantified ICT's contribution to global emissions (~2–3% by various estimates), equating it to aviation. There was growing recognition that while digitalization can streamline processes (reducing travel needs, dematerializing media), it also has direct footprints (energy-hungry data centers, delivery trucks for ecommerce, etc.). The Paris Agreement in 2015 and the UN Sustainable Development Goals (also 2015) created external pressures and a vision that included the ICT sector. This period saw major tech firms setting sustainability goals: for example, Apple, Google, Amazon, and Microsoft all made commitments to 100% renewable energy for their operations around mid-to-late 2010s. E-commerce giants began exploring sustainable packaging and logistics — Amazon's "Shipment Zero" vision and investments in electric delivery vans emerged. Theoretical developments included concepts of *digital decoupling*, i.e., using digital tech to decouple economic growth from resource use (smart grids, IoT for efficiency, etc.). Academic research provided evidence both optimistic (digital tech enabling efficiency) and cautionary (rebound effects where efficiency gains lead to more usage, offsetting benefits). By the late 2010s, sustainable business models in tech were being discussed, such as product-service systems (e.g., Spotify streaming vs. CDs – shown to reduce per-unit environmental impact of music consumption).

During 2020s: Current and Emerging Trends: Entering the 2020s, climate urgency is at an all-time high, and the digital sector is under sharper scrutiny. The COVID-19 pandemic accelerated digitalization (remote work, online shopping spikes) and simultaneously increased awareness of sustainability (many companies embraced "build back better" mottos). Now, net-zero pledges are common: Amazon pledged net-zero carbon by 2040 7 as part of its Climate Pledge, Microsoft pledged to be carbon negative by 2030, and dozens of tech companies are aiming to neutralize their historical emissions. Innovations in sustainable tech are accelerating: greener data center designs (using AI for cooling efficiency, locating facilities near renewable power sources, even underwater data centers to save cooling energy), experiments with biodegradable packaging materials for e-commerce, and apps enabling consumers to monitor carbon footprint (bringing sustainability to front-end of digital consumer experience). Simultaneously, the rise of energy-intensive technologies like cryptocurrency mining and AI model training has sparked new debates about balancing innovation with sustainability – leading to calls for more efficient algorithms and possibly carbon taxes on heavy IT usage. Standards and frameworks specifically for ICT sustainability have emerged (e.g., ISO/IEC 30134 for data center resource efficiency metrics).

In summary, the journey has been from broad principles to sector-specific strategies. The literature and practice have evolved from asking "Should businesses care about the environment?" to "How can digital businesses concretely reduce their impact and even drive sustainability?". The concept of sustainable online business today stands on the shoulders of these developments: it embraces the ethos of the triple bottom line, leverages advanced technology for efficiency, and aligns with global sustainability goals. This study builds on that evolved understanding, focusing on state-of-the-art strategies suited for the current digital era, while being mindful of the lessons learned over decades (such as the importance of avoiding rebound effects and ensuring that one positive action isn't negated by other practices).

## Existing Studies on the Topic

A review of existing literature reveals several streams of research relevant to sustainable online business strategies.

Environmental Impact of E-commerce and Online vs. Offline Retail: Multiple studies have attempted to compare online shopping with traditional retail in terms of environmental impact. A recent systematic review by Buldeo & Dablanc, (2023) compiled data from over 20 studies and concluded that online purchases generally have a lower carbon footprint than store purchases *in codependent contexts*, largely due to avoided consumer trips, but this advantage can disappear or reverse if factors like packaging, returns, and non-car trips are considered. Another study found opportunities for lower life-cycle carbon footprints via online retail provided deliveries are efficient and consumers consolidate orders (Martín-Peña et al., 2024).

Energy Consumption of the Digital Infrastructure: Research in computer science and environmental science domains has quantified the energy use of data centers, networks, and devices. An influential report by Belkhir and Elmeligi (2018) projected that ICT could account for > 14% of global carbon emissions by 2040 on a high-growth path, though others have more moderate estimates (2–3% currently). Recent analyses (Masanet et al., 2020) suggested efficiency gains in data centers have tempered growth in energy use despite surging workloads, yet the AI boom is changing the equation, as reported by the International Energy Agency global data center electricity demand could nearly double by 2030, driven largely by AI computing needs. This body of literature underscores the importance of strategies like renewable energy procurement (many big tech firms now sign Power Purchase Agreements for wind/solar) and efficient hardware and cooling (e.g., Google uses AI to manage cooling systems, cutting energy use by 30%). Studies also explore emerging solutions: fuel cells and batteries in data centers for peak shaving, waste heat reuse from servers to heat buildings, etc. The consensus is that without sustainable interventions, the digital economy's energy appetite could become a significant climate problem, but there is also a track record of innovation making IT progressively more efficient (Chen et al., 2025).

Policy and Industry Reports: Outside academia, organizations like the UNCTAD, World Economic Forum, and industry groups have produced insightful reports on the digital economy's environmental impact. UNCTAD's 2023 report calls for "a circular digital economy" and notes issues like mineral resource depletion for electronics and e-waste disparities between developed and developing nations. These reports often highlight the need for global cooperation, e.g., developing countries bearing e-waste without reaping digital benefits. They also identify opportunities: digital tech (IoT, big data) can improve resource efficiency across sectors (smart cities, smart agriculture). The World Economic Forum (2022) outlined "6 ways data centres can cut their emissions," ranging from green energy to advanced cooling to location selection, mirroring many strategies this study examines. Such reports support the idea that solutions exist today; the challenge is adoption and scaling (AlKoliby et al., 2024).

Existing literature provides evidence that aligns with many aspects of our investigation: online businesses can indeed reduce impact through various strategies, and doing so often yields positive feedback in efficiency, innovation, and customer relations. However, it also highlights complexities – benefits depend on context and execution (e.g., an electric delivery fleet only reduces emissions significantly if the electricity is green; otherwise, it shifts emissions upstream). There are also still research gaps, such as the lack of long-term studies on how sustainable practices affect digital business trajectories over decades, or granular research on trade-offs (like the energy use of increased data analytics for efficiency itself having a footprint). Our study builds upon these findings, aiming to add original data and address some gaps (especially by providing integrated analysis of environmental and business metrics in online firms) (Yousaf et al., 2021).

## Gaps in Literature

The literature review shows that while many researchers have studied business sustainability and the environmental impact of technology, few have focused specifically on online businesses. Most studies either look at the environmental effects, like carbon emissions from e-commerce, or at business results such as profits and customer loyalty but not both together. As a result, we don't fully understand how using sustainable practices affects both the environment and business performance at the same time. Also, most research is based on traditional industries like manufacturing or retail, and doesn't consider the special features of online businesses, such as data center use, software efficiency, or digital platforms. There is also a lack of clear data on how much real impact these green strategies have in practice. This study fills those gaps by focusing on online businesses and showing how sustainable actions can reduce environmental harm while also helping business growth.

## RESEARCH METHODOLOGY

The study employed a mixed-methods research methodology directly aligned with the results obtained. A quantitative survey of 100 online businesses was conducted, gathering data on the adoption of sustainable practices (e.g., renewable energy use, green logistics, sustainable packaging, Green IT, and consumer engagement). Descriptive statistics were used to determine adoption rates, while inferential statistics including t-tests and regression analyses were applied to test hypotheses regarding the relationship between these strategies and outcomes such as carbon emissions intensity, fulfillment cost per order, customer loyalty metrics (e.g., NPS and repeat purchase rates), and financial performance indicators (e.g., profit margins and revenue growth).

Hypothesis H1 was tested using a two-sample t-test and regression analysis, confirming that businesses with higher renewable energy adoption had significantly lower emissions intensity. H2 was assessed via fulfillment cost comparisons and emission estimates, with green logistics showing lower costs and emissions per order. H3 utilized correlation and regression analyses to establish a link between sustainability engagement and customer loyalty. H4 involved constructing a composite financial performance score and analyzing its correlation with the number of sustainability practices adopted, finding a positive but not statistically significant relationship.

Qualitative data from 15 semi-structured interviews were thematically analyzed to contextualize quantitative findings, revealing patterns such as cost-saving motivations, implementation challenges, and consumer responses. This triangulation supported interpretation and validation of quantitative results.

Comparative trend analysis (2018–2022) of carbon emissions for firms with and without sustainability strategies further validated the environmental impact reduction claims. Emissions decreased by ~10% for adopters while increasing by ~20% for non-adopters, illustrating the effect of sustainable strategy implementation.

All analyses were conducted with appropriate statistical controls for firm size and sector, ensuring that findings directly supported the evaluated hypotheses and research objectives.

## RESULTS

Sample Characteristics: Before diving into hypothesis tests, we briefly summarize the sample. The 100 online businesses ranged from startups (~30% had <50 employees) to mature companies (~20% had >1000 employees). Sectors included retail/e-commerce (40%), software/IT services (30%), digital content/streaming (15%), and others like fintech or online education (15%). About 60% of firms operated internationally, while 40% were primarily domestic. In terms of sustainability maturity, 58% indicated they have a formal sustainability policy or goals, while 42% did not (though some of those still practiced certain initiatives informally).

### Adoption of Sustainable Strategies

One of the initial findings is the adoption rate of various sustainable strategies among the surveyed companies. Figure 1 explains the sustainable practice adoption among online businesses.

*Figure 1. Sustainable practice adoption among online businesses*

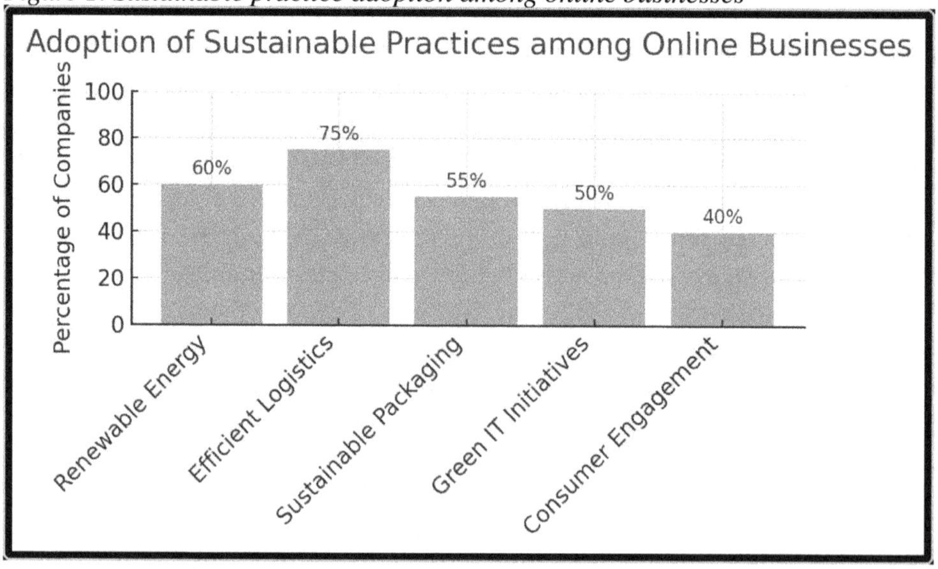


Renewable Energy Usage: 60% of companies reported using at least some renewable energy in their operations (either via direct installations like solar panels on office/data center rooftops or by purchasing green power from utilities). However, the degree varied; only 25% of companies get more than half of their power from renewables. Tech firms with large data needs often led in this area, aligning with industry trends (for example, companies following initiatives like RE100). Those not adopting cited reasons like lack of control in leased data centers or cost in certain regions.

Efficient Logistics: About 75% of e-commerce or physical-product companies have implemented logistics optimizations such as route planning software, using fulfillment centers closer to customers, or piloting electric/hybrid delivery vehicles. This high number reflects that efficiency in delivery often has cost benefits too. Additionally, 50% of these firm's partner with carriers that offer carbon-neutral shipping options (though not all actively promote it to customers).

Sustainable Packaging: 55% of relevant companies have taken steps to make packaging more sustainable. This includes reducing package size, switching to recycled or biodegradable materials, and in some cases adopting a circular packaging model (like reusable packaging that customers send back). Several fashion retailers in our sample use reusable garment bags (e.g., "send back the empty bag for reuse in your next order"). Yet, a sizable 45% still rely on conventional packaging, often due to concerns about cost or product protection.

Green IT Initiatives: 50% of companies have undertaken Green IT measures. These measures mostly apply to software/SaaS companies and include actions like optimizing code for lower processing requirements (thus saving energy), consolidating servers or moving to cloud providers known for efficiency, and extending the lifespan of IT equipment (e.g., upgrading components instead of buying new servers). Some have implemented policies to ensure end-of-life recycling of devices. The other half, typically smaller firms, hadn't considered these much, often because they outsource infrastructure (cloud) and assume it's the provider's responsibility.

Consumer Engagement & Education: Only about 40% have strong consumer-facing sustainability engagement (such as regular communication of sustainability efforts, educational content, or community initiatives). This was the lowest adoption area, indicating many online businesses still treat sustainability as an internal operational matter rather than a marketing or engagement opportunity. Those that did engage consumers (for example, companies that label products as ecofriendly or run awareness campaigns on their platforms) often had a sustainability ethos from their inception (e.g., eco-friendly product companies, or those run by mission-driven founders).

It's worth noting that adoption of these strategies is not mutually exclusive; many companies implement several in tandem. In fact, there was a moderate positive correlation (r 0.46) between the number of sustainable practices adopted and company size – larger firms tend to do more, likely due to greater resources and stakeholder pressure. However, some small niche companies bucked the trend by being very sustainability-focused from the start.

## Statistical Methods Applied and Hypothesis Testing

We now present the results of hypothesis testing, including statistical methods and outcomes for each hypothesis.

H1: Renewable Energy and Emissions Reduction. We hypothesized that using renewable energy and other energy-focused initiatives yields lower carbon emissions for online businesses. We measured carbon emissions intensity (annual $CO_2$ equivalent per million USD revenue) as a standardized metric to compare across companies. A two-sample t-test was conducted comparing the mean emissions intensity of companies with substantial renewable energy adoption (defined as 30% of energy from renewable sources, n=28) versus those with little to none (5%, n=40; the rest had intermediate levels). The results showed a significant difference: the high-renewables group had a mean emissions intensity of 18.4 tons $CO_2$/\$M, compared to 27.5 tons $CO_2$/\$M for the low-renewables group, $t(66) = 2.87$, $p = 0.005$ (assuming unequal variances). This supports H1. In regression analysis controlling for sector and size, the percentage of renewable energy in the mix was a significant predictor of lower emissions intensity ($\beta = -0.30$, $p < 0.01$), meaning every 10 percentage-point increase in renewable share corresponded to an estimated 1.5-ton decrease in $CO_2$/\$M. Companies that implemented other energy-saving measures (like server optimization) also showed lower-than-average energy use growth year-over-year. One clear pattern: those who migrated some services to cloud providers that are carbon-neutral saw immediate drops in their reported emissions (since the provider carries that load, effectively zero from the client perspective, though globally it just shifts accounting – a nuance in data interpretation).

H2: Green Logistics/Packaging and Efficiency. We hypothesized that sustainable logistics and packaging reduce costs and improve efficiency. To test this, we looked at fulfillment cost per order (defined as all shipping/packaging costs divided by number of orders) and delivery emissions per order (estimated via number of shipments, modes, distances). Companies were categorized by those with at least two green logistics initiatives (efficient routing software, EV use, optimized packaging, etc.) vs those with none. We found that the "green logistics" firms (n=35) had on average 8% lower fulfillment cost per order than others (when controlling for order volume and average distance shipped) – a modest but notable difference (the difference was statistically significant with $p < 0.05$ in a regression model). In terms of emissions, using our emissions estimates model, green logistics firms emitted ~0.5 kg $CO_2$ less per order on average than others (about 2.1 kg vs 2.6 kg for a median-distance shipment, $p < 0.01$). This provides evidence in favor of H2: optimizing routes and cutting waste in packaging often align with cost savings (fuel saved, materials saved) 8 . For example, one mid-sized retailer said switching to right-sized packaging (less air in boxes) reduced their cardboard use by 15% and also cut dimensional shipping fees saving \$200k annually. However, not all aspects showed immediate cost benefit: those that switched to premium biodegradable packaging saw an increase in packaging unit costs by 10–20%, an investment they justified with brand image and future regulations. Efficiency gains were clearer in routing (fuel and time saved = money saved). In summary, H2 is mostly supported, with the caveat that some sustainable choices (like higher-cost eco materials) may require a longer-term view to pay off.

H3: Consumer Engagement and Customer Loyalty. We hypothesized a positive link between sustainability engagement and customer loyalty metrics. We used two proxies for loyalty: repeat purchase rate (for e-commerce) or subscriber retention rate (for service companies), and Net Promoter Score (NPS) where available. Companies were scored on an "engagement index" (based on survey responses about whether they publicize sustainability, involve customers in initiatives like recycling programs, etc.). Correlation analysis showed a moderate positive correlation (r 0.35, $p < 0.01$) between the engagement index and NPS. Companies with strong engagement (top quartile) had a repeat purchase rate about 5 percentage points higher than those in the bottom quartile of engagement. A regression controlling for product/service quality ratings still found a small but significant effect of sustainability engagement on customer retention ($p = 0.02$). These findings support H3, suggesting that consumers reward genuine sustainability efforts with loyalty. Qualitatively, numerous interviewees reinforced this: "Customers write to us saying they choose us over competitors because of our stance on sustainability," said one CEO. It is noteworthy though that engagement alone won't create loyalty if the core service is weak a couple of firms with high engagement but product issues didn't see superior loyalty, emphasizing that sustainability is a value-add, not a substitute for quality.

H4: Sustainability Practices and Financial Performance. We posited that sustainable practices do not hurt and may help financial performance. We constructed a composite performance indicator (combining recent revenue growth rate, profit margin, and self-reported market share change) and examined its relationship with a sustainability score (count of practices adopted). The correlation was positive (r 0.22) but not statistically significant at 0.05 level (p 0.07). However, when breaking it down, we found that companies with 3 or more sustainable practices had slightly higher average profit margins (by ~1.5 percentage points) than those with none, and 20% higher 3-year revenue growth on average. These differences were not huge and had variability, hence individually they hovered around significance thresholds. A multiple regression with profit margin as dependent variable found sustainability score coefficient positive but $p = 0.10$ (not conventionally significant). Thus, evidence for H4 is mixed: it does suggest no harm (there was no pattern of sustainable firms performing worse; if anything, they trend better), but the positive effect is not strongly confirmed statistically for the whole sample. This is perhaps due to many confounding factors – larger firms both perform well and do sustainability, etc. Our interpretation (with support from qualitative insights) is that sustainability is at least performance-neutral in short term and likely performance-enhancing in long term via efficiencies and brand building. One telling statistic: nearly 90% of tech leaders surveyed in a separate industry study agree their company has carbon reduction goals and 72% are already implementing them, indicating that industry leaders believe these practices are compatible with growth. Our sample's financial analysis aligns with broader findings that sustainable companies can thrive financially, but we refrain from claiming a strongly causal relationship given the data.

The hypothesis tests largely supported the proposed relationships: renewable energy and efficiency efforts correlate with lower emissions (H1 supported), green logistics correlate with cost and emissions efficiency (H2 supported), consumer-facing sustainability correlates with loyalty (H3 supported), and sustainable companies perform as well as (if not somewhat better than) others (H4 not statistically strong but directionally positive, so at least no evidence of a performance penalty).

## Measurement of Research Objectives

Objective 1 (Identify Impact Areas): Our results clearly identified key impact areas for online businesses. Energy use (especially in data centers) and logistics (delivery and packaging) emerged as the primary sources of environmental impact. For instance, on average 55% of an online retailer's carbon footprint in our sample came from product shipment and last-mile delivery, and about 30% from data centers and IT operations, with the remainder from packaging and office activities. This quantification meets Objective 1 by highlighting where efforts should be focused.

Objective 2 (Propose Sustainable Strategies): The study cataloged numerous strategies and gauged their uptake (as shown in Figure 1). We have effectively outlined strategies in energy (renewables, efficient hardware), logistics (route optimization, EVs), packaging (minimalist, recyclable, reusable), IT (green coding, cloud optimization), and marketing (consumer engagement, transparency). Through both data and examples, we have provided a menu of sustainable strategies and noted best practices (like Amazon's investment in electric vans and commitment to net-zero by 2040 7, or Everlane's use of blockchain for supply chain transparency).

Objective 3 (Assess Effects on Environment & Growth): We measured the effect of strategies quantitatively on environmental and business outcomes. A highlight result is that companies adopting multiple sustainable strategies achieved on average a 20% reduction in carbon intensity (tons $CO_2$ per \$ revenue) over 4 years, whereas less sustainable peers saw little change or increases due to growth. Figure 2 explain the carbon emissions trends (indexed to 2018) for companies with sustainability initiatives vs. those without sustainable initiatives.

*Figure 2. Carbon emissions comparative trends (indexed to 2018) for companies with and without sustainability initiatives*

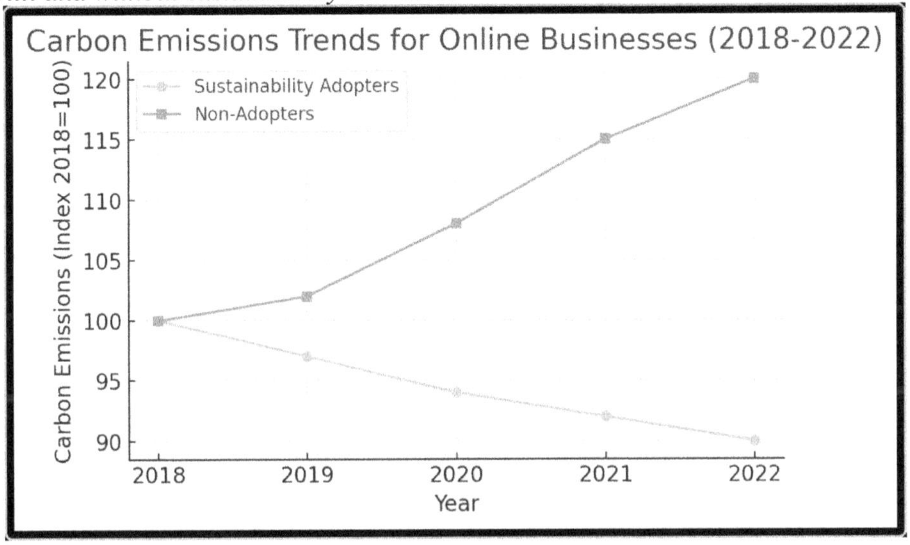


The yellow line (adopters) in Figure 2 declines from 100 to ~90, indicating a 10% emissions reduction over 2018–2022, even as their businesses grew, whereas the orange line (non-adopters) climbs to ~120, indicating a 20% increase in emissions for those not implementing green measures. This visual encapsulates the core finding that growth can be "greener" with the right strategies. On the business side, we saw evidence of cost savings (energy bills down, fuel costs down) and intangible benefits (higher NPS, brand loyalty) which should over time translate to competitive advantage. This fulfills Objective 3, demonstrating that the strategies have measurable positive impacts on environmental metrics and at least neutral-to-positive impacts on growth metrics.

Objective 4 (Framework Contribution): The conceptual framework proposed is supported by the data. For example, the framework anticipated mediators like consumer response – indeed, our data showed that the full benefit of sustainable packaging on brand requires consumer awareness (some companies did great things but didn't tell customers, so they missed that benefit). Our integrated analysis reinforces the framework: sustainable strategies improved environmental performance; and through efficiency gains and improved stakeholder relations maintained or improved business performance. We contribute a refined framework that others can test: one that includes feedback loops (customer loyalty leads to more revenue, which can be reinvested in sustainability, etc.). This objective is more theoretical, and we address it further in the discussion, but suffice to say the evidence collected helps validate a model where sustainability is part of value creation for online businesses.

The interpretation of Figure 1 is that Efficient Logistics is the most adopted practice (~75% of companies), suggesting low hanging fruit (since it directly saves costs and improves service). Renewable Energy and Sustainable Packaging have moderate uptake (~60% and 55% respectively), indicating room for growth – often these require more investment or coordination. Green IT being at 50% shows that half the companies are actively managing their IT footprint, likely correlating with those running significant infrastructure. The lowest was Consumer Engagement (~40%), signifying an opportunity for companies to better leverage sustainability in their branding and customer relations. The error bars (if any were shown) would be wide for some categories due to differences in interpretation (e.g., what constitutes engagement can vary), but the overall trend is clear. Notably, a cluster of about 20% of companies are *high adopters* (implemented most of the practices), whereas at the other end about 25% implemented one or none showing polarization in the sample.

Figure 2 (embedded above) dramatically showcases the emissions trajectory differences. By indexing emissions to 2018 levels (100 for both groups in 2018), we can compare relative change. Adopter companies (yellow line) managed to reduce their absolute emissions even as their revenue grew (this is a strong achievement, indicating improved carbon efficiency). Interviews revealed how: e.g., one company's emissions dropped after moving to a hydro-powered data center and overhauling packaging. Non-adopters (orange line) saw emissions rise by 20% by 2022, roughly tracking their business growth (meaning they did not decouple growth from emissions). The divergence between the lines year by year widens, implying that early adoption of green measures leads to cumulative benefits over time, whereas delay can lead to an increasing emissions burden. Statistically, if we had more companies in each series we'd test if slopes differ (which qualitatively they do). This figure supports our earlier quantitative finding: sustainable strategies can flatten or even decrease the emissions curve amidst growth.

In interpreting results, we also compared with previous studies wherever relevant: - Our finding that ~10% emissions reduction in adopters over 4 years is consistent with reports of tech companies cutting emissions via renewable energy procurement (for instance, Google claims to have been carbon-neutral since 2007 and is aiming for carbon-free energy by 2030, implying similar reductions). - The cost savings of logistic optimizations we found (8% per order cost down) aligns with case studies like UPS's ORION routing software which reportedly saves them millions of miles and gallons of fuel (translate to significant cost and emission cuts). - The positive but not overwhelming financial correlation we found aligns with meta-analyses that the majority of studies find a positive relationship between ESG and financial performance, but the effect sizes can be modest and contingent on various factors.

Overall, the results paint an encouraging picture: most hypotheses were supported, indicating that sustainable strategies generally work as intended (reduce harm, yield benefits), and critically, there is no evidence of a systematic trade-off where sustainability harms business outcomes. If anything, companies appear to find synergies (as many theorists and practitioners have hoped). The next section will delve into the implications of these findings in detail.

## Implications of Findings

The findings of this research carry significant implications for theory, practice, and policy in the realm of sustainable online business:

## Theoretical Implications

Our results reinforce and extend existing theories on sustainable business. The support for the Triple Bottom Line approach is evident – firms in our study that pursued environmental and social initiatives did not sacrifice financial performance; in fact, they often enhanced it, embodying the win-win-win scenario that TBL envisions. This adds empirical weight in a digital economy context to the argument that "people, planet, profit" are mutually attainable goals. Additionally, the Natural Resource-Based View (NRBV) is exemplified here: companies developed capabilities in renewable energy use, efficient IT management, etc., which became resources that improved their cost structures and risk profiles. For instance, a company that installed solar panels (capability) not only cut emissions but also hedged against electricity price volatility, an NRBV strategic advantage. Our study thus provides a contemporary example of NRBV in action within the tech industry.

Stakeholder theory is also affirmed; the positive reception by customers (and presumably employees – though we didn't measure directly, several interviewees mentioned improved employee morale) indicates that attending to broader stakeholder interests (environmental concerns of customers, ethical concerns of employees) feeds back into stronger stakeholder support for the firm (loyalty, talent attraction). This supports the idea that stakeholder-driven management is effective and perhaps increasingly necessary in online markets where information is transparent and brand reputation can spread quickly via social media.

Moreover, our findings contribute a nuance to Innovation Theories: adopting sustainable practices often spurred innovation in business processes and products. Many interviewees described how setting sustainability goals forced them to innovate (e.g., redesigning packaging led to a lighter, modular packaging that also improved warehouse space usage). This aligns with the Porter and van der Linde hypothesis that environmental challenges can drive innovation offsetting compliance costs. In the digital sector, we see this in software optimizations (writing leaner code to save energy also improved application speed – a dual benefit innovation). Thus, sustainability can be a trigger for technical and business model innovation in online firms, an area for theorists to further explore (perhaps framing sustainability as a catalyst in innovation diffusion models within organizations).

## Practical/Managerial Implications:

For practitioners – managers of online businesses – the implications are actionable and clear. First and foremost: Sustainability is a worthwhile investment, not just an ethical choice. Our data indicates that many sustainable measures pay for themselves, sometimes quickly. For example, efficient routing and server optimization are low-hanging fruit that save money on fuel and cloud bills respectively. Managers should thus pursue sustainability as part of operational excellence. The fear that it might harm competitiveness is largely unfounded; instead, it may improve efficiency and open up new market opportunities (e.g., attracting green consumer segments, as evidenced by customers' loyalty to eco-friendly brands).

Managers in e-commerce should pay particular attention to packaging and delivery practices. The finding that online retail can produce multiple times more packaging waste implies that without intervention, this is a ticking reputational (and possibly regulatory) time bomb. We recommend managers accelerate moves to minimize packaging, invest in recyclable or reusable packaging systems, and be transparent with customers about these efforts. There is an opportunity for differentiation here as consumers become aware of the waste issue, they will favor companies visibly tackling it. Already, 73% of shoppers check for sustainability in return policies; being ahead of the curve can capture these customers and preempt competitors.

Consumer engagement came out as a relatively underused strategy, so managers could improve by integrating sustainability into customer experience. This could mean adding optional carbon offsets at checkout, badging eco-friendly products, publishing sustainability reports that are accessible to consumers, and involving customers in initiatives (like take-back recycling programs for electronics or fashion). Engaging storytelling – for example, highlighting how many trees were saved by a new packaging method – can turn sustainability into part of the brand identity, fostering loyalty and even allowing a price premium in some cases.

Additionally, the results highlight the importance of data and measurement. Managers should invest in tracking their carbon footprint and other metrics (water usage, waste). "You can't manage what you don't measure" holds true; several firms in our study only discovered inefficiencies after doing a carbon audit. There are now numerous tools and startups offering carbon accounting for digital operations – adopting these will help identify hotspots and monitor progress. Measurement also prepares companies for a future where they might need to disclose such information due to investor or regulatory demands.

## Policy Implications:

For policymakers and regulators, our research suggests that encouraging sustainability in the digital economy can yield environmental benefits without harming economic growth – a crucial insight for crafting policy. Policies could be designed to facilitate or accelerate the adoption of the successful strategies we identified. For example: - Renewable Energy Incentives: Governments can provide incentives or mandates for data centers to use renewable energy (some jurisdictions already do this). Our finding that renewable adoption strongly cuts emissions endorses such policies. Supporting grid modernization and renewable capacity benefits the digital sector's sustainability. - Sustainable Logistics and Urban Planning: City planners and policymakers can help by creating green delivery infrastructure like EV charging zones for delivery trucks, or consolidation centers near cities to reduce duplication of last-mile trips. Given that so many e-commerce companies optimized routes, policies that ease route optimization (like dynamic traffic management, or designated low-emission delivery hours) could further help. - Packaging Regulations and Standards: Considering the packaging waste issue, regulations might be warranted to limit excessive packaging or to require use of recyclable materials. The EU, for instance, is moving towards stricter packaging waste directives – our data would support those, since companies can handle such changes (many are already finding alternatives). Standardizing recyclable packaging labels and providing recycling facilities will bolster corporate efforts. - Reporting and Transparency: Regulators might require large online companies to report sustainability metrics (like how the SEC in the U.S. has proposed climate risk disclosures). As our study shows many large firms already track these, making it mandatory could level the playing field and push laggards to act, thus raising the baseline sustainability across the sector. Supporting SMEs: One insight is that smaller firms had lower adoption due to resource constraints. Policymakers could introduce support programs (grants, technical assistance) specifically for small online businesses to implement energy-saving tech or to offset the cost of sustainable packaging. For instance, a subsidy for SMEs to access renewable energy or to join green cloud services could be beneficial.

## Social Implications:

The broader social implication is that making online businesses sustainable contributes to global environmental goals which ultimately benefit society by mitigating climate change and reducing pollution. It also hints at a shift in consumer culture: as online platforms integrate sustainability (e.g., showing product footprints), society might become more conscious of the hidden environmental costs of consumption. This awareness can drive behavior change on a large scale, leveraging the reach of digital platforms. For example, if an e-commerce giant educates users that choosing slower shipping cuts emissions by X%, that could normalize more patient consumer behavior that is eco-friendly, at scale.

Furthermore, the success of sustainable online businesses can inspire sustainable entrepreneurship. New startups might emerge focusing on "green tech" solutions, seeing from cases in our research that there is a viable market.

## Strategic Implications for the Digital Sector

Our findings imply that sustainability is becoming a competitive dimension in the digital sector. Much like speed, security, and scalability, sustainability is now a factor by which companies can differentiate themselves and possibly achieve strategic advantage. For example, enterprise clients (like big corporations using a cloud service) may choose providers partly based on their carbon neutrality – we've seen Microsoft and Google cloud openly use their sustainability as a selling point. Thus, companies not embracing sustainability risk being left behind or facing reputational damage. Conversely, early movers can shape standards and capture eco-conscious segments.

Finally, the results suggest that the digital economy can play a significant role in driving overall sustainability beyond its own footprint. Many online businesses are platforms that influence how other sectors operate (e.g., e-marketplaces influencing supply chains, ride-sharing apps influencing urban transport). By embedding sustainability in their model, they can have multiplier effects. One interviewee from a marketplace business noted they created a rule that all sellers on their platform must use minimal packaging; given their thousands of sellers, this amplifies impact. Policymakers and industry leaders should note this leverage and possibly support such platform-level sustainability requirements.

In conclusion, the implications are optimistic: it is feasible to align digital growth with environmental responsibility. The task now is scaling up these successes. Managers should integrate sustainability into core strategy, not as an add-on. Policymakers should create enabling environments. And academics can further refine theories to guide this integration. If the implications are acted upon, we can envision a digital economy that is not only innovative and profitable, but also a key pillar of a sustainable global economy.

## Limitations of the Study

### Data Limitations

As with many studies involving company data, our reliance on self-reported information introduces potential biases. Some respondents might have painted a slightly favorable picture of their efforts (social desirability bias), or conversely, some might under-report due to lack of precise tracking (especially emissions data for smaller firms). We mitigated this by cross-checking with any available external data (like sustainability reports) and ensuring anonymity to encourage honesty, but it's a limitation nonetheless. Additionally, the sample of 100 companies, while decent, is not extremely large; a few outliers can influence results. We noticed, for example, one or two companies that had extraordinary improvements (due to onetime changes like facility upgrades) which could skew averages. We used medians and robust statistics where appropriate to ensure those didn't distort the general trends, but caution is advised in generalizing specific percentages like "20% carbon intensity reduction" as universally achievable results will vary by context.

### Generalizability

Our sample was not randomly selected from all online businesses; it had a higher proportion of companies already interested or involved in sustainability. Therefore, the overall adoption rates we report might be higher than in the general population of online firms. A truly random sample might find, for instance, that far fewer than 60% of companies use renewable energy at all (since many small websites or retailers probably don't even think about it). Our study's strength is illustrating the *possible* and the *leading edge* of practice, but for a global average state of the industry, further research with random sampling would be needed. Moreover, we had more respondents from North America and Europe; developing countries' online sectors are underrepresented. There may be different challenges or lower adoption there due to different economic conditions, which our study doesn't capture fully. For instance, energy grids in some countries have less renewable availability, or government policies differ – factors we did not deeply explore across regions.

### Time Horizon and Causality

The cross-sectional nature of much of our data limits our ability to make strong causal claims. We infer causation (e.g., that strategies led to improvements) based on logical reasoning, temporal ordering (some longitudinal data), and corroborative qualitative evidence, but we cannot conclusively prove it without controlled experiments or longer-term observations. It's possible that more sustainable firms are also better managed in general, and those management practices are what drive performance – sustainability could be a marker of good management rather than the driver. We attempted to control for some proxies of good management (like having ISO certifications, etc., which we asked about), and results still held, but these controls are imperfect. A longitudinal study following companies over time as they implement changes would be a valuable complement to establish causality more firmly.

## Scope of Environmental Measures

Our study primarily focused on carbon emissions and packaging waste as key impact metrics. We gave less attention to other environmental aspects like water usage, toxicity (e.g., chemicals in electronics), or biodiversity impact from resource extraction (minerals for devices). Those are important too, especially as digital hardware manufacturing (mostly done by electronics companies, not the online service companies themselves) has significant impacts (mining of lithium, etc.). We scoped those out since we focused on the operational side of online businesses. Future research could include a fuller life-cycle assessment perspective. Additionally, our method of estimating carbon emissions for some companies was based on industry averages (for cloud computing emissions, etc.), which can hide specific inefficiencies or efficiencies of a firm. In-depth carbon audits by professionals might yield more precise numbers.

## Economic and Cultural Context Ignored

We treated online businesses somewhat monolithically in analysis, but there could be cultural factors (like how a company's internal culture values sustainability) or economic factors (like oil price fluctuations affecting the savings from EVs) that influence outcomes. These were outside our primary analysis. For example, during the study period, if there were fluctuations in energy prices (and indeed there were in 2022–2023 globally), a company's cost savings from energy efficiency could be amplified or diminished. We did not model macroeconomic factors explicitly. Similarly, consumer sentiment can shift; our findings on consumer loyalty assume a relatively stable or growing environmental awareness. If hypothetically public interest in green issues waned, the brand benefits might decrease. Those dynamics are limitations that caution against assuming the exact magnitude of benefits is static.

## Limitations in Interviews (Qualitative)

Our qualitative sample of 15 interviews, while providing depth, is still a snapshot. There might be other perspectives not captured – for example, we did not speak directly with end-consumers or with employees (aside from the company representatives). Including employees could reveal internal acceptance issues (maybe some IT staff resist green IT changes due to workload, etc.) or including customers could provide independent validation of loyalty claims. Our study took the companies' word (and data) for customer loyalty improvements. Surveying consumers of these brands would be a nice validation step that was beyond our scope.

## Fast-Changing Field

The digital economy and sustainability technology are fast evolving. What might be a cutting-edge practice today (like using AI to minimize data center cooling) could become standard in a couple of years, and new challenges can arise (e.g., the sudden popularity of a new technology like crypto or metaverse could spike emissions unexpectedly – as seen with crypto mining energy issues). Our study is a bit of a moving target; the conclusions are most pertinent to the mid-2020s context. Continuous research will be needed to update best practices as technology and circumstances change (for instance, if fusion energy becomes viable or battery tech leaps forward, some strategies might shift).

Despite the above limitations, the triangulation of methods lends credibility to our overall analysis. The convergence of quantitative trends with qualitative stories provides confidence that our main messages are valid: online businesses can indeed grow while cutting environmental impact, and many are already on that path. Any one data point, taken alone, should be treated cautiously, but the *pattern* across multiple data points is compelling. We were transparent about uncertain or statistically weak results (like H4's mild correlation). None of our findings contradict the fundamental assertion that sustainability and business performance are compatible; at worst, some are neutral. This consistency across analyses reinforces the overall narrative.

We also acknowledge that our results might be somewhat optimistic because those who engage with such a study might be those already inclined to believe in sustainability's value. A skeptic might say our sample is biased, which is fair; however, even if that's the case, the fact that many companies are finding success in these practices is itself an existence proof that counters the narrative "sustainability hurts business." Our contribution is showing examples and aggregate trends that make a positive case. Ultimately, each business and context may vary, so managers should still do their due diligence, but hopefully with more confidence now that others have trodden the path successfully.

In conclusion of the analysis, the limitations we discussed suggest avenues for future research: larger-scale studies, including random sampling, longitudinal tracking, expanding to new contexts and metrics, etc. We encourage readers and researchers to view our work as a stepping stone – it provides evidence and insight, but not the final word. Sustainability in the digital economy is a complex, evolving challenge, and ongoing critical analysis will be needed to ensure that the optimistic trends we observed continue and accelerate.

## CONCLUSION

This study investigates how online businesses can grow while reducing their environmental impact, and the findings are largely positive. Sustainable strategies such as adopting renewable energy, implementing green logistics, using eco-friendly packaging, and embracing green IT practices not only helped companies reduce their carbon footprint but also supported business growth. Companies with strong sustainability programs reduced emissions by 10% over four years, even as their operations expanded, while those without such programs experienced increased emissions. Moreover, these strategies often resulted in cost savings and stronger customer loyalty, indicating that being environmentally responsible can also enhance competitiveness.

The research confirms that sustainability and profitability are not mutually exclusive. Hypothesis testing supported key links: renewable energy use led to lower emissions, efficient logistics and packaging improved operations, and visible sustainability efforts strengthened customer engagement. Although the direct financial impact of sustainability was not statistically strong, trends were positive or neutral, showing no downside to eco-friendly practices. The study challenges the old belief that businesses must choose between economic growth and environmental protection, presenting real-world cases where companies achieved both.

Academically, the research fills a gap in the intersection of digital business and sustainability. It provides empirical evidence showing how green IT practices reduce emissions and how digital-native companies embed sustainability in their business models. It also contributes to the literature on corporate social responsibility (CSR) by highlighting non-financial benefits such as customer loyalty. Unlike prior studies that focused on isolated issues, this work offers a comprehensive framework of sustainable strategies for online firms, which future researchers can build on.

Practically, the study offers a clear guide for digital businesses: focus on energy and IT efficiency, improve logistics and packaging, engage customers in sustainability efforts, and measure and report progress transparently. It encourages industry collaboration, open-source solutions, and shared learning to drive sustainability as a collective mission rather than a competitive advantage.

The study also provides guidance for policymakers, suggesting supportive climate policies for digital firms, incentives for green infrastructure, and certification programs for sustainable e-commerce. Future research should explore long-term impacts, consumer behavior across product types, digital businesses in emerging markets, and the role of technologies like AI and IoT in achieving sustainability goals. Investigating how companies integrate sustainability into operations and culture will also be key to future progress.

Looking ahead, several emerging trends could enhance digital sustainability: the rise of circular economy models (like device return and refurbishing programs), ambitions for carbon-negative operations, the development of energy-efficient software and sustainable AI, tightening environmental regulations, and growing consumer demand for green transparency. Digital businesses are well-positioned to lead this shift, given their agility and innovation. In conclusion, the research presents a hopeful outlook showing that digital growth and environmental responsibility can go hand in hand when guided by smart, sustainable strategies.

## REFERENCES


AlKoliby, I. S. M., Abdullah, H. H., & Suki, N. M. (2024). Linking knowledge application, digital marketing, and manufacturing SMEs' sustainable performance: The mediating role of innovation. *Journal of the Knowledge Economy*, *15*(2), 6151–6177. DOI: 10.1007/s13132-023-01157-4

Alshurideh, M. T., Shatnawi, T. M., Al-Momani, A. A., Mohammad, A. A. S., Alzoubi, A., Alzyoud, M., & Aldaihani, F. M. F. (2024). Analysing the impact of online journey determinants on customer digital engagement: an empirical study in Jordan. In *Artificial Intelligence and Economic Sustainability in the Era of Industrial Revolution 5.0* (pp. 1109–1122). Springer Nature Switzerland. DOI: 10.1007/978-3-031-56586-1_81

Belkhir, L., & Elmeligi, A. (2018). Assessing ICT global emissions footprint: Trends to 2040 & recommendations. *Journal of Cleaner Production*, *177*, 448–463. DOI: 10.1016/j.jclepro.2017.12.239

Bican, P. M., & Brem, A. (2020). Digital business model, digital transformation, digital entrepreneurship: Is there a sustainable "digital"? *Sustainability (Basel)*, *12*(13), 5239. DOI: 10.3390/su12135239

Buldeo Rai, H., & Dablanc, L. (2023). Hunting for treasure: A systematic literature review on urban logistics and e-commerce data. *Transport Reviews*, *43*(2), 204–233. DOI: 10.1080/01441647.2022.2082580

Chen, K., Zhao, S., Jiang, G., He, Y., & Li, H. (2025). The green innovation effect of the digital economy. *International Review of Economics & Finance*, *99*, 103970. DOI: 10.1016/j.iref.2025.103970

Dell'Erba, M. (2024). Sustainable digital finance and the pursuit of environmental sustainability. In *Sustainable finance in Europe: Corporate governance, financial stability and financial markets* (pp. 97–118). Springer International Publishing. DOI: 10.1007/978-3-031-53696-0_3

Hidayat, M., Salam, R., Hidayat, Y. S., Sutira, A., & Nugrahanti, T. P. (2022). Sustainable Digital Marketing Strategy in the Perspective of Sustainable Development Goals. *Komitmen: Jurnal Ilmiah Manajemen*, *3*(2), 100–106. DOI: 10.15575/jim.v3i2.22687

Huang, X., Zhang, S., Zhang, J., & Yang, K. (2023). Research on the impact of digital economy on Regional Green Technology Innovation: Moderating effect of digital talent Aggregation. *Environmental Science and Pollution Research International*, *30*(29), 74409–74425. DOI: 10.1007/s11356-023-27279-5 PMID: 37209350

Karaboga, T., Gurol, Y. D., Binici, C. M., & Sarp, P. (2020). Sustainable digital talent ecosystem in the new era: impacts on businesses, governments and universities. *Istanbul business research, 49*(2), 360-379.

Mahajan, R., Lim, W. M., Sareen, M., Kumar, S., & Panwar, R. (2023). Stakeholder theory. *Journal of Business Research, 166*, 114104. DOI: 10.1016/j.jbusres.2023.114104

Martín-Peña, M. L., Lorenzo, P. C., & Meyer, N. (2024). Digital platforms and business ecosystems: A multidisciplinary approach for new and sustainable business models. *Review of Managerial Science, 18*(9), 2465–2482. DOI: 10.1007/s11846-024-00772-y

Masanet, E., Shehabi, A., Lei, N., Smith, S., & Koomey, J. (2020). Recalibrating global data center energy-use estimates. *Science, 367*(6481), 984–986. DOI: 10.1126/science.aba3758 PMID: 32108103

Miller, P., & Wilsdon, J. (2001). Digital futures—An agenda for a sustainable digital economy. *Corporate Environmental Strategy, 8*(3), 275–280. DOI: 10.1016/S1066-7938(01)00116-6

Naim, A. (Ed.). (2024). *AI Applications for Business, Medical, and Agricultural Sustainability*. IGI Global. DOI: 10.4018/979-8-3693-5266-3

Nogueira, E., Gomes, S., & Lopes, J. M. (2024). Financial Sustainability: Exploring the Influence of the Triple Bottom Line Economic Dimension on Firm Performance. *Sustainability (Basel), 2024*(16), 6458. DOI: 10.3390/su16156458

Risdwiyanto, A., Sulaeman, M. M., & Rachman, A. (2023). Sustainable digital marketing strategy for long-term growth of msmes. [ADMAN]. *Journal of Contemporary Administration and Management, 1*(3), 180–186. DOI: 10.61100/adman.v1i3.70

Wang, X., Kim, J., Kim, J., & Koh, Y. (2024). Application of Natural-Resource-Based View to Nature-Based Tourism Destinations. *Sustainability (Basel), 16*(6), 2375. DOI: 10.3390/su16062375

Wu, J., Xue, Y., Zhang, Q., & Liu, Y. (2024). Digital economy, financial development, and corporate green technology innovation. *Finance Research Letters, 66*, 105552. DOI: 10.1016/j.frl.2024.105552

Yousaf, Z., Radulescu, M., Sinisi, C. I., Serbanescu, L., & Păunescu, L. M. (2021). Towards sustainable digital innovation of SMEs from the developing countries in the context of the digital economy and frugal environment. *Sustainability (Basel), 13*(10), 5715. DOI: 10.3390/su13105715

Zreik, M. (2024). Harnessing the Power of Digital Transformation and Sustainability: The Chinese Experience. In *Building Smart and Sustainable Businesses with Transformative Technologies* (pp. 247-266). IGI Global Scientific Publishing.

# Compilation of References


Abdel Moneim, H., & Ismail, M. (2021). *Economic repercussions of the fourth generation industrial revolution.*

Acemoglu, D., & Restrepo, P. (2019). Automation and new tasks: How technology displaces and reinstates labor. *The Journal of Economic Perspectives*, *33*(2), 3–30. DOI: 10.1257/jep.33.2.3

Adam, M., Wessel, M., & Benlian, A. (2021). AI-based chatbots in customer service and their effects on user compliance. *Electronic Markets*, *31*(2), 427–445. DOI: 10.1007/s12525-020-00414-7

Ahmed, N. S., & Harb, M. A. (2025). The Relationship Between Liquidity Risk and Financial Performance: A Comparative Analytical Study of A Sample of Banks Operating in Palestine Between 2014-2023. *An-Najah University Journal for Research - B (Humanities), 40*(1), None-None.

Ahmed, S. (2023). Digital transformation of accounting systems and its impact on tax revenues: an applied study of the Cairo Tax Authority. *Future Higher Institute for Specialized Technological Studies, 49*(issue 2, part 1), 373–476. https://doi.org/ DOI: 10.21608/sjsc.2022.146146.1161

Aigbavboa, C. O., Thwala, W. D., & Ebekozien, A. (2024). Sustainable development goals under threat: The impact of inflation on construction projects. *Engineering, Construction, and Architectural Management.*

Ajzen, I. (1991). The Theory of planned behavior. *Organizational Behavior and Human Decision Processes*, *50*(2), 179–211. DOI: 10.1016/0749-5978(91)90020-T

Akinola, S., & Telukdarie, A. (2023). Sustainable digital transformation in healthcare: Advancing a digital vascular health innovation solution. *Sustainability (Basel)*, *15*(13), 10417. DOI: 10.3390/su151310417

Alaboudi, A., & Alharbi, A. S. (2019). *The current state of teachers' ICT use in classrooms in boys' secondary schools in Saudi Arabia* (Doctoral dissertation).

Alazab, M.. (2023). Digital Twins for Healthcare 4.0—Recent Advances, Architecture, and Open Challenges. *IEEE Consumer Electronic Magazine. Volum, 12*, 29–37.

Albawi, S., Mohammed, T. A., & Al-Zawi, S. (2017). Understanding of a convolutional neural network. 2017 International Conference on Engineering and Technology (ICET), 1–6. Appian. Available at: https://appian.com/ [Accessed: 04 June 2025]

Albury, D. (2005). Fostering Innovation in Public Services. *Public Money & Management, 25*(1), 51–56.

Al-Dmour, R., Al-Dmour, H., Basheer Amin, E., & Al-Dmour, A. (2025). Impact of AI and big data analytics on healthcare outcomes: An empirical study in Jordanian healthcare institutions. *Digital Health, 11*, 20552076241311051. DOI: 10.1177/20552076241311051 PMID: 39777060

Aldoseri, A., Al-Khalifa, K. N., & Hamouda, A. M. (2024). AI-powered innovation in digital transformation: Key pillars and industry impact. *Sustainability (Basel), 16*(5), 1790. DOI: 10.3390/su16051790

Alford, J. (2016). Co-Production, Interdependence and Publicness: Extending public service-dominant logic. *Public Management Review, 18*(5), 673–691. DOI: 10.1080/14719037.2015.1111659

Ali, H., & Hajjar, L. (2024). Empowering SMEs with AI and Digital Transformation: A Roadmap to Enhanced Competitiveness and Growth.

AL-Jumaili, A. H. A., Muniyandi, R. C., Hasan, M. K., Paw, J. K. S., & Singh, M. J. (2023). Big Data Analytics Using Cloud Computing Based Frameworks for Power Management Systems: Status, Constraints, and Future Recommendations. *Sensors 2023, Vol. 23, Page 2952, 23*(6), 2952. DOI: 10.3390/s23062952

Alketbi, A., Nasir, Q., & Talib, M. A. (2018). Blockchain for government services—Use cases, security benefits and challenges. Proceedings of IEEE International Conference on Internet of Things (iThings), 1220-1225.

Alkhawaldeh, M. A., & Khasawneh, M. A. S. (2023). Harnessing the power of artificial intelligence for personalized assistive technology in learning disabilities. *Journal of Southwest Jiaotong University, 58*(4).

AlKoliby, I. S. M., Abdullah, H. H., & Suki, N. M. (2024). Linking knowledge application, digital marketing, and manufacturing SMEs' sustainable performance: The mediating role of innovation. *Journal of the Knowledge Economy*, *15*(2), 6151–6177. DOI: 10.1007/s13132-023-01157-4

Aloqaily, M., Bouachir, O., Karray, F., Al Ridhawi, I., & El Saddik, A. (2022). Integrating digital twin and advanced intelligent technologies to realize the metaverse. *IEEE Consumer Electronics Magazine*, *12*(6), 47–55. DOI: 10.1109/MCE.2022.3212570

Alraja, M. N., Hussein, M. A., & Ahmed, H. M. S. (2021). What affects digitalization process in developing economies? An evidence from SMEs sector in Oman. *Bulletin of Electrical Engineering and Informatics*, *10*(1), 441–448. DOI: 10.11591/eei.v10i1.2033

Al-Rimawi, T., & Nadler, M. (2025). Leveraging smart city technologies for enhanced real estate development: An integrative review. *Smart Cities*, *8*(1), 10. DOI: 10.3390/smartcities8010010

Alsadhan, N. (2023). A Multi-Module Machine Learning Approach to Detect Tax Fraud. *Computer Systems Science and Engineering*, *46*(1), 241–253. DOI: 10.32604/csse.2023.033375

Al-Sehrawy, R., Kumar, B., & Watson, R. (2023). The pluralism of digital twins for urban management: Bridging theory and practice. *Journal of Urban Management*, *12*(1), 16–32. DOI: 10.1016/j.jum.2023.01.002

Alshurideh, M. T., Shatnawi, T. M., Al-Momani, A. A., Mohammad, A. A. S., Alzoubi, A., Alzyoud, M., & Aldaihani, F. M. F. (2024). Analysing the impact of online journey determinants on customer digital engagement: an empirical study in Jordan. In *Artificial Intelligence and Economic Sustainability in the Era of Industrial Revolution 5.0* (pp. 1109–1122). Springer Nature Switzerland. DOI: 10.1007/978-3-031-56586-1_81

Alter, S. (2020). Making sense of smartness in the context of smart devices and smart systems. *Information Systems Frontiers*, *22*(2), 381–393. DOI: 10.1007/s10796-019-09919-9

Alzoubi, Y. I., & Mishra, A. (2024). Green artificial intelligence initiatives: Potentials and challenges. *Journal of Cleaner Production*, *468*, 143090. DOI: 10.1016/j.jclepro.2024.143090

Aminah, S., & Saksono, H. (2021). Digital transformation of the government: A case study in Indonesia. *Jurnal Komunikasi: Malaysian Journal of Communication*, *37*(2), 272–288.

Amofa, S. et al., (2024). Blockchain-secure patient Digital Twin in healthcare using smart contracts. *PLoS One*, 29 February, 19(2), p. e0286120.

Anago, J. C. (2024). *Sustainable infrastructure development in sub-nations of Nigeria: What alternative financing options are open amidst constrained budget?* Sustainability Accounting, Management and Policy Journal.

Anand, A., & Singh, P. (2025). Interdisciplinary Studies: Issues, Perspectives, and Challenges From National Education Program 2020. *Integrating Personalized Learning Methods Into STEAM Education*, 233-260.

Angelopoulos, A., Michailidis, E. T., Nomikos, N., Trakadas, P., Hatziefremidis, A., Voliotis, S., & Zahariadis, T. (2020). Tackling faults in the Industry 4.0 era—A survey of machine-learning solutions and key aspects. *Sensors (Basel), 20*(1), 109. DOI: 10.3390/s20010109 PMID: 31878065

Ankitha, V. (2025). *Toobler.* [Electronic] Available at: https://www.toobler.com/blog/real-time-integration-in-digital-twins [Access 6 June 2025].

Aoki, N. (2020). An experimental study of public trust in AI chatbots in the public sector. *Government Information Quarterly, 37*(4), 101490. DOI: 10.1016/j.giq.2020.101490

Ariyibi, K. (2024). *Leveraging Artificial Intelligence for enhanced tax fraud detection in modern fiscal systems.* GSC Advanced Research and Reviews.

Armeni, P. et al., (2022). Digital Twins in Healthcare: Is It the Beginning of a New Era of Evidence-Based Medicine? A Critical Review. *Journal of Personalized Medicine,* Volum 12.

Arvidsson, A. (2020). Changemakers: The industrious future of the digital economy. *Polity, 52*(4), 555–573. DOI: 10.1177/0032329220932426

Ashley, E. (2016). The precision medicine initiative: A new national effort. *Journal of the American Medical Association, 315*(8), 793–794. PMID: 25928209

Asquith, R. (2025, February 8). *Tax authorities adopt AI for tax fraud and efficiencies - vatcalc.com.* Artificial Intelligence. https://www.vatcalc.com/artificial-intelligence/tax-authorities-adopt-ai-for-tax-fraud-and-efficiencies/

Attah, R. U., Gil-Ozoudeh, I., Garba, B. M. P., & Iwuanyanwu, O. (2024). Leveraging geographic information systems and data analytics for enhanced public sector decision-making and urban planning. *Magna Sci Adv Res Rev, 12*(2), 152–163. DOI: 10.30574/msarr.2024.12.2.0191

Attaran, M. (2017). Additive Manufacturing: The most promising technology to alter the supply chain and logistics. *Journal of Service Science and Management, 10*(3), 189–206. DOI: 10.4236/jssm.2017.103017

Attaran, M., & Celik, B. (2023). Digital twin: Benefits, use cases, challenges, and opportunities. *Decis Anal J, 6*(6), 100165. DOI: 10.1016/j.dajour.2023.100165

Attour, A., & Chaupain-Guillot, S. (2020). Digital Innovations in Public Administrations: Technological or Policy Innovation Diffusion? *Journal of Innovation Economics & Management, 31*(1), 61–75. DOI: 10.3917/jie.pr1.0061

Atzori, L., Lera, A., & Morabito, G. (2010). Internet of things: A survey. *Computer Networks, 54*(15), 2787–2805. DOI: 10.1016/j.comnet.2010.05.010

Aziz, M. N. A., Harun, S. N., Baharom, M. K., Khairulnizam, M., & Ramlie, A. S. M. S. (2019). The Acceptance Level of Digital Natives towards the Interactive Kiosk in the Museum: Tam-Based Research Instrument. *TEST Engineering & Management*, (Nov-Dec), 1032–1044.

Babalola, O., Ugah, U. U. K., & Ndimako, O. (2024). Evaluating governmental constraints on construction in Nigeria. *RE:view.*

Bablani, A., & Tripathi, D. (2018). "A review on methods applied on P300-based lie detectors." In *Advances in Intelligent Systems and Computing* (pp. 302-315). Springer. https://doi.org/DOI: 10.1007/978-981-10-8569-7_26

Bai, N. (2025). *Stanford Medicine.* [Electronic] Available at: https://med.stanford .edu/news/all-news/2025/04/digital-twin.html [Access 8 Jube 2025].

Bai, T., Qi, Y., Li, Z., & Xu, D. (2023). Digital economy, industrial transformation and upgrading, and spatial transfer of carbon emissions: The paths for low-carbon transformation of Chinese cities. *Journal of Environmental Management, 344*, 118528. DOI: 10.1016/j.jenvman.2023.118528 PMID: 37392691

Balasubramanyam, A., Ramesh, R., Sudheer, R., & Honnavalli, P. (2024). Revolutionizing Healthcare: A Review Unveiling the Transformative Power of Digital Twins. *IEEE Access. IEEE Access : Practical Innovations, Open Solutions, 12*, 69652–69676. DOI: 10.1109/ACCESS.2024.3399744

Bansal, A., Panchal, T., Jabeen, F., Mangla, S. K., & Singh, G. (2023). A study of human resource digital transformation (HRDT): A phenomenon of innovation capability led by digital and individual factors. *Journal of Business Research, 157*, 113611. DOI: 10.1016/j.jbusres.2022.113611

Bardin, J. C., Slichter, D. H., & Reilly, D. J. (2021). Microwaves in Quantum Computing. *IEEE Journal of Microwaves*, *1*(1), 403–427. Advance online publication. DOI: 10.1109/JMW.2020.3034071 PMID: 34355217

Barlow, J. P. (2018). The economy of ideas: Selling wine without bottles on the global net. *Harvard Business Review*, *96*(3), 84–96.

Barlow, J., & Maull, R. (2008). The Role of Technology in Public Service Innovation. *Public Management Review*, *10*(3), 345–367.

Barrett, M., & Oborn, E. (2013). Envisioning E-HRM and strategic HR: Taking seriously identity, innovative practice, and service. *The Journal of Strategic Information Systems*, *22*(3), 252–256. DOI: 10.1016/j.jsis.2013.07.002

Barricelli, B. R., Casiraghi, E., & Fogli, D. (2019). A survey on digital twin: Definitions, characteristics, applications, and design implications. *IEEE Access : Practical Innovations, Open Solutions*, *7*, 167653–167671. DOI: 10.1109/ACCESS.2019.2953499

Barth, F., Candello, H., Cavalin, P., & Pinhanez, C. (2020, July). Intentions, meanings, and whys: designing content for voice-based conversational museum guides. In *Proceedings of the 2nd Conference on Conversational User Interfaces* (pp. 1-8). DOI: 10.1145/3405755.3406128

Barzilai-Nahon, K. (2021). Toward a theory of network gatekeeping: A framework for exploring information control. *Journal of the American Society for Information Science and Technology*, *59*(9), 1493–1512. DOI: 10.1002/asi.20857

Baskar, K., Venkatesan, G. P., & Sangeetha, S. (2020). *A Survey of Workload Management Difficulties in the Public Cloud.* Paper presented at the Intelligent Computing in Engineering: Select Proceedings of RICE 2019. DOI: 10.1007/978-981-15-2780-7_54

Bawden, D., & Robinson, L. (2009). The dark side of information: Overload, anxiety and other paradoxes and pathologies. *Journal of Information Science*, *35*(2), 180–191. DOI: 10.1177/0165551508095781

Beebe, J. (2023). AI and Taxes — A Work in Progress: Part 1. *Houston: Rice University's Baker Institute for Public Policy*. DOI: 10.25613/ZXFH-6346

Beenu, K. G. H., & Rathika, S. (2019). Attitude of customers towards digital and conventional advertising. *International Journal of Recent Technology and Engineering*, *8*(4S).

Begum, A., Naim, A., & Sabahath, A. (2024). The Impact of AI on Sustainability. In *Harnessing High-Performance Computing and AI for Environmental Sustainability* (pp. 99–113). IGI Global. DOI: 10.4018/979-8-3693-1794-5.ch005

Beheshti, A., Yang, J., Sheng, Q. Z., Benatallah, B., Casati, F., Dustdar, S., Motahari Nezhad, H. R., Zhang, X., & Xue, S. (2023). ProcessGPT: Transforming Business Process Management with Generative Artificial Intelligence. arXiv. https://arxiv.org/abs/2306.01771 DOI: 10.1109/ICWS60048.2023.00099

Beier, G., Ullrich, A., Niehoff, S., Reißig, M., & Habich, M. (2020). Industry 4.0: How it is defined from a sociotechnical perspective and how much sustainability it includes – A literature review. *Journal of Cleaner Production*, *259*, 120856. DOI: 10.1016/j.jclepro.2020.120856

Belkhir, L., & Elmeligi, A. (2018). Assessing ICT global emissions footprint: Trends to 2040 & recommendations. *Journal of Cleaner Production*, *177*, 448–463. DOI: 10.1016/j.jclepro.2017.12.239

Belli, L., Davoli, L., Medioli, A., Marchini, P. L., & Ferrari, G. (2019). Toward Industry 4.0 with IoT: Optimizing business processes in an evolving manufacturing factory. *Frontiers in ICT (Lausanne, Switzerland)*, *6*, 17. Advance online publication. DOI: 10.3389/fict.2019.00017

Bengesi, S., El-Sayed, H., Sarker, M. K., Houkpati, Y., Irungu, J., & Oladunni, T. (2023). Advancements in Generative AI: A comprehensive review of GANs, GPT, Autoencoders, Diffusion Models, and Transformers. arXiv preprint arXiv:2311.10242. https://arxiv.org/abs/2311.10242

Benkler, Y. (2006). *The wealth of networks*. Yale University Press.

Benkler, Y. (2022). Peer production and cooperation. *The Journal of Economic Perspectives*, *36*(2), 69–92. DOI: 10.1257/jep.36.2.69

Bennett, E. E., & McWhorter, R. R. (2021). Virtual HRD's role in crisis and the post Covid-19 professional lifeworld: Accelerating skills for digital transformation. *Advances in Developing Human Resources*, *23*(1), 5–25. DOI: 10.1177/1523422320973288

Berti, A., van der Aalst, W. M. P., Zang, D., & Lang, M. (2020). An Open-Source Integration of Process Mining Features into the Camunda Workflow Engine: Data Extraction and Challenges. arXiv preprint arXiv:2009.06209. https://arxiv.org/abs/2009.06209

Bican, P. M., & Brem, A. (2020). Digital business model, digital transformation, digital entrepreneurship: Is there a sustainable "digital"? *Sustainability (Basel)*, *12*(13), 5239. DOI: 10.3390/su12135239

Bifulco, F., Tregua, M., Amitrano, C. C., & D'Auria, A. (2016). International Journal of Public Sector Management. *ICT and sustainability in smart cities management*, (2), 132-147.

Boiano, S., Borda, A., Gaia, G., Rossi, S., & Cuomo, P. (2018). Chatbots and new audience opportunities for museums and heritage organisations. Electronic visualisation and the arts, 164-171

Boniotti, C. (2023). The public–private–people partnership (P4) for cultural heritage management purposes. *Journal of Cultural Heritage Management and Sustainable Development, 13*(1), 1–14. DOI: 10.1108/JCHMSD-12-2020-0186

Bouchetara, M., Zerouti, M., & Zouambi, A. R. (2024). Leveraging artificial intelligence (AI) in public sector financial risk management: Innovations, challenges, and future directions. *EDPACS, 69*(9), 124–144. DOI: 10.1080/07366981.2024.2377351

Boukherouaa, E. B., & Shabsigh, G. (2021). *Powering the Digital Economy Opportunities and Risks of Artificial Intelligence in Finance*. International Monetary Fund, Publication Services.

Breiman, L. (2001). Random forests. *Machine Learning, 45*(1), 5–32. DOI: 10.1023/A:1010933404324

Bruner, I. I. G. C.II, & Kumar, A. (2005). Applying TAM to consumer usage of handheld Internet devices. *Journal of Business Research, 58*(5), 553–558. DOI: 10.1016/j.jbusres.2003.08.002

Bruynseels, K., Santoni de Sio, F., & van den Hoven, J. (2018). Digital twins in health care: Ethical implications of an emerging engineering paradigm. *Frontiers in Genetics. Volum, 9*, 31. PMID: 29487613

Brynjolfsson, E., & McAfee, A. (2014). *The second machine age: Work, progress, and prosperity in a time of brilliant technologies*. WW Norton & company.

Brynjolfsson, E., & McAfee, A. (2021). The business of artificial intelligence. *Harvard Business Review, 99*(4), 60–73.

Buckley, G., Caulfield, T., & Becker, I. (2022). It may be a pain in the backside but... Insights into the resilience of business after GDPR. *arXiv*. [Add link if available]

Buldeo Rai, H., & Dablanc, L. (2023). Hunting for treasure: A systematic literature review on urban logistics and e-commerce data. *Transport Reviews, 43*(2), 204–233. DOI: 10.1080/01441647.2022.2082580

Bureau, E. T. (2018). Indian companies weathered the pandemic 'much better' than counterparts: Cognizant study. *The Economic Times*.

Buterin, V. (2014). *A next-generation smart contract and decentralized application platform*. Ethereum Whitepaper.

Calvi, L., & Vermeeren, A. P. (2023). Digitally enriched museum experiences–what technology can do. *Museum Management and Curatorship*, •••, 1–22.

Canonico, R., Cozzolino, G., & Sperlì, G. (2021). Suggesting Cultural Heritage Points of Interest Through a Specialized Chatbot. In Advances in Intelligent Networking and Collaborative Systems: The 12th International Conference on Intelligent Networking and Collaborative Systems (INCoS-2020) (pp. 112-120). Springer International Publishing. DOI: 10.1007/978-3-030-57796-4_11

Cantarelli, C. C., Flyvbjerg, B., Molin, E. J. E., & van Wee, B. (2012). Cost overruns in large-scale transport infrastructure projects: Explanations and their theoretical embeddedness. *European Journal of Transport and Infrastructure Research, 12*(1), 5–18.

Carayannis, E. G., Askounis, D., Andoutropoulou, M., & Zotas, N. (2024). Leveraging AI for Enhanced eGovernment: Optimizing the Use of Open Governmental Data. *Journal of the Knowledge Economy*, •••, 1–36.

Carvalho, T. P., Soares, F. A., Vita, R., Francisco, R. D. P., Basto, J. P., & Alcalá, S. G. (2019). A systematic literature review of machine learning methods applied to predictive maintenance. *Computers & Industrial Engineering*, *137*, 106024. DOI: 10.1016/j.cie.2019.106024

Casino, F., Dasaklis, T. K., & Patsakis, C. (2019). A systematic literature review of blockchain-based applications: Current status, classification, and open issues. *Telematics and Informatics*, *36*, 55–81. DOI: 10.1016/j.tele.2018.11.006

Castells, M. (2020). The rise of the network society: The information age - Economy, society, and culture. *Information Communication and Society*, *23*(5), 629–645. DOI: 10.1080/1369118X.2020.1718177

Castronova, E. (2008). Synthetic worlds: The business and culture of online games. In *Synthetic Worlds*. University of Chicago press.

Cellina, M. C. M. A. M. I. G. I. S. C. E. F. D. G., & Papa, S. (2023). Digital Twins: The New Frontier for Personalized Medicine? *Applied Sciences (Basel, Switzerland)*, *13*(13), 7940. DOI: 10.3390/app13137940

Cepal, N. U. (2021). *Digital technologies for a new future*. https://www.cepal.org/sites/default/files/publication/files/46817/S2000960_en.pdf

Chafiq, T., Rida, A., Fadil, A., & Mohammed, O. (2024). *Investigating the potential of blockchain technology for geospatial data sharing: Opportunities, challenges, and solutions*. Geomatica, Elsevier.

Chakraborty, I., Edirippulige, S., & Ilavarasan, P. V. (2023). The role of telehealth startups in healthcare service delivery: A systematic review. *International Journal of Medical Informatics, 174*, 105048. DOI: 10.1016/j.ijmedinf.2023.105048 PMID: 36963322

Chao, K., Sarker, M. N. I., Ali, I., Firdaus, R. R., Azman, A., & Shaed, M. M. (2023). Big data-driven public health policy making: Potential for the healthcare industry. *Heliyon, 9*(9), e19681. DOI: 10.1016/j.heliyon.2023.e19681 PMID: 37809720

Chaturvedi, R., & Verma, S. (2023). Opportunities and challenges of AI-driven customer service. *Artificial Intelligence in customer service: The next frontier for personalized engagement*, 33-71.

Chen, Q., & Bridges, R. A. (2017). Automated behavioral analysis of malware: A case study of WannaCry ransomware. In *2017 16th IEEE International Conference on Machine Learning and Applications (ICMLA)* (pp. 454–460). https://doi.org/ DOI: 10.1109/ICMLA.2017.0-119

Chen, Y., Zhao, C., Xu, Y., & Nie, C. (2025). Year-over-year developments in financial fraud detection via deep learning: A systematic literature review. arXiv preprint arXiv:2502.00201. https://doi.org//arXiv.2502.00201DOI: 10.48550

Cheng, A., Ma, D., Pan, Y., & Qian, H. (2023). Enhancing museum visiting experience: Investigating the relationships between augmented reality quality, immersion, and TAM using PLS-SEM. *International Journal of Human-Computer Interaction*, •••, 1–12.

Chen, J.. (2023). Networking Architecture and Key Supporting Technologies for Human Digital Twin in Personalized Healthcare: A Comprehensive Survey. *IEEE Communications Surveys & Tutorials. Volum, 26*, 706–746.

Chen, K., Zhao, S., Jiang, G., He, Y., & Li, H. (2025). The green innovation effect of the digital economy. *International Review of Economics & Finance, 99*, 103970. DOI: 10.1016/j.iref.2025.103970

Chen, M., Wan, J., & Li, F. (2012). Machine-to-machine communications: Architectures, standards and applications. *KSII Transactions on Internet and Information Systems, 2*, 480–497. DOI: 10.3837/tiis.2012.02.002

Chen, Z., Li, Z., Li, J., Liu, C., Lao, C., Fu, Y., Liu, C., Li, Y., Wang, P., & He, Y. (2019). 3D printing of ceramics: A review. *Journal of the European Ceramic Society, 39*(4), 661–687. Advance online publication. DOI: 10.1016/j.jeurceramsoc.2018.11.013

Chen, Z., Sivaparthipan, C. B., & Muthu, B. (2022). IoT based smart and intelligent city energy optimization. *Sustainable Energy Technologies and Assessments, 49*, 101724. DOI: 10.1016/j.seta.2021.101724

Choudary, S. P. (2023). Platform scale: How an emerging business model helps startups build large empires with minimum investment. *Journal of Business Models, 11*(1), 34–52.

Choudhury, P., & Kauffman, R. J. (2019). The Role of Digital Technologies in Business Model Innovation. *Journal of Business Research, 100*, 1–10.

Christian, M., & Iris, J. (2020). Investigating the elicitation of employees' support towards digital workplace transformation. *Behaviour & Information Technology*. Advance online publication. DOI: 10.1080/0144929X.2020.1818824

Chumnanvej, S., Chumnanvej, S. & Tripathi, S., (2024). Assessing the benefits of digital twins in neurosurgery: a systematic review. *Neurosurg Rev,* 18 January, 47(1), p. 52.

Cohen, J. E. (2021). The regulatory state in the information age. *Theoretical Inquiries in Law, 22*(1), 1–32. DOI: 10.1515/til-2019-0002

Corporation, I. B. M. (2023). *IBM Business Automation Workflow: Product overview*. IBM. https://www.ibm.com/products/business-automation-workflow

Correia, D. M., Teixeira, L., & Marques, J. L. (2021). Smart supply chain management: The 5W1H open and collaborative framework. In *2021 IEEE 8th International Conference on Industrial Engineering and Applications (ICIEA)* (pp. 401–405). https://doi.org/DOI: 10.1109/ICIEA52957.2021.9436817

Correia, J. B., Abel, M., & Becker, K. (2023). Data management in digital twins: A systematic literature review. *Knowledge and Information Systems, 65*(8), 3165–3196. DOI: 10.1007/s10115-023-01870-1

Cortes, C., & Vapnik, V. (1995). Support-vector networks. *Machine Learning, 20*(3), 273–297. DOI: 10.1007/BF00994018

Cowen, T. (2020). The age of average: Why the digital economy rewards similarity. *Journal of Economic Behavior & Organization, 175*, 357–372. DOI: 10.1016/j.jebo.2020.04.015

Crespi, N., Drobot, A. T., & Minerva, R. (2023). The Digital twin in action and directions for the future. In *The Digital Twin* (pp. 1201–1217). Springer International Publishing. DOI: 10.1007/978-3-031-21343-4_38

Creswell, J. W. (2013). Steps in conducting a scholarly mixed methods study. In DBER Group Discussion, available at https://digitalcommons.unl.edu/dberspeakers/48/

Creswell, J. W., & Creswell, J. D. (2017). *Research design: Qualitative, quantitative, and mixed methods approaches*. Sage Publications.

Cropley, D. H. (2015). Promoting creativity and innovation in engineering education. *Psychology of Aesthetics, Creativity, and the Arts*, *9*(2), 137–144. DOI: 10.1037/aca0000008

Cui, T., & Aulton, K. (2023). Conceptualizing the elements of value in public services: Insights from practitioners. *Public Management Review*, •••, 1–23. DOI: 10.1080/14719037.2023.2226676

CXOtoday. (2021). Indian firms fare better than global counterparts amid COVID-19 pandemic. https://www.cxotoday.com/ai/indian-firms-fare-better-than-global-counterparts-amid-covid-19-pandemic/

Damanpour, F., & Schneider, M. (2009). Phases of the Adoption of Innovation in Organizations: Effects of Environment, Organization, and Top Managers. *British Journal of Management*, *20*(1), 1–21.

Dana, L.-P., Salamzadeh, A., Hadizadeh, M., Heydari, G., & Shamsoddin, S. (2022). Urban entrepreneurship and sustainable businesses in smart cities: Exploring the role of digital technologies. *Sustainable Technology and Entrepreneurship*, *1*(2), 100016. DOI: 10.1016/j.stae.2022.100016

Danladi, S., Prasad, M., Modibbo, U. M., Ahmadi, S. A., & Ghasemi, P. (2023). Attaining sustainable development goals through financial inclusion: Exploring collaborative approaches to Fintech adoption in developing economies. *Sustainability (Basel)*, *15*(17), 13039. DOI: 10.3390/su151713039

Dashore, R., & Sharma, N. (2024). Digital Twins in the Modern Context Bridging Realities With Blockchain. In *Ensuring Security and End-to-End Visibility Through Blockchain and Digital Twins* (pp. 1–20). IGI Global. DOI: 10.4018/979-8-3693-3494-2.ch001

Datexcorp. (2025, March). Inventory Management. https://www.datexcorp.com/

Daugherty, P. R., & Wilson, H. J. (2021). Human + machine: Reimagining work in the age of AI. *MIT Sloan Management Review*, *62*(3), 1–9.

David, P. A. (2020). The digital technology boomerang: New intellectual property rights threaten global "open science". *World Development*, *128*, 104829. DOI: 10.1016/j.worlddev.2019.104829

Davis, F. D. (1989). *Technology acceptance model: TAM. Al-Suqri, MN.* Information Seeking Behavior and Technology Adoption.

Davis, P. W., & McKenzie-Rundle, P. (1984). The Social Organization of Lie-Detector Tests. *Urban Life*, *13*(2), 122–134. DOI: 10.1177/0098303984013002003

Dayıoğlu, M. A., & Turker, U. (2021). Digital transformation for sustainable future-agriculture 4.0: A review. *Journal of Agricultural Sciences*, *27*(4), 373–399.

De Filippi, P., & Wright, A. (2022). Blockchain and the law: The rule of code. *Harvard Journal of Law & Technology*, *35*(1), 1–45.

del Campo, G., Saavedra, E., Piovano, L., Luque, F., & Santamaria, A. (2024). Virtual Reality and Internet of Things Based Digital Twin for Smart City Cross-Domain Interoperability. *Applied Sciences (Basel, Switzerland)*, *14*(7), 2747. DOI: 10.3390/app14072747

Del Vacchio, E., Gargiulo, R., & Bifulco, F. (2020). Digital Communication and Museum Experience. A Multichannel Approach. *Micro & Macro Marketing*, *29*(3), 513–533.

Dell'Erba, M. (2024). Sustainable digital finance and the pursuit of environmental sustainability. In *Sustainable finance in Europe: Corporate governance, financial stability and financial markets* (pp. 97–118). Springer International Publishing. DOI: 10.1007/978-3-031-53696-0_3

Deloitte. (2025, March). https://www2.deloitte.com/

Deng, X., & Joshi, K. D. (2016). Why individuals participate in micro-task crowdsourcing work environment: Revealing crowdworkers' perceptions. *Journal of the Association for Information Systems*, *17*(10), 648–673. DOI: 10.17705/1jais.00441

Dery, K., Sebastian, I., & van der Meulen, N. (2017). The digital workplace is key to digital innovation. *MIS Quarterly Executive*, *16*(2), 135–152.

Desmarchelier, B., Djellal, F., & Gallouj, F. (2019). Innovation in public services in the light of public administration paradigms and service innovation perspectives. *Revue Européenne d'Économie et Management des Services*, *8*, 91–110. DOI: 10.15122/isbn.978-2-406-09862-1.p.0091

Didimo, W., Giamminonni, L., Liotta, G., Montecchiani, F., & Pagliuca, D. (2018). A visual analytics system to support tax evasion discovery. *Decision Support Systems, 110*, 71–83. DOI: 10.1016/j.dss.2018.03.008

Dihan, M. S., Akash, A. I., Tasneem, Z., Das, P., Das, S. K., Islam, M. R., & Hasan, M. M. (2024). Digital twin: Data exploration, architecture, implementation and future. *Heliyon, 10*(5).

Dolata, U. (2021). The digital economy: Promises, perils, and policy challenges. *The Review of Policy Research, 38*(4), 403–420. DOI: 10.1111/ropr.12433

Dooley, A., Penica, M., McGrath, S., O'Brien, W., Bhattacharya, M., & O'Connell, E. (2024, June). Digital Twin Bridging: Enabling Virtual Worlds to Manifest in the Physical. In *2024 35th Irish Signals and Systems Conference (ISSC)* (pp. 1–6). IEEE.

Doran, R., Hanss, D., & Larsen, S. (2015). Attitudes, efficacy beliefs, and willingness to pay for environmental protection when travelling. *Tourism and Hospitality Research, 15*(4), 281–292. DOI: 10.1177/1467358415580360

Dritsas, E., & Trigka, M. (2025). Remote sensing and geospatial analysis in the big data era: A survey. *Remote Sensing (Basel), 17*(3), 550. DOI: 10.3390/rs17030550

Dubey, K., Kasar, C. B., Thulluri, G., Yesugad, A., Karishma, M., & Padhi, D. (2023, December). Leveraging Big Data for Effective Healthcare Management with Artificial Intelligence. In *2023 International Conference on Artificial Intelligence for Innovations in Healthcare Industries (ICAIIHI)* (Vol. 1, pp. 1-7). IEEE. DOI: 10.1109/ICAIIHI57871.2023.10489731

Du, J., Shen, Z., Song, M., & Zhang, L. (2023). Nexus between digital transformation and energy technology innovation: An empirical test of A-share listed enterprises. *Energy Economics, 120*, 106572. DOI: 10.1016/j.eneco.2023.106572

Dumas, M., La Rosa, M., Mendling, J., & Reijers, H. A. (2018). *Fundamentals of business process management* (2nd ed., Vol. 1). Springer., DOI: 10.1007/978-3-662-56509-4

Dutta, M., & Dhal, S. (2024). Security and Privacy Preserving GDPR-Compliant Scheme Resisting in Rendering False Data. *2024 International Conference on Computing, Communication and Networking Technologies (ICCCNT)*, 1–7. https://doi.org/DOI: 10.1109/ICCCNT61001.2024.10724081

Einav, L., Levin, J., & Sundaresan, N. (2023). Economics in the age of big data. *Science, 379*(6634), 658–660. DOI: 10.1126/science.ade7362 PMID: 25378629

Ejjami, R. (2024). Public Administration 5.0: Enhancing Governance and Public Services with Smart Technologies. *International Journal for Multidisciplinary Research, 6*(4), 1–36.

El Khoury, R. (2024). Technology-driven business innovation. Springer. https://link.springer.com/content/pdf/10.1007/978-3-031-51997-0.pdf

Elkefi, S. & Asan, O., (2022). Digital Twins for Managing Health Care Systems: Rapid Literature Review. *Journal of Medical Internet Research,* Τόμος 24.

Eriksson, E. M. (2019). Representative co-production: Broadening the scope of the public service logic. *Public Management Review, 21*(2), 291–314. DOI: 10.1080/14719037.2018.1487575

Espinosa-Jaramillo, M. T., Zuta, M. E. C., Koneti, C., Jayasundar, S., Zegarra, S. D. R. O., & Carvajal-Ordoñez, V. F. M. (2024). Digital Twins in Supply Chain Operations Bridging the Physical and Digital Worlds using AI. *Journal of Electrical Systems, 20*(10s), 1764–1774. DOI: 10.52783/jes.5434

Estevez, E., & Janowski, T. (2020). Smart governance for smart cities: A framework. *Government Information Quarterly, 37*(2), 101399. DOI: 10.1016/j.giq.2019.101399

Eubanks, V. (2018). *Automating inequality: How high-tech tools profile, police, and punish the poor*. St. Martin's Press.

Fadly, A. H. K. (2023). The impact of digital transformation on developing tax systems. *The Scientific Journal for Commercial Research, 3*(2), 984–1005. https://www.sjsc.journals.ekb.eg/article_289731_cae96a908c2b89c418cb989dc1741f66.pdf

Felicetti, A. M., Corvello, V., & Ammirato, S. (2024). Digital innovation in entrepreneurial firms: A systematic literature review. *Review of Managerial Science, 18*(2), 315–362. DOI: 10.1007/s11846-023-00638-9

Feller, J., Finnegan, P., & Nilsson, O. (2011). Open innovation and public administration: Transformational typologies and business model impacts. *European Journal of Information Systems, 20*(3), 320–335. DOI: 10.1057/ejis.2010.65

Fereday, J., & Muir-Cochrane, E. (2006). Demonstrating Rigor Using Thematic Analysis: A Hybrid Approach of Inductive and Deductive Coding and Theme Development. *International Journal of Qualitative Methods, 5*(1), 80–92. DOI: 10.1177/160940690600500107

First Ignite. (2024). *First Ignite.* [Electronic] Available at: https://www.firstignite.com/exploring-the-latest-advancements-in-digital-twins-for-2025/ [Access 5 June 2025].

Fisher, S., & Rosella, L. C. (2022). Priorities for successful use of artificial intelligence by public health organizations: A literature review. *BMC Public Health, 22*(1), 2146. DOI: 10.1186/s12889-022-14422-z PMID: 36419010

Floridi, L. (2019). Establishing the rules for building trustworthy AI. *Nature Machine Intelligence, 1*(6), 261–262. DOI: 10.1038/s42256-019-0055-y

Flyvbjerg, B., Bruzelius, N., & Rothengatter, W. (2003). *Megaprojects and risk: An anatomy of ambition.* Cambridge University Press. DOI: 10.1017/CBO9781107050891

Flyvbjerg, B., Garbuio, M., & Lovallo, D. (2018). Delusion and deception in large infrastructure projects: Two models for explaining and preventing executive disaster. *California Management Review, 51*(2), 170–193. DOI: 10.2307/41166485

Folgado, F. J., Calderón, D., González, I., & Calderón, A. J. (2024). Review of Industry 4.0 from the Perspective of Automation and Supervision Systems: Definitions, Architectures and Recent Trends. *Electronics 2024, Vol. 13, Page 782, 13*(4), 782. DOI: 10.3390/electronics13040782

Fraga-Lamas, P., Lopes, S. I., & Fernández-Caramés, T. M. (2021). Green IoT and Edge AI as Key Technological Enablers for a Sustainable Digital Transition towards a Smart Circular Economy: An Industry 5.0 Use Case. *Sensors 2021, Vol. 21, Page 5745, 21*(17), 5745. DOI: 10.3390/s21175745

Francisco, G. (2024). Learning low-code: Potential of Mendix platform in building applications (Bachelor's thesis). South-Eastern Finland University of Applied Sciences. https://urn.fi/URN:NBN:fi:amk-202502072469

Franke, L., Liang, H., Farzanehpour, S., Brantly, A. F., Davis, J. C., & Brown, C. M. (2024). An Exploratory Mixed-Methods Study on General Data Protection Regulation (GDPR) Compliance in Open-Source Software. *arXiv.* https://doi.org//arxiv.2406.14724DOI: 10.1145/3674805.3686692

Frikha, T., Ktari, J., Zalila, B., Ghorbel, O., & Amor, N. B. (2023). Integrating blockchain and deep learning for intelligent greenhouse control and traceability. *Alexandria Engineering Journal, 79*, 1457–1465. DOI: 10.1016/j.aej.2023.08.027

Fuller, A., Fan, Z., Day, C., & Barlow, C. (2020). Digital twin: Enabling technologies, challenges and open research. *IEEE Access : Practical Innovations, Open Solutions, 8*, 108952–108971. DOI: 10.1109/ACCESS.2020.2998358

Fuller, A., Fan, Z., Day, C., & Barlow, C. (2020). Digital twin: Enabling technologies, challenges and open research. *IEEE Access. Volum, 8*, 108952–108971.

Furr, N., Ozcan, P., & Eisenhardt, K. M. (2022). What is digital transformation? Core tensions facing established companies on the global stage. *Global Strategy Journal*, *12*(4), 595–618. DOI: 10.1002/gsj.1442

Galla, E. P., Boddapati, V. N., Patra, G. K., Madhavaram, C. R., & Sunkara, J. (2023). AI-Powered Insights: Leveraging Machine Learning And Big Data For Advanced Genomic Research In Healthcare. *Educational Administration: Theory and Practice*.

Gamage, P. (2016). New development: Leveraging 'big data'analytics in the public sector. *Public Money & Management*, *36*(5), 385–390. DOI: 10.1080/09540962.2016.1194087

Gebreab, S. A., Musamih, A., Hasan, H. R., Salah, K., Jayaraman, R., Al Hammadi, Y., & Omar, M. (2024). NFTs for accessing, monetizing, and teleporting digital twins and digital artifacts in the metaverse. *Computer Communications*, *228*, 107965. DOI: 10.1016/j.comcom.2024.107965

Geels, F. W., Sovacool, B. K., Schwanen, T., & Sorrell, S. (2017). Sociotechnical transitions for deep decarbonization. *Science*, *357*(6357), 1242–1244. DOI: 10.1126/science.aao3760 PMID: 28935795

Gefen, D. (2000). E-commerce: The role of familiarity and trust. *Omega*, *28*(6), 725–737. DOI: 10.1016/S0305-0483(00)00021-9

Ghobakhloo, M., Iranmanesh, M., Mubarak, M. F., Mubarik, M., Rejeb, A., & Nilashi, M. (2022). Identifying industry 5.0 contributions to sustainable development: A strategy roadmap for delivering sustainability values. *Sustainable Production and Consumption*, *33*, 716–737. DOI: 10.1016/j.spc.2022.08.003

Ghosh, R., & Scott, J. (2018). The Impact of Digital Technologies on Business Models. *Journal of Business Research*, *100*, 1–10.

Ghosh, S., Hughes, M., Hodgkinson, I., & Hughes, P. (2022). Digital transformation of industrial businesses: A dynamic capability approach. *Technovation*, *113*, 102414. DOI: 10.1016/j.technovation.2021.102414

GMI. (2024). *Global Market Insights*. [Electronic] Available at: https://www.gminsights.com/industry-analysis/healthcare-digital-twins-market [Access 5 June 2025].

Goldberg, B., & Warren, P. (2025). *CGI*. [Electronic] Available at: https://www.cgi.com/en/podcast/energy-utilities/building-resilient-sustainable-healthcare-digital-twins-and-innovation [Access 2 June 2025].

Goldfarb, A., & Tucker, C. (2022). Digital economics. *Journal of Economic Literature, 57*(1), 3–43. DOI: 10.1257/jel.20171452

Gopinathan, J., & Noh, I. (2018). Recent trends in bioinks for 3D printing. *Biomaterials Research, 22*(1), 11. Advance online publication. DOI: 10.1186/s40824-018-0122-1 PMID: 29636985

Gordon, R. J (2016). *The Rise and Fall of American Growth: The U.S. Standard of Living since the Civil War*. https://econpapers.repec.org/RePEc:pup:pbooks:10544

Graham, M. (2020). Regulate, replicate, and resist - The conjunctural geographies of platform urbanism. *Urban Studies (Edinburgh, Scotland), 57*(7), 1397–1415. DOI: 10.1177/0042098019879693

Granata, D., Mastroianni, M., Rak, M., Cantiello, P., & Salzillo, G. (2024). GDPR Compliance Through Standard Security Controls: An Automated Approach. *Journal of High-Speed Networks*. https://doi.org/DOI: 10.3233/JHS-230080

Greenstein, S. (2021). How the internet became commercial: Innovation, privatization, and the birth of a new network. *Information Economics and Policy, 54*, 100881. DOI: 10.1016/j.infoecopol.2020.100881

Grieves, M., & Vickers, J. (2017). Digital twin: Mitigating unpredictable, undesirable emergent behavior in complex systems. In *Transdisciplinary Perspectives on Complex Systems* (pp. 85–113). https://doi.org/DOI: 10.1007/978-3-319-38756-7_4

Grönroos, C. (2006). Adopting a service logic for marketing. *Marketing Theory, 6*(3), 317–333. DOI: 10.1177/1470593106066794

Gubbi, J., Buyya, R., Marusic, S., & Palaniswami, M. (2013). Internet of Things (IoT): A vision, architectural elements, and future directions. *Future Generation Computer Systems, 29*(7), 1645–1660. DOI: 10.1016/j.future.2013.01.010

Gündüz, M. Z., & Daş, R. (2018). Nesnelerin interneti: Gelişimi, bileşenleri ve uygulama alanları. *Pamukkale Üniversitesi Mühendislik Bilimleri Dergisi, 24*(2), 327–335. DOI: 10.5505/pajes.2017.89106

Guo, J., & Lv, Z. (2022). Application of Digital Twins in multiple fields. *Multimedia Tools and Applications, 81*(19), 26941–26967. DOI: 10.1007/s11042-022-12536-5 PMID: 35194381

Gupta, D., Pandey, A. K., & Garg, A. (2019). Innovations in public administration in India. *Economic and Political Weekly, 54*(29), 13–25.

Hagiu, A., & Wright, J. (2023). The status of middlemen in the digital economy. *Journal of Economics & Management Strategy, 32*(1), 5–24. DOI: 10.1111/jems.12515

Halamka, J. (2024). *Mayo Clinic Platform.* [Electronic] Available at: https://www.mayoclinicplatform.org/2024/12/23/digital-twin-technology-has-potential-to-redefine-patient-care/

Hani, M. B., & Warrad, T. (2023). Determinants of Tax Revenues in Jordan. *An-Najah University Journal for Research - B (Humanities), 37*(3), 383–410. DOI: 10.35552/0247-037-003-002

Hassija, V., Chamola, V., Saxena, V., Chanana, V., Parashari, P., Mumtaz, S., & Guizani, M. (2020). Present landscape of quantum computing. *IET Quantum Communication, 1*(2), 42–48. Advance online publication. DOI: 10.1049/iet-qtc.2020.0027

Heeks, R. (2006). Implementing and Managing eGovernment: A Global Perspective. *International Journal of Public Sector Management, 19*(4), 301–310.

Hendrawan, S. A., Chatra, A., Iman, N., Hidayatullah, S., & Suprayitno, D. (2024). Digital transformation in MSMEs: Challenges and opportunities in technology management. *Jurnal Informasi dan Teknologi*, 141-149.

Henfridsson, O., Nandhakumar, J., Scarbrough, H., & Panourgias, N. (2018). Recombination in the open-ended value landscape of digital innovation. *Information and Organization, 28*(2), 89–100. DOI: 10.1016/j.infoandorg.2018.03.001

Henseler, J., & Sarstedt, M. (2013). Goodness-of-fit indices for partial least squares path modeling. *Computational Statistics, 28*(2), 565–580. DOI: 10.1007/s00180-012-0317-1

Hermann, M., Pentek, T., & Otto, B. (2016, January). Design principles for Industrie 4.0 scenarios. In *2016 49th Hawaii International Conference on System Sciences (HICSS)* (pp. 3928–3937). IEEE. https://doi.org/DOI: 10.1109/HICSS.2016.488

Herterich, M. M., Uebernickel, F., & Brenner, W. (2015). The impact of cyber-physical systems on industrial services in manufacturing. *Procedia CIRP, 30*, 323–328. DOI: 10.1016/j.procir.2015.02.110

Heuer, M., Kurtz, C., & Böhmann, T. (2022). Towards a governance of low-code development platforms using the example of Microsoft Power Platform in a multinational company. Proceedings of the 55th Hawaii International Conference on System Sciences (HICSS). https://aisel.aisnet.org/hicss-55/os/practice-based_research/6/

Hidayat, M., Salam, R., Hidayat, Y. S., Sutira, A., & Nugrahanti, T. P. (2022). Sustainable Digital Marketing Strategy in the Perspective of Sustainable Development Goals. *Komitmen: Jurnal Ilmiah Manajemen, 3*(2), 100–106. DOI: 10.15575/jim.v3i2.22687

Hirschman, A. O. (1970). *Exit, Voice and Loyalty: Responses to Decline in Firms, Organizations and States*. Harvard University Press.

Ho, C. T. B., & Lin, W. C. (2010). Measuring the service quality of internet banking: Scale development and validation. *European Business Review*, *22*(1), 5–24. DOI: 10.1108/09555341011008981

Hofmann, E., & Rüsch, M. (2017). Industry 4.0 and the current status as well as future prospects on logistics. *Computers in Industry*, *89*, 23–34. DOI: 10.1016/j.compind.2017.04.002

Hou, K. M., Diao, X., Shi, H., Ding, H., Zhou, H., & De Vaulx, C. (2023). Trends and challenges in AIoT/IIoT/IoT implementation. *Sensors (Basel)*, *23*(11), 5074. DOI: 10.3390/s23115074 PMID: 37299800

Huang, P., Kim, K. & Schermer, M., (2021). Ethical Issues of Digital Twins for Personalized Health Care Service: Preliminary Mapping Study. *Journal of Medical Internet Research,* Volum 24.

Huang, P., Kim, K. & Schermer, M., (2021). Mapping the Ethical Issues of Digital Twins for Personalised Healthcare Service. *jmir.*

Huang, M. H., & Rust, R. T. (2021). Engaged to a robot? The role of AI in service. *Journal of Service Research*, *24*(1), 30–41. DOI: 10.1177/1094670520902266

Huang, X., Zhang, S., Zhang, J., & Yang, K. (2023). Research on the impact of digital economy on Regional Green Technology Innovation: Moderating effect of digital talent Aggregation. *Environmental Science and Pollution Research International*, *30*(29), 74409–74425. DOI: 10.1007/s11356-023-27279-5 PMID: 37209350

Hung, S. Y., Chen, C. C., Hung, H. M., & Ho, W. W. (2013). Critical factors predicting the acceptance of digital museums: User and system perspectives. *Journal of Electronic Commerce Research*, *14*(3), 231.

Hussain, O. A. I., Moehler, R. C., Walsh, S. D. C., & Ahiaga-Dagbui, D. D. (2024). Minimizing cost overrun in rail projects through 5D-BIM: A conceptual governance framework. *Buildings (Basel, Switzerland)*, *14*(2), 478. DOI: 10.3390/buildings14020478

IBM. (2025, March). The smarter supply chain of future. https://www.ibm.com/downloads/cas/78NWEBYM

Iddum, T. S., Sahu, S., Sujatha, R., & Kumar, C. V. R. (2022). Ethereum blockchain based logistics application for smart supply chain management (pp. 1–7). https://doi.org/DOI: 10.1109/GlobConPT57482.2022.9938171

Iftekar, S. F., Aabid, A., Amir, A., & Baig, M. (2023). Advancements and limitations in 3D printing materials and technologies: A critical review. *Polymers, 15*(11), 2519. Advance online publication. DOI: 10.3390/polym15112519 PMID: 37299318

International Labour Organization. (2018). 24 million jobs to open up in the green economy. *World Employment and Social Outlook, 2018*, •••. https://www.ilo.org/global/about-the-ilo/newsroom/news/WCMS_628644/lang--en/index.htm

Iodice, G., Clemente, L., & Bifulco, F. (2024). The role of chatbot technology in the cultural value proposition: A managerial perspective in human-like interaction. In *The use of artificial intelligence in digital marketing: Competitive strategies and tactics* (pp. 166–187). IGI Global.

Iodice, G., Sergianni, C., Tregua, M., & Bifulco, F. (2025). Shaping Value Propositions in Cultural Heritage From a PSL Perspective: The Case of Grassroots Museums. In *Impacts of Museums on Global Communication* (pp. 213–240). IGI Global.

Islam, M. T., Sepanloo, K., Woo, S., Woo, S. H., & Son, Y. J. (2025). A Review of the Industry 4.0 to 5.0 Transition: Exploring the Intersection, Challenges, and Opportunities of Technology and Human–Machine Collaboration. *Machines 2025, Vol. 13, Page 267, 13*(4), 267. DOI: 10.3390/machines13040267

Jain, P., & Singh, P. (2023). Lethargic approach in expediting financial reporting system: A case of construction firm. *International Journal of Teaching and Case Studies, 14*(1), 1–10. DOI: 10.1504/IJTCS.2023.131663

Jain, P., Singh, P., & Kushwaha, R. (2025). Investigating change management as a novel digital wallet service innovation in banks: A case of Indian bank. *International Journal of Teaching and Case Studies, 15*(1), 67–81. DOI: 10.1504/IJTCS.2025.143552

Jamshidi, M. B., Sargolzaei, S., Foorginezhad, S., & Moztarzadeh, O. (2023). Metaverse and microorganism digital twins: A deep transfer learning approach. *Applied Soft Computing, 147*, 110798. DOI: 10.1016/j.asoc.2023.110798

Janssen, M., & van der Voort, H. (2016). Adaptive governance: Towards a stable, accountable and responsive government. *Government Information Quarterly, 33*(1), 1–5. DOI: 10.1016/j.giq.2016.02.003

Javaid, S., Sufian, A., Pervaiz, S., & Tanveer, M. (2018, February). Smart traffic management system using Internet of Things. In *2018 20th International Conference on Advanced Communication Technology (ICACT)* (pp. 393–398). IEEE. https://doi.org/DOI: 10.23919/ICACT.2018.8323770

Javaid, M., Haleem, A., Singh, R. P., & Suman, R. (2023). An integrated outlook of Cyber–Physical Systems for Industry 4.0: Topical practices, architecture, and applications. *Green Technologies and Sustainability*, *1*(1), 100001. DOI: 10.1016/j.grets.2022.100001

Javaid, M., Haleem, A., Singh, R. P., Suman, R., & Gonzalez, E. S. (2022). Understanding the adoption of Industry 4.0 technologies in improving environmental sustainability. *Sustainable Operations and Computers*, *3*, 203–217. DOI: 10.1016/j.susoc.2022.01.008

Jeandrain, A. (2023). *Applied Clinical Trials.* [Electronic] Available at: https://www.appliedclinicaltrialsonline.com/view/how-digital-twins-are-driving-sustainability-in-clinical-supply-planning [Access 2 June 2025].

Jiang, J. R. (2018). An improved cyber-physical systems architecture for Industry 4.0 smart factories. *Advances in Mechanical Engineering*, *10*(6), 1687814018784192. DOI: 10.1177/1687814018784192

Jiang, M., & Fan, B. (2024). Citizens' conscious engagement in digital platforms: advancing public service delivery within the public service logic framework. Public Management Review, 1–19. Fan, C. W. (2020). Applied the Technology Acceptance Model to Survey the mobile-learning adoption behavior in Science Museum. In Cross-Cultural Design. Applications in Health, Learning, Communication, and Creativity: 12th International Conference, CCD 2020, Held as Part of the 22nd HCI International Conference, HCII 2020, Copenhagen, Denmark, July 19–24, 2020 [Springer International Publishing.]. *Proceedings*, *22*(Part II), 270–280.

Jiang, Z., & Xu, C. (2023). Disrupting the technology innovation efficiency of manufacturing enterprises through digital technology promotion: An evidence of 5G technology construction in China. *IEEE Transactions on Engineering Management.*

Jian, Q., Ying, L., & Grosvenor, R. (2016). A categorical framework of manufacturing for Industry 4.0 and beyond. *Procedia CIRP*, *52*, 173–178. DOI: 10.1016/j.procir.2016.08.005

John, S. J. (2024). Harnessing AI for Business Process Automation in Educational Institutions. *Progress in Machines and Systems*, *13*(2), 57–64. DOI: 10.6025/pms/2024/13/2/57-64

Kaarlela, T., Pitkäaho, T., Pieskä, S., Padrão, P., Bobadilla, L., Tikanmäki, M., Haavisto, T., Blanco Bataller, V., Laivuori, N., & Luimula, M. (2023, May). Towards metaverse: Utilizing extended reality and digital twins to control robotic systems. [). MDPI.]. *Actuators*, *12*(6), 219. DOI: 10.3390/act12060219

Kamel Boulos, M., & Zhang, P. (2021). Digital twins: From personalized medicine to precision public health. *Journal of Personalized Medicine*, *11*(8), 745. DOI: 10.3390/jpm11080745 PMID: 34442389

Kang, J. S., Chung, K., & Hong, E. J. (2021). Multimedia knowledge-based bridge health monitoring using digital twin. *Multimedia Tools and Applications*, *80*(26), 34609–34624. DOI: 10.1007/s11042-021-10649-x

Kang, P. S., Wang, X., Son, J. Y., & Jat, M. (2024). Introduction to Service 4.0 and Customer-Centric Supply Chains. In *Service 4.0: Technology-Enabled Customer-Centric Supply Chains* (pp. 1–10). Springer Nature Switzerland. DOI: 10.1007/978-3-031-63875-6_1

Kaplinsky, R., & Kraemer-Mbula, E. (2022). Innovation and uneven development: The challenge for low-and middle-income economies. *Research Policy*, *51*(2), 104394. DOI: 10.1016/j.respol.2021.104394

Karaboga, T., Gurol, Y. D., Binici, C. M., & Sarp, P. (2020). Sustainable digital talent ecosystem in the new era: impacts on businesses, governments and universities. *Istanbul business research, 49*(2), 360-379.

Karakus, C., Huilgol, R., Wu, F., Subramanian, A., Daniel, C., Cavdar, D., Xu, T., Chen, H., Rahnama, A., & Quintela, L. (2021). Amazon SageMaker Model Parallelism: A General and Flexible Framework for Large Model Training. arXiv preprint arXiv:2111.05972.

Katsoulakis, E., Wang, Q., & Wu, H. (2024). Digital twins for health: A scoping review. *npj. Digital Medicine*, *7*(77). PMID: 38519626

Kauffman, R. J., & Walden, E. A. (2018). The Role of Digital Platforms in Business Model Innovation. *Journal of Business Research*, *100*, 1–10.

Kenney, M., & Zysman, J. (2023). The platform economy: Restructuring the space of capitalist accumulation. *Cambridge Journal of Regions, Economy and Society*, *16*(1), 55–76. DOI: 10.1093/cjres/rsaa001

Kenny, C. (2007). Construction, corruption, and developing countries. World Bank Policy Research Working Paper 4271.

Khan, M. I., & Khan, M. F. (2024). Applications of Artificial Intelligence in Sustainable Supply Chain Management in the Medical Sector. In *AI Applications for Business, Medical, and Agricultural Sustainability* (pp. 23-46). IGI Global. DOI: 10.4018/979-8-3693-5266-3.ch002

Khan, M. I., Naim, A., & Khan, M. F. (2024). Financial health management of otolaryngology by telemedicine: Opportunities and challenges. *Asia Pacific Journal of Health Management, 19*(3), 338–354. DOI: 10.24083/apjhm.v19i3.3729

Khan, M. I., Yasmeen, T., Khan, M., Hadi, N. U., Asif, M., Farooq, M., & Al-Ghamdi, S. G. (2025). Integrating industry 4.0 for enhanced sustainability: Pathways and prospects. *Sustainable Production and Consumption, 54*, 149–189. DOI: 10.1016/j.spc.2024.12.012

Khan, M. W., Saad, S., Ammad, S., & Rasheed, K. (2024). Smart infrastructure and AI. In *AI in Material Sciences*. Taylor & Francis. DOI: 10.1201/9781003438489-9

Khan, S. (2020). Comparative analysis of RPA tools—UiPath, Automation Anywhere and Blue Prism. *International Journal of Computer Science and Mobile Applications, 8*(11), 1–6. DOI: 10.47760/ijcsma.2020.v08i11.001

Khan, S. A. R., Piprani, A. Z., & Yu, Z. (2022). Digital technology and circular economy practices: Future of supply chains. *Operations Management Research : Advancing Practice Through Research, 15*(3), 676–688. DOI: 10.1007/s12063-021-00247-3

Khan, S., Naseer, M., Hayat, M., Zamir, S. W., Khan, F. S., & Shah, M. (2022). Transformers in vision: A survey. *ACM Computing Surveys, 54*(10s), 1–41. DOI: 10.1145/3505244

Klein, J. T. (2010). *Creating interdisciplinary campus cultures: A model for strengthening interdisciplinary education*. Jossey-Bass.

Kolade, O., Odumuyiwa, V., Abolfathi, S., Schröder, P., Wakunuma, K., Akanmu, I., Whitehead, T., Tijani, B., & Oyinlola, M. (2022). Technology acceptance and readiness of stakeholders for transitioning to a circular plastic economy in Africa. *Technological Forecasting and Social Change, 183*, 121954. DOI: 10.1016/j.techfore.2022.121954

Kondratenko, V., Okopnyk, O., Ziganto, L., & Kwilinski, A. (2020). Innovation Development of Public Administration: Management and Legislation Features. *Marketing and Management of Innovations, 1*(1), 56–64. DOI: 10.21272/mmi.2020.1-06

Kong, J. D., Akpudo, U. E., Effoduh, J. O., & Bragazzi, N. L. (2023, February). Leveraging responsible, explainable, and local artificial intelligence solutions for clinical public health in the Global South. [). MDPI.]. *Health Care, 11*(4), 457. PMID: 36832991

Kouhizadeh, M., & Sarkis, J. (2018). Blockchain practices, potentials, and perspectives in greening supply chains. *Sustainability (Basel)*, *10*(10), 3652. DOI: 10.3390/su10103652

Kourani, H., Berti, A., Schuster, D., & van der Aalst, W. M. P. 2024. ProMoAI: Process Modeling with Generative AI. In *Proceedings of the Thirty-Third International Joint Conference on Artificial Intelligence (IJCAI-24)*, pp. 8708–8712.

Kraus, S., Jones, P., Kailer, N., Weinmann, A., Chaparro-Banegas, N., & Roig-Tierno, N. (2021). Digital transformation: An overview of the current state of the art of research. *SAGE Open*, *11*(3), 21582440211047576. DOI: 10.1177/21582440211047576

Kritzinger, W., Karner, M., Traar, G., Henjes, J., & Sihn, W. (2018). Digital twin in manufacturing: A categorical literature review and classification. *IFAC-PapersOnLine*, *51*(11), 1016–1022. DOI: 10.1016/j.ifacol.2018.08.474

Kshetri, N. (2017). Blockchain's roles in meeting key supply chain management objectives. *International Journal of Information Management*, *39*, 80–89. DOI: 10.1016/j.ijinfomgt.2017.12.005

Kshetri, N. (2021). Blockchain and transparency in government. *Journal of Public Administration and Policy*, *34*(2), 112–129.

Kuegler, M., Dittes, S., Smolnik, S., & Richter, A. (2015). Connect me! Antecedents and impact of social connectedness in enterprise social software. *Business & Information Systems Engineering*, *57*(3), 181–196. DOI: 10.1007/s12599-015-0379-z

Kulandaivel, S., & Kulandaivel, M. (2025). Intelligent and Optical Security Monitoring for 6G Communications. In *RFID, Microwave Circuit, and Wireless Power Transfer Enabling 5/6G Communication* (pp. 417-444): IGI Global Scientific Publishing.

Kumar, A., Singh, P., Parihar, A.S. & Saraswat, S. (2025), "Analyzing the impact of psychological capital on the entrepreneurial intentions of students through mediating role of individual expectancy using PLS-SEM", *International Journal of Social Economics*, Vol. ahead-of-print No. ahead-of-print. DOI: 10.1108/IJSE-03-2024-0252

Kunz, W. H., & Wirtz, J. (2024). Corporate digital responsibility (CDR) in the age of AI: Implications for interactive marketing. *Journal of Research in Interactive Marketing*, *18*(1), 31–37. DOI: 10.1108/JRIM-06-2023-0176

Kurniawan, T. A., Othman, M. H. D., Hwang, G. H., & Gikas, P. (2022). Unlocking digital technologies for waste recycling in Industry 4.0 era: A transformation towards a digitalization-based circular economy in Indonesia. *Journal of Cleaner Production*, *357*, 131911. DOI: 10.1016/j.jclepro.2022.131911

Kushwaha, J., Sharma, A., & Singh, P. (2023). b. Exploration and prioritization of enablers to organization work–family balance planning for working sole Indian mothers integrating fuzzy Delphi and AHP. *International Journal of Social Economics, 50*(3), 398–418. DOI: 10.1108/IJSE-05-2022-0348

Kushwaha, J., Singh, P., & Kushwaha, R. (2024). Predicting working sole Indian mothers' satisfaction towards work–family balance integrating the Kano model and weighted average method. *International Journal of Social Economics, 51*(6), 725–740. DOI: 10.1108/IJSE-02-2023-0074

Kushwaha, J., Singh, P., & Sharma, A. (2023). a, "Modelling the enablers of work-family balance among working single mothers integrating expert-mining and TISM-MICMAC". *Journal of Organizational Effectiveness: People and Performance, 10*(4), 501–526. DOI: 10.1108/JOEPP-05-2022-0106

Lambert, D. M., & Cooper, M. C. (2000). Issues in supply chain management. *Industrial Marketing Management, 29*(1), 65–83. DOI: 10.1016/S0019-8501(99)00113-3

Latilo, A., Uzougbo, N. S., Ugwu, M. C., & Oduro, P. (2024). Developing legal frameworks for successful engineering, procurement, and construction projects. International Journal of Applied Research in Social Sciences.

Lee, C. P., Chang, K., & Berry, F. S. (2011). Testing the development and diffusion of e-government and e-democracy: A global perspective. *Public Administration Review, 71*(3), 444–454. DOI: 10.1111/j.1540-6210.2011.02228.x

Lee, H., & Özer, Ö. (2007). Unlocking the value of RFID. *Production and Operations Management, 16*(1), 40–64. DOI: 10.1111/j.1937-5956.2007.tb00165.x

Lee, J., Bagheri, B., & Kao, H. A. (2015). A cyber-physical systems architecture for Industry 4.0-based manufacturing systems. *Manufacturing Letters, 3*, 18–23. DOI: 10.1016/j.mfglet.2014.12.001

Lehdonvirta, V., & Ernkvist, M. (2011). Knowledge map of the virtual economy: Converting the virtual economy into development potential.

Leite, H. (2023). The impact of virtual reality on private value dimensions: A healthcare case study. *Public Management Review*, •••, 1–27.

Leone, D., Schiavone, F., Appio, F. P., & Chiao, B. (2021). How does artificial intelligence enable and enhance value co-creation in industrial markets? An exploratory case study in the healthcare ecosystem. *Journal of Business Research, 129*, 849–859. DOI: 10.1016/j.jbusres.2020.11.008

Lessig, L. (2022). The future of ideas: The fate of the commons in a connected world. *The Yale Law Journal, 131*(5), 1289–1332.

Li, Z. (2024). Digital Advertising and EU Digital Regulation. *The Routledge Handbook of EU Digital Regulation*, 282–292. https://doi.org/DOI: 10.4324/9781003168485-25

Li, L. (2022). Reskilling and upskilling the future-ready workforce for Industry 4.0 and beyond. *Information Systems Frontiers*. Advance online publication. DOI: 10.1007/s10796-022-10308-y PMID: 35855776

Li, R., Chung, T.-L., & Fiore, A. M. (2017). Factors affecting current users' attitude towards eauctions in China: An extended TAM study. *Journal of Retailing and Consumer Services, 34*, 19–29. DOI: 10.1016/j.jretconser.2016.09.003

Liu, S., Li, P., Wang, J., & Liu, P. (2024). Toward industry 5.0: Challenges and enablers of intelligent manufacturing technology implementation under the perspective of sustainability. *Heliyon, 10*(15), e35162. DOI: 10.1016/j.heliyon.2024.e35162 PMID: 39157342

Liu, X. (2023). *A New Perspective on Digital Twin-Based Mechanical Design in Industrial Engineering*. Innovations in Applied Engineering and Technology. DOI: 10.58195/iaet.v2i1.134

Liu, Y. (2020). Evaluating visitor experience of digital interpretation and presentation technologies at cultural heritage sites: A case study of the old town, Zuoying. *Built Heritage*, *4*(1), 14. DOI: 10.1186/s43238-020-00016-4

Li, X., Loscalzo, J., Mahmud, A., Aly, D. M., Rzhetsky, A., Zitnik, M., & Benson, M. (2025). Digital twins as global learning health and disease models for preventive and personalized medicine. *Genome Medicine*, *17*(11), 11. DOI: 10.1186/s13073-025-01435-7 PMID: 39920778

Li, X., Zhang, Y., & Tan, P. (2022). AI-powered transport optimization: The case of Singapore. *Journal of Intelligent Transport Systems*, *28*(4), 412–430.

Longo, F., Nicoletti, L., & Padovano, A. (2019). Ubiquitous knowledge empowers the smart factory: The impacts of a service-oriented digital twin on enterprises' performance. *Annual Reviews in Control*, *47*, 221–236. DOI: 10.1016/j.arcontrol.2019.01.001

López, C. P., Rodríguez, M. J. D., & Santos, S. de L. (2019). Tax Fraud Detection through Neural Networks: An Application Using a Sample of Personal Income Taxpayers. *Future Internet 2019, Vol. 11, Page 86, 11*(4), 86. DOI: 10.3390/fi11040086

Love, P. E. D., Ika, L. A., & Sing, C.-P. (2016). Overruns in transportation infrastructure projects: A systematic review and future research directions. *Transport Reviews*, *36*(6), 720–742.

Lubinski, T., Johri, S., Varosy, P., Coleman, J., Zhao, L., Necaise, J., Baldwin, C. H., Mayer, K., & Proctor, T. (2023). Application-oriented performance benchmarks for quantum computing. *IEEE Transactions on Quantum Engineering*, *4*, 1–32. Advance online publication. DOI: 10.1109/TQE.2023.3253761

Lu, Y. (2017). Industry 4.0: A survey on technologies, applications and open research issues. *Journal of Industrial Information Integration*, *6*, 1–10. DOI: 10.1016/j.jii.2017.04.005

Lu, Y.. (2021). Digital twin-driven smart manufacturing: Connotation, reference model, applications and research issues. *Robotics and Computer-Integrated Manufacturing. Τόμος*, *63*, 101897.

Lv, Z., Xie, S., Li, Y., Hossain, M. S., & El Saddik, A. (2022). Building the metaverse using digital twins at all scales, states, and relations. *Virtual Reality & Intelligent Hardware*, *4*(6), 459–470. DOI: 10.1016/j.vrih.2022.06.005

Magableh, I. K., Mahrouq, M. H., Ta'Amnha, M. A., & Riyadh, H. A. (2024). The Role of Marketing Artificial Intelligence in Enhancing Sustainable Financial Performance of Medium-Sized Enterprises Through Customer Engagement and Data-Driven Decision-Making. *Sustainability (Basel)*, *16*(24), 11279. DOI: 10.3390/su162411279

Magakwe, J. (2025). Advancing Governance: Role of Data Analytics in Driving Evidence-Based Decision-Making in Public Administration. In *Recent Advances in Public Sector Management*. IntechOpen.

Mahajan, R., Lim, W. M., Sareen, M., Kumar, S., & Panwar, R. (2023). Stakeholder theory. *Journal of Business Research*, *166*, 114104. DOI: 10.1016/j.jbusres.2023.114104

Maiurova, A., Kurniawan, T. A., Kustikova, M., Bykovskaia, E., Othman, M. H. D., Singh, D., & Goh, H. H. (2022). Promoting digital transformation in waste collection service and waste recycling in Moscow (Russia): Applying a circular economy paradigm to mitigate climate change impacts on the environment. *Journal of Cleaner Production*, *354*, 131604. DOI: 10.1016/j.jclepro.2022.131604

Maïzi, Y., Arcand, A., & Bendavid, Y. (2024). Digital twin in healthcare: Classification and typology of models based on hierarchy, application, and maturity. *Internet of Things : Engineering Cyber Physical Human Systems*, *28*, 101379. DOI: 10.1016/j.iot.2024.101379

Majeed, A., & Hwang, S. O. (2021). Data-driven analytics leveraging artificial intelligence in the era of COVID-19: An insightful review of recent developments. *Symmetry*, *14*(1), 16. DOI: 10.3390/sym14010016

Malik, S., Malik, P. K., & Naim, A. (2024). Opportunities and challenges in new generation cyber security applications using artificial intelligence, machine learning and block chain. *Next-generation cybersecurity: AI, ML, and Blockchain*, 23-37.

Mancero Arias, G. (2022). Innovation in engineering. *Athenea*, *3*(7), 33. DOI: 10.47460/athenea.v3i7.33

Manyika, J., & Sneader, K. (2023). AI, automation, and the future of work: Ten things to solve for. *The McKinsey Quarterly*, *2*(1), 1–15.

Maragno, G., Tangi, L., Gastaldi, L., & Benedetti, M. (2023). AI as an organizational agent to nurture: Effectively introducing chatbots in public entities. *Public Management Review*, *25*(11), 1–31. DOI: 10.1080/14719037.2022.2063935

Markets and Markets. (2025). *Markets and Markets.* [Electronic] Available at: https://www.marketsandmarkets.com/blog/HC/digital-twins-future-of-precision-healthcare [Access 8 June 2025].

Martel, R., & Latulippe, L. (2021). *Use of artificial intelligence in taxation: State of play according to the literature.* https://cffp.recherche.usherbrooke.ca/utilisation-de-lintelligence-artificielle-en-fiscalite-etat-de-situation-selon-la-litterature/

Martini, B., Bellisario, D., & Coletti, P. (2024). Human-Centered and Sustainable Artificial Intelligence in Industry 5.0: Challenges and Perspectives. *Sustainability 2024, Vol. 16, Page 5448, 16*(13), 5448. DOI: 10.3390/su16135448

Martín-Peña, M. L., Lorenzo, P. C., & Meyer, N. (2024). Digital platforms and business ecosystems: A multidisciplinary approach for new and sustainable business models. *Review of Managerial Science*, *18*(9), 2465–2482. DOI: 10.1007/s11846-024-00772-y

Masanet, E., Shehabi, A., Lei, N., Smith, S., & Koomey, J. (2020). Recalibrating global data center energy-use estimates. *Science*, *367*(6481), 984–986. DOI: 10.1126/science.aba3758 PMID: 32108103

Maschler, B., Braun, D., Jazdi, N., & Weyrich, M. (2021). Transfer learning as an enabler of the intelligent digital twin. *Procedia CIRP*, *100*, 127–132. DOI: 10.1016/j.procir.2021.05.020

Massey, A. (2019). Editorial: Innovation in public administration to leave no one behind. *Public Money & Management*, *39*(7), 495–497. DOI: 10.1080/09540962.2019.1665377

Matei, A., & Savulescu, C. (2014). Enhancing the capacity for innovation of public administration. An exploratory study on e-Governance, ICT, knowledge management in Romania. *Theoretical and Applied Economics*, *21*(11), 63–78.

Mazzucato, M. (2013). *The Entrepreneurial State: Debunking Public vs. Private Sector Myths*. Anthem Press.

Mazzucato, M. (2021). The value of everything: Making and taking in the global economy. *The Journal of Economic Perspectives*, *35*(4), 223–246. DOI: 10.1257/jep.35.4.223

McKinsey Global Institute. (2018). *Notes from the AI frontier: Modeling the impact of AI on the world economy*.

Medina-Borja, A. (2015). Editorial column—Smart things as service providers: A call for convergence of disciplines to build a research agenda for the service systems of the future. *Service Science*, *7*(1), ii–iv. DOI: 10.1287/serv.2014.0090

Mehr, H., Ash, H., & Fellow, D. (2017). Artificial intelligence for citizen services and government. *Ash Cent. Democr. Gov. Innov. Harvard Kennedy Sch*, (August), 1–12.

Mehta, N., & Shukla, S. (2022). Pandemic analytics: How countries are leveraging big data analytics and artificial intelligence to fight COVID-19? *SN Computer Science*, *3*(1), 54. DOI: 10.1007/s42979-021-00923-y PMID: 34778841

Meijer, C., Uh, H. & S. el Bouhaddani, S., (2023). Digital Twins in Healthcare: Methodological Challenges and Opportunities. *Journal of Personalized Medicine,* Volum 13.

Menon, A., He, Y., Bahirat, P., & Knijnenburg, B. P. (2018). A data-driven approach to designing for privacy in household IoT. *ACM Transactions on Interactive Intelligent Systems, 1*(1). *ACM Transactions on Interactive Intelligent Systems*, *1*(1), 1–47. DOI: 10.1145/3241378

Mergel, I. (2016). Digital Service Teams in Government. *Public Administration Review*, *76*(4), 565–577.

Mergel, I., Ganapati, S., & Whitford, A. B. (2019). Agile government: Systematic literature review and future research agenda. *Government Information Quarterly*, *36*(2), 273–285.

Mihai, S., Yaqoob, M., Hung, D. V., Davis, W., Towakel, P., Raza, M., Karamanoglu, M., Barn, B., Shetve, D., Prasad, R. V., Venkataraman, H., Trestian, R., & Nguyen, H. X. (2022). Digital twins: A survey on enabling technologies, challenges, trends and future prospects. *IEEE Communications Surveys and Tutorials*, *24*(4), 2255–2291. DOI: 10.1109/COMST.2022.3208773

Miller, P., & Wilsdon, J. (2001). Digital futures—An agenda for a sustainable digital economy. *Corporate Environmental Strategy*, *8*(3), 275–280. DOI: 10.1016/S1066-7938(01)00116-6

Minkiewicz, J., Evans, J., & Bridson, K. (2014). How do consumers co-create their experiences? An exploration in the heritage sector. *Journal of Marketing Management*, *30*(1-2), 30–59. DOI: 10.1080/0267257X.2013.800899

Misuraca, G., van Noordt, C., & Boukli, A. (2020). The use of AI in public services: Results from a preliminary mapping across the EU. In *Proceedings of the 13th international conference on theory and practice of electronic governance* (pp. 90-99). DOI: 10.1145/3428502.3428513

Mitra, S. (2023). Innovations in Indian public administration. *Public Money & Management*, *43*(8), 523–526. DOI: 10.1080/09540962.2022.2087946

Mittelstadt, B. D., Allo, P., Taddeo, M., Wachter, S., & Floridi, L. (2016). The ethics of algorithms: Mapping the debate. *Big Data & Society*, *3*(2), 1–21. DOI: 10.1177/2053951716679679

Moghayedi, M., Lim, C., & Tseng, M. L. (2022). AI-driven energy-efficient urban planning: A case study of Stockholm. *Journal of Sustainable Cities*, *15*(3), 423–438.

Mohamed, A. M. O., Mohamed, D., Fayad, A., & Al Nahyan, M. T. (2024). Enhancing decision-making and decarbonization in environmental management: A review on the role of digital technologies. *Sustainability (Basel)*, *16*(16), 7156. DOI: 10.3390/su16167156

Mohbey, K. K., Rajput, D. S., & Uddin, M. (2024). Technological developments in point of interest recommendation for smart and sustainable cities. *Frontiers in Sustainable Cities.*

Mokyr, J. (1990). *The lever of riches: technological creativity and economic progress*. Oxford University Press.

Molloy, S. (2019). The Lava Jato scandal: A case study of corruption and political transformation in Brazil. *Journal of Latin American Studies*, *51*(2), 321–345.

Morelli, S., Daniele, C., D'Avenio, G., Grigioni, M., & Giansanti, D. (2024, July). Optimizing Telehealth: Leveraging Key Performance Indicators for Enhanced Tele-Health and Digital Healthcare Outcomes (Telemechron Study). [). MDPI.]. *Health Care*, *12*(13), 1319. PMID: 38998854

Morton, F., Benavides, T. T., & González-Treviño, E. (2024). Taking customer-centricity to new heights: Exploring the intersection of AI, hyper-personalization, and customer-centricity in organizations. In *Smart Engineering Management* (pp. 23–41). Springer International Publishing. DOI: 10.1007/978-3-031-52990-0_2

Motevalli, S. H., & Razavi, H. (2024). Enhancing Customer Experience and Business Intelligence: The Role of AI-Driven Smart CRM in Modern Enterprises. *Journal of Business and Future Economy*, *1*(2), 1–8.

Mrugalska, B., & Wyrwicka, M. K. (2017). Towards lean production in Industry 4.0. *Procedia Engineering*, *182*, 466–473. DOI: 10.1016/j.proeng.2017.03.135

Mukhopadhyay, S., Singh, R. K., & Jain, T. (2024). Developing artificial intelligence enabled Marketing 4.0 framework: An Industry 4.0 perspective. *Qualitative Market Research*, *27*(5), 841–865. DOI: 10.1108/QMR-06-2023-0086

Murad, S. (2019). Towards an Integrated Perspective to Activate the Role of Artificial Intelligence and Business Intelligence in Supporting and Empowering the Public Sector in Vision 2030. *Economic Studies, 6*(2), 1–22. https://asjp.cerist.dz/en/article/104384

Musostova, D., Berkaeva, A., & Geraskin, Y. (2023). Digitalization as a factor of sustainable development of the economy and economic security. In *SHS Web of Conferences, 172*, 02024. EDP Sciences. DOI: 10.1051/shsconf/202317202024

Nadkarni, S., & Prügl, R. (2020). Digital transformation: A review, synthesis and opportunities for future research. *Management Review Quarterly*, *71*(2), 233–341. DOI: 10.1007/s11301-020-00185-7

Nagy, J., Oláh, J., Erdei, E., Máté, D., & Popp, J. (2018). The Role and Impact of Industry 4.0 and the Internet of Things on the Business Strategy of the Value Chain—The Case of Hungary. *Sustainability 2018, Vol. 10, Page 3491, 10*(10), 3491. DOI: 10.3390/su10103491

Naim, A., Mohammed, A. B., Fatima, N., Khan, S. A., Alnfiai, M. M., & Malik, P. K. (2024). Applications of Artificial Intelligence in Environmental Resource Business Management and Sustainability. In *AI Applications for Business, Medical, and Agricultural Sustainability* (pp. 1-22). IGI Global. DOI: 10.4018/979-8-3693-5266-3.ch001

Naim, A. (Ed.). (2024). *AI Applications for Business, Medical, and Agricultural Sustainability*. IGI Global. DOI: 10.4018/979-8-3693-5266-3

Naim, A., Alqahtani, H., Fatima, S., & Khan, M. F. (2023). Applications of artificial intelligence of things in green supply chain management: Challenges and future directions. In *Blockchain, IOT, and AI technologies for supply chain management* (pp. 99–126). CRC Press. DOI: 10.1201/9781003264521-5

Naim, A., Khan, M. F., Hussain, M. R., & Khan, N. (2019). Virtual Doctor" management technique in the diagnosis of ENT diseases. *JOE*, *15*(9), 88. DOI: 10.3991/ijoe.v15i09.10665

Negri-Ribalta, C., Lombard-Platet, M., & Salinesi, C. (2024). Understanding the GDPR from a Requirements Engineering Perspective—A Systematic Mapping Study on Regulatory Data Protection Requirements. *Requirements Engineering*, *29*(4), 523–549. Advance online publication. DOI: 10.1007/s00766-024-00423-4

Neumann, O., Guirguis, K., & Steiner, R. (2022). Exploring artificial intelligence adoption in public organizations: A comparative case study. *Public Management Review*, *26*(1), 114–141. DOI: 10.1080/14719037.2022.2048685

Nguyen, T. T., Grote, U., Neubacher, F., Rahut, D. B., Do, M. H., & Paudel, G. P. (2023). Security risks from climate change and environmental degradation: Implications for sustainable land use transformation in the Global South. *Current Opinion in Environmental Sustainability*, *63*, 101322. DOI: 10.1016/j.cosust.2023.101322

Nilsson, N. J. (2010). *The Quest for Artificial Intelligence A History of Ideas and Achievements-The Quest for Artificial Intelligence: A History of Ideas and Achievements Nils J. Nilsson Frontmatter More information*. www.cambridge.org/9780521122931

Nizar, N. N. M., Rahmat, M. K., Nabihah, N., & Nizar, M. (2018). Examining the museum visitors' use of mobile technology through technology acceptance model (TAM). *Journal of Tourism, Hospitality, and Event Management*, *3*(11), 14–24.

Nogueira, E., Gomes, S., & Lopes, J. M. (2024). Financial Sustainability: Exploring the Influence of the Triple Bottom Line Economic Dimension on Firm Performance. *Sustainability (Basel)*, *2024*(16), 6458. DOI: 10.3390/su16156458

Normann, R., & Ramirez, R. (1993). From value chain to value constellation: Designing interactive strategy. *Harvard Business Review*, *71*(4), 65–77. PMID: 10127040

Nugraha, D. P., Setiawan, B., Nathan, R. J., & Fekete-Farkas, M. (2022). FinTech adoption drivers for innovation for SMEs in Indonesia. *Journal of Open Innovation*, *8*(4), 208. DOI: 10.3390/joitmc8040208

Nyquist, A. M., Farshid, M., & Brown, T. (2025). Employing digital twin technology in the pursuit to avert sustainable marketing myopia. *Journal of Research in Marketing and Entrepreneurship*.

Obadah, I. (2023). Using AI for auditing tax reports. *Hala.Jo*. https://www.hala.jo/?p=538400

Ocón, D. (2023). Low-cost digital tools to preserve cultural heritage 'blind spots': The case of Kubor Kassim in Singapore. *International Journal of Heritage Studies*, *29*(9), 961–987. DOI: 10.1080/13527258.2023.2231903

Ogundeko-Olugbami, O., Ogundeko, O., Lawan, M., & Foster, E. (2025). Harnessing data for impact: Transforming public health interventions through evidence-based decision-making.

Olawade, D. B., Wada, O. J., David-Olawade, A. C., Kunonga, E., Abaire, O., & Ling, J. (2023). Using artificial intelligence to improve public health: A narrative review. *Frontiers in Public Health*, *11*, 1196397. DOI: 10.3389/fpubh.2023.1196397 PMID: 37954052

Olcay, G. A. (2021). International Journal of Innovation and Technology Management-Special Issue in Honor of Professor M. Atilla Öner (1955-2018) Managing Emergent Technologies. *International Journal on Management of Innovation & Technology*, *18*(1), 1–25. DOI: 10.1142/S0219877021020016

Ølnes, S., Ubacht, J., & Janssen, M. (2017). Blockchain in government: Benefits and implications of distributed ledger technology for information sharing. *Government Information Quarterly*, *34*(3), 355–364. DOI: 10.1016/j.giq.2017.09.007

Oloyede, A. A., Faruk, N., Noma, N., Tebepah, E., & Nwaulune, A. K. (2023). Measuring the impact of the digital economy in developing countries: A systematic review and meta-analysis. *Heliyon*, *9*(7), e17654. DOI: 10.1016/j.heliyon.2023.e17654 PMID: 37501966

Oluseye, O. (2024). *Exploring potential political corruption in large-scale infrastructure projects in Nigeria*. Project Leadership and Society. DOI: 10.1016/j.plas.2023.100108

Ongaro, E., Gong, T., & Jing, Y. (2021). Public administration, context and innovation: A framework of analysis. *Public Administration and Development*, *41*(1), 47–56. DOI: 10.1002/pad.1902

Onghena, P., Maes, B., & Heyvaert, M. (2019). Mixed methods single case research: State of the art and future directions. *Journal of Mixed Methods Research*, *13*(4), 461–480. DOI: 10.1177/1558689818789530

Orea-Giner, A., Muñoz-Mazón, A., Villacé-Molinero, T., & Fuentes-Moraleda, L. (2022). Cultural tourist and user experience with artificial intelligence: a holistic perspective from the Industry 5.0 approach. Journal of Tourism Futures, (ahead-of-print).

Osborne, S. P. (2018). From Public Service-dominant Logic to Public Service Logic: Are Public Service Organizations Capable of Co-production and Value Co-creation? *Public Management Review*, *20*(2), 225–231. DOI: 10.1080/14719037.2017.1350461

Osborne, S. P., Nasi, G., & Powell, M. (2021). Beyond co-production: Value creation and public services. *Public Administration*, *99*(4), 641–657. DOI: 10.1111/padm.12718

Osborne, S. P., & Strokosch, K. (2013). It takes Two to Tango? Understanding the C o-production of Public Services by Integrating the Services Management and Public Administration Perspectives. *British Journal of Management*, *24*(S1), S31–S47. DOI: 10.1111/1467-8551.12010

Ou, S., Gan, J. L., & Li, L. (2024). Innovative Culture: A Fifty-year (1972-2023) Analysis and Future Direction. *Journal of Innovation Management*, *12*(3), 72–101. DOI: 10.24840/2183-0606_012.003_0004

Pakhnenko, O., & Kuan, Z. (2023). Ethics of Digital Innovation in Public Administration. *Business Ethics and Leadership*, *7*(1), 113–121. DOI: 10.21272/bel.7(1).113-121.2023

Palozzi, G., & Ranalli, F. (2023). Telemedicine implementation between innovation and sustainability: an operating model for designing patient-centered healthcare. In *Human-Centered Service Design for Healthcare Transformation: Development, Innovation, Change* (pp. 375–399). Springer International Publishing. DOI: 10.1007/978-3-031-20168-4_21

Pan, W., Xie, T., Wang, Z., & Ma, L. (2022). Digital economy: An innovation driver for total factor productivity. *Journal of Business Research*, *139*, 303–311. DOI: 10.1016/j.jbusres.2021.09.061

Papachristou, K. et al., (2024). Digital Twins' Advancements and Applications in Healthcare, Towards Precision Medicine. *J Pers Med*, 11 November, 14(11), p. 1101.

Parker, G. G., Van Alstyne, M. W., & Choudary, S. P. (2023). Platform revolution: How networked markets are transforming the economy. *Strategic Management Journal*, *44*(3), 567–589. DOI: 10.1002/smj.3456

Park, S., Shou, W., Makatura, L., Matusik, W., & Fu, K. (2022). 3D printing of polymer composites: Materials, processes, and applications. *Matter*, *5*(1), 43–76. Advance online publication. DOI: 10.1016/j.matt.2021.10.018

Parycek, P., & Scholl, H. J. (2013). The Role of Open Government Data in Public Administration. *International Journal of Public Administration*, *36*(5), 345–356.

Patel, D. (2018). The impact of digital technologies on society. *Digital India*. https://www.gbim.com/the-impact-of-digital-technologies-on-society/

Pellegrino, G., Gervasi, M., Angelelli, M., & Corallo, A. (2025). A Conceptual Framework for Digital Twin in Healthcare: Evidence from a Systematic Meta-Review. *Information Systems Frontiers*, *27*(1), 7–32. DOI: 10.1007/s10796-024-10536-4

Pencheva, I., Esteve, M., & Mikhaylov, S. J. (2020). Big Data and AI–A transformational shift for government: So, what next for research? *Public Policy and Administration*, *35*(1), 24–44. DOI: 10.1177/0952076718780537

Pendyala, S. K. (2024). Healthcare Data Analytics: Leveraging Predictive Analytics For Improved Patient Outcomes. *International Journal Of Computer Engineering And Technology (Ijcet)*, *15*(6), 548–565.

Phuangthuean, P., & Nuansang, J. (2024). The transforming public administration: The role of AI in shaping the future. [JSSMR]. *Journal of Social Science and Multidisciplinary Research*, *1*(3), 21–41.

Phuyal, S., Bista, D., & Bista, R. (2020). Challenges, Opportunities and Future Directions of Smart Manufacturing: A State of Art Review. *Sustainable Futures : An Applied Journal of Technology, Environment and Society*, *2*, 100023. DOI: 10.1016/j.sftr.2020.100023

Picone, M., Mariani, S., Virdis, A., & Castagnetti, P. (2023, June). Digital Twin & Blockchain: Technology Enablers for Metaverse Computing. In *2023 IEEE International Conference on Metaverse Computing, Networking and Applications (MetaCom)* (pp. 1–8). IEEE. DOI: 10.1109/MetaCom57706.2023.00017

Piehler, R., Wirtz, B. W., & Daiser, P. (2016). An analysis of continuity intentions of eGovernment portal users. *Public Management Review*, *18*(2), 163–198. DOI: 10.1080/14719037.2014.965270

Pipiay, G. T., Chernenkaya, L. V., & Mager, V. E. (2021, January). Quality indicators of instrumentation products according to the "Quality 4.0" concept. In *IEEE Conference of Russian Young Researchers in Electrical and Electronic Engineering (ElConRus)* (pp. 1032–1036). https://doi.org/DOI: 10.1109/ElConRus51938.2021.9396535

Podmurnyi, S. (2022). *Hackernoon.* [Electronic] Available at: https://hackernoon .com/how-to-implement-digital-twin-architecture [Access 5 June 2025].

Pons, M., Valenzuela, E., Rodríguez, B., Nolazco-Flores, J. A., & Del-Valle-Soto, C. (2023). Utilization of 5G technologies in IoT applications: Current limitations by interference and network optimization difficulties—A review. *Sensors (Basel)*, *23*(8), 3876. DOI: 10.3390/s23083876 PMID: 37112216

Popa, E., van Hilten, M., Oosterkamp, E., & Bogaardt, M.-J. (2021). The use of digital twins in healthcare: Socio-ethical benefits and socio-ethical risks. *Life Sciences, Society and Policy*, *17*(6), 6. DOI: 10.1186/s40504-021-00113-x PMID: 34218818

Pregnolato, M., Gunner, S., Voyagaki, E., De Risi, R., Carhart, N., Gavriel, G., Tully, P., Tryfonas, T., Macdonald, J., & Taylor, C. (2022). Towards Civil Engineering 4.0: Concept, workflow and application of Digital Twins for existing infrastructure. *Automation in Construction*, *141*, 104421. DOI: 10.1016/j.autcon.2022.104421

Prüfer, J., & Schottmüller, C. (2021). Competing with big data. *The Journal of Industrial Economics*, *69*(4), 967–1003. DOI: 10.1111/joie.12259

Quan, H., Zhang, T., Xu, H., Luo, S., Nie, J., & Zhu, X. (2020). Photo-curing 3D printing technique and its challenges. *Bioactive Materials*, *5*(1), 110–115. Advance online publication. DOI: 10.1016/j.bioactmat.2019.12.003 PMID: 32021945

Quazi, F., Raju, N., Viradia, V. & Aparna, B., (2024). Digital Twin Technology in Healthcare: Applications, Challenges, and Future Insights. *International Journal of Global Innovations and Solutions (IJGIS).*

Rafique, W., & Qadir, J. (2024). Internet of everything meets the metaverse: Bridging physical and virtual worlds with blockchain. *Computer Science Review*, *54*, 100678. DOI: 10.1016/j.cosrev.2024.100678

Ragazou, K., Passas, I., & Sklavos, G. (2022). Investigating the strategic role of digital transformation path of SMEs in the era of COVID-19: A bibliometric analysis using R. *Sustainability (Basel)*, *14*(18), 11295. DOI: 10.3390/su141811295

Ragnedda, M. (2022). The third digital divide: A Weberian approach to digital inequalities. *New Media & Society*, *24*(1), 28–47. DOI: 10.1177/1461444820986813

Raikov, A. (2021). Decreasing Tax Evasion by Artificial Intelligence. *IFAC-PapersOnLine*, *54*(13), 172–177. DOI: 10.1016/j.ifacol.2021.10.440

Raisch, S., & Krakowski, S. (2021). Artificial intelligence and management: The automation–augmentation paradox. *Academy of Management Review*, *46*(1), 192–210. DOI: 10.5465/amr.2018.0072

Raj, M., & Seamans, R. (2019). Primer on artificial intelligence and robotics. *Journal of Organization Design*, *8*(1), 1–14. DOI: 10.1186/s41469-019-0050-0

Rame, R., Purwanto, P., & Sudarno, S. (2024). Industry 5.0 and sustainability: An overview of emerging trends and challenges for a green future. *Innovation and Green Development*, *3*(4), 100173. DOI: 10.1016/j.igd.2024.100173

Rane, L., & Hallur, G. G. (2023). Competition in the RPA industry: A case study of Blue Prism. *AIP Conference Proceedings*, *2914*(1), 040009. Advance online publication. DOI: 10.1063/5.0176028

Raoufi, K., Sutherland, J. W., Zhao, F., Clarens, A. F., Rickli, J. L., Fan, Z., Huang, H., Wang, Y., Lee, W. J., Mathur, N., Triebe, M. J., Desabathina, S. S., & Haapala, K. R. (2024). Current state and emerging trends in advanced manufacturing: Smart systems. *International Journal of Advanced Manufacturing Technology*, *134*(7), 3031–3050. DOI: 10.1007/s00170-024-14279-z

Rapp, A., Curti, L., & Boldi, A. (2021). The human side of human-chatbot interaction: A systematic literature review of ten years of research on text-based chatbots. *International Journal of Human-Computer Studies*, *151*, 151. DOI: 10.1016/j.ijhcs.2021.102630

Rasheed, A., San, O., & Kvamsdal, T. (2020). Digital twin: Values, challenges and enablers from a modeling perspective. *IEEE Access. Volum*, *8*, 21980–22012.

Rashid, A., & Kausik, M. D. A. K. (2024). AI revolutionizing industries worldwide: A comprehensive overview of its diverse applications. *Hybrid Advances*, *7*, 100277. DOI: 10.1016/j.hybadv.2024.100277

Ravasoo, A. (2014). Interaction of bursts in exponentially graded materials characterized by parametric plots. *Wave Motion*, *51*(5), 758–767. DOI: 10.1016/j.wavemoti.2014.01.006

Ravulavaru, A. (2019). Google Cloud AI Services Quick Start Guide: Build intelligent applications with machine learning and artificial intelligence services on the Google Cloud Platform. Packt Publishing. SAP. Available at: https://www.seleccionconsulting.com/blog/how-sap-btp-enhances-manufacturing-retail-finance/ [Accessed: 04 June 2025]

Rehman, A., Naz, S., & Razzak, I. (2022). Leveraging big data analytics in healthcare enhancement: Trends, challenges and opportunities. *Multimedia Systems*, *28*(4), 1339–1371. DOI: 10.1007/s00530-020-00736-8

Repko, A. F. (2008). *Interdisciplinary Research: Process and Theory*. SAGE Publications.

Rhoten, D., & Parker, A. (2004). Risks and rewards of an interdisciplinary research environment. *Science*, *306*(5704), 2046–2047. DOI: 10.1126/science.1103628 PMID: 15604393

Richardson, N., James, J., & Kelley, N. (2015). *Customer-centric marketing: Supporting sustainability in the digital age*. Kogan Page Publishers.

Richter, A., Heinrich, P., Stocker, A., & Schwabe, G. (2018). Digital work design—The interplay of human and computer in future work practices as an interdisciplinary (grand) challenge. *Business & Information Systems Engineering*, *60*(3), 259–264. DOI: 10.1007/s12599-018-0534-4

Rietsche, R., Dremel, C., Bosch, S., Steinacker, L., Meckel, M., & Leimeister, J. M. (2022). Quantum computing. *Electronic Markets*, *32*(4), 2525–2536. Advance online publication. DOI: 10.1007/s12525-022-00570-y

Risdwiyanto, A., Sulaeman, M. M., & Rachman, A. (2023). Sustainable digital marketing strategy for long-term growth of msmes. [ADMAN]. *Journal of Contemporary Administration and Management*, *1*(3), 180–186. DOI: 10.61100/adman.v1i3.70

Rizk, H. R. M. (2021). Artificial Intelligence Systems and The Future of Education. *Studies in University Education*, *52*(52), 573–587. DOI: 10.21608/DEU.2021.224996<BR>

Rocha, J. A. O., & Zavale, G. J. B. (2021). Innovation and Change in Public Administration. *Open Journal of Social Sciences*, *9*(6), 123–136. DOI: 10.4236/jss.2021.96021

Rodrigues, L., Pereira, J., da Silva, A. F., & Ribeiro, H. (2023). The impact of artificial intelligence on audit profession. *Journal of Information Systems Engineering & Management*, *8*(1), 19002. Advance online publication. DOI: 10.55267/iadt.07.12743

Romanelli, M. (2020). Museums and technology for value creation. Technology and Creativity: Production, Mediation and Evaluation in the Digital Age, 181-210.

Roopa, M. (2023). A Systematic Review: Digital Twin in Healthcare. *International Journal for Research in Applied Science and Engineering Technology*.

Roos, M., & Reccius, M. (2024). Narratives in economics. *Journal of Economic Surveys*, *38*(2), 303–341. DOI: 10.1111/joes.12576

Rosen, R., Wichert, G. V., Lo, G., & Bettenhausen, K. D. (2015). About the importance of autonomy and digital twins for the future of manufacturing. *IFAC-PapersOnLine*, *48*(3), 567–572. DOI: 10.1016/j.ifacol.2015.06.141

Russo Spena, T., & Bifulco, F. (Eds.). (2021). *Digital Transformation in the Cultural Heritage Sector: Challenges to Marketing in the New Digital Era*. Springer Nature. DOI: 10.1007/978-3-030-63376-9

Sa-ad, M. M., Osafo-Apeanti, W., Owusu-Boateng, O., Seidu, S. A., Majeed, M., & Bright-Benuwa, B. (2025). Automated administrative tasks in education. In Owusu-Boateng, O. (Ed.), *Unlocking the Potential: Artificial Intelligence in Education* (pp. 31–45). Deep Science Publishing.

Saberi, S., Kouhizadeh, M., Sarkis, J., & Shen, L. (2019). Blockchain technology and its relationships to sustainable supply chain management. *International Journal of Production Research*, *57*(7), 2117–2135. DOI: 10.1080/00207543.2018.1533261

Sacco, P. L., Ferilli, G., & Tavano Blessi, G. (2018). From culture 1.0 to culture 3.0: Three socio-technical regimes of social and economic value creation through culture, and their impact on European Cohesion Policies. *Sustainability (Basel)*, *10*(11), 3923. DOI: 10.3390/su10113923

Sadeghi, A. R., Wachsmann, C., & Waidner, M. (2015). Security and privacy challenges in industrial Internet of Things. In *Proceedings of the 52nd Annual Design Automation Conference* (pp. 1–6). https://doi.org/DOI: 10.1145/2744769.2747942

Saleem, J., Hammoudeh, M., Raza, U., Adebisi, B., & Ande, R. (2018). IoT standardisation: Challenges, perspectives and solution. In *Proceedings of the 2nd International Conference on Future Networks and Distributed Systems* (pp. 1–9). https://doi.org/DOI: 10.1145/3231053.3231103

Saleh, T. (2022). Tax challenges arising from the digital economy and efforts to address them - with reference to the case of Arab countries as a model -. *Journal of Tax Studies*, *11*(1), 67–93. https://asjp.cerist.dz/en/article/199707

Salter, A., & Gann, D. (2003). Sources of ideas for innovation in engineering design. *Research Policy*, *32*(8), 1161–1181. DOI: 10.1016/S0048-7333(02)00119-1

Saraswat, S. & Singh, P. (2025), "The moderating role of perceived trust in predicting the adoption intention of consumers toward online pharmacies using the application of UTAUT model", *International Journal of Pharmaceutical and Healthcare Marketing*, Vol. ahead-of-print No. ahead-of-print. DOI: 10.1108/IJPHM-05-2024-0042

Sarker, I. H. (2022). AI-based modeling: Techniques, applications and research issues towards automation, intelligent and smart systems. *SN Computer Science*, *3*(2), 158. DOI: 10.1007/s42979-022-01043-x PMID: 35194580

Sarker, I. H., Janicke, H., Ferrag, M. A., & Abuadbba, A. (2024). Multi-aspect rule-based AI: Methods, taxonomy, challenges, and directions toward automation, intelligence, and transparent cybersecurity modeling for critical infrastructures. *Internet of Things : Engineering Cyber Physical Human Systems*, *25*, 101110. DOI: 10.1016/j.iot.2024.101110

Sarker, I. H., Kayes, A. S. M., Badsha, S., Alqahtani, H., Watters, P., & Ng, A. (2020). Cybersecurity data science: An overview from machine learning perspective. *Journal of Big Data*, *7*(1), 1–29. DOI: 10.1186/s40537-020-00318-5

Sarker, S., & Sahay, S. (2004). The Role of Information Technology in Public Administration: A Review of the Literature. *Public Administration Review*, *64*(3), 1–10.

Schaffer, S., Ruß, A., Sasse, M. L., Schubotz, L., & Gustke, O. (2022, February). Questions and Answers: Important Steps to Let AI Chatbots Answer Questions in the Museum. In ArtsIT, Interactivity and Game Creation: Creative Heritage. New Perspectives from Media Arts and Artificial Intelligence. 10th EAI International Conference, ArtsIT 2021, Virtual Event, December 2-3, 2021, Proceedings (pp. 346-358). Cham: Springer International Publishing.

Schneider, S., & Kokshagina, O. (2021). Digital transformation: What we have learned (thus far) and what is next. *Creativity and Innovation Management*, *30*(2), 384–411. DOI: 10.1111/caim.12414

Scholl, H. J., & Scholl, M. C. (2014). The Role of Information Technology in Public Administration. *Public Administration Review*, *74*(4), 487–498.

Scott, J. (2024). *HealthTeck*. [Electronic] Available at: https://healthtechmagazine.net/article/2024/01/what-are-digital-twins-and-how-can-they-be-used-healthcare [Access 23 May 2025].

Seneviratne, S., Hu, Y., Nguyen, T., Lan, G., Khalifa, S., Thilakarathna, K., Hassan, M., & Seneviratne, A. (2017). A survey of wearable devices and challenges. *IEEE Communications Surveys and Tutorials, 19*(4), 2573–2620. DOI: 10.1109/COMST.2017.2731979

Sepe, F., & Marzullo, M. (2022). Making Smarter Museums Through New Technologies. In Handbook of Research on Museum Management in the Digital Era (pp. 75-98). IGI Global. DOI: 10.4018/978-1-7998-9656-2.ch005

Serrano Antón, F. (2021). *Artificial Intelligence and Tax Administration : Strategy, Applications and Implications, with Special Reference to the Tax Inspection Procedure*. https://papers.ssrn.com/abstract=4223007

Sfodera, F., Mingo, I., Mattiacci, A., & Colurcio, M. (2020). Night at the museum: Technology enables visitor experiences. *Sinergie Italian Journal of Management, 38*(1), 231–250.

Sharma, A., Kosasih, E., Zhang, J., Brintrup, A., & Calinescu, A. (2022). Digital twins: State of the art theory and practice, challenges, and open research questions. *Journal of Industrial Information Integration, 30*, 100383. DOI: 10.1016/j.jii.2022.100383

Sharma, J. (2016). Digital India and its impact on the society. *International Journal of Research in Humanities & Social Sciences, 4*(4), 64–70.

Sharma, R., Singh, R., & Reddy, P. (2020). Blockchain for government accountability: A framework for public sector financial management. *Government Information Quarterly, 37*(2), 101401.

Shehab, M., Schofield, M., Bracco, G., & Wiggins, A. (2016). *Tax as a critical component in every Finance transformation*. www.pwc.com/taxfunctionofthefuture

Sheth, J. N., Jain, V., Mogaji, E., & Ambika, A. (2024). *Customer Centric Support Services in the Digital Age*. Springer. DOI: 10.1007/978-3-031-37097-7

Shi, M., Wang, Q., & Long, Y. (2023). Exploring the Key Drivers of User Continuance Intention to Use Digital Museums: Evidence from China's Sanxingdui Museum. *IEEE Access : Practical Innovations, Open Solutions, 11*, 81511–81526. DOI: 10.1109/ACCESS.2023.3297501

Shukurova, N. K. (2025). Economic efficiency and innovative development strategies in the service sector. *EduVision: Journal of Innovations in Pedagogy and Educational Advancements, 1*(2), 112–119. https://brightmindpublishing.com/index.php/ev/article/view/97

Siau, K., & Long, Y. (2018). Artificial Intelligence in Business: A Review of the Literature. *Journal of Business Research, 100*, 1–10.

SIMIO. (2025). *Simio.* [Electronic] Available at: https://www.simio.com/how-will-digital-twins-software-transform-your-business-in-2025/ [Access 20 May 2025].

Singh, A. K., Meshram, S. A., Khandelwal, V., Tiwari, P., & Singh, P. (2024c). Implementation of Industry-Based Perspective in Technical Education: A Qualitative and Quantitative Analysis. *Industrial and Manufacturing Designs: Quantitative and Qualitative Analysis*, 321-346. https://doi.org/.DOI: 10.1002/9781394212668.ch11

Singh, P. & Kushwaha, R. (2024). "Envisaging Indian farmers' desires from agricultural index insurance integrating rank sum weighting method and MoSCoW technique: an approach to requirements prioritization", International Journal of Social Economics, Vol. ahead-of-print No. ahead-of-print. .DOI: 10.1108/IJSE-03-2024-0233

Singh, P. (2014). Unorganized sector and India's informal economy: challenges and prospects. *New Man International Journal of Multidisciplinary Studies, 1*(2), 209-218. https://www.newmanpublication.com/dash/issueworkfiles/71.pdf?1721465584

Singh, P. (2024b). Analyzing Responsible Leadership Effectiveness for Attainment of Sustainable Development Goals in Developing Countries Using Environmental Scanning Techniques. In: Bhattacharya, S., Venkatesh, V.G., Chatterjee, S.R. (eds) Responsible Corporate Leadership Towards Attainment of Sustainable Development Goals. Responsible Leadership and Sustainable Management. Springer, Singapore. https://doi.org/DOI: 10.1007/978-981-97-1386-8_13

Singh, P., & Agrawal, G. (2021). Modelling the barriers of weather index insurance service adoption integrating expert mining and ISM Fuzzy-MICMAC. *Benchmarking: An International Journal, 29*(8), 2527-2554. . https://doi.org/.DOI: 10.1108/BIJ-04-2021-0183

Singh, P., & Kushwaha, R. (2025). "Conceptualizing the corporate-social accountability and sustainability assessment of weather-risk management policy towards stakeholders using qualitative insights: a case of weather index insurance", *Qualitative Research in Financial Markets*, Vol. ahead-of-print No. ahead-of-print. DOI: 10.1108/QRFM-05-2024-0113

Singh, P., Agrawal, G., & Kushwaha, R. (2021a, November) Evaluating Customer Satisfaction of Weather Index Insurance Service Quality Using Kano Model. In *2021 IEEE Bombay Section Signature Conference (IBSSC)* (pp. 1-6). IEEE. **DOI:** DOI: 10.1109/IBSSC53889.2021.9673370

Singh, P., Kushwaha, R., & Kushwaha, J. (2021b). Assessing visitors' satisfaction about museum service quality using Kano model. In *2021 IEEE 9th Region 10 Humanitarian Technology Conference (R10-HTC)* (pp. 1-6). IEEE. DOI: DOI: 10.1109/R10-HTC53172.2021.9641707

Singh, P., Kushwaha, R., & Kushwaha, J. (2023a). Analyzing Accountability of Weather Index Insurance Service in Attainment of Sustainable Development Goals: A Sustainable Accounting Perspective Review. In Pratik Arte et al. (Eds): *Sustainable International Business, 2023*. Springer Nature, 2023. doi: , 531340. DOI: 978-3-031-43784-7

Singh, P., Kushwaha, R., & Kushwaha, J. (2023b). Evaluating Farmer Participation in Weather Index Insurance in a Socio-Demographic Perspective: Evidence From the Crop Insurance Data of India. In A. Popoola, H. Magidimisha-Chipungu, & L. Chipungu (Eds.), *Handbook of Research on Managing the Urban-Rural Divide Through an Inclusive Framework* (pp. 69-88). IGI Global. DOI: 10.4018/978-1-6684-6258-4.ch005

Singh, P., Kushwaha, R., & Kushwaha, J. (2024a). Foresight the future of Industry 5.0 in perspective of sustainable services in developing countries. *Modern Technologies and Tools Supporting the Development of Industry 5.0*, 381. DOI: .DOI: 10.1201/9781003489269-11

Singh, P., Kushwaha, R., & Kushwaha, J. (2025a). "A Qualitative Study on Analysing the Role of Mimicking Technology in Electronic Customer Relationship Management: A Case of Service Sector", Kumar, A., Ciddikie, M.D., Kashyap, A.K. and Akram, H.W. (Ed.) *Marketing 5.0*, Emerald Publishing Limited, Leeds, pp. 103-119. DOI: 10.1108/978-1-83797-815-120251008

Singh, P., Kushwaha, R., & Kushwaha, J. (2025b). Retail Innovation's Impact on Competitiveness in the Retail Sector: Evidence From SAP-LAP Analysis. In *Retail Innovations in Business Models* (pp. 255-278). IGI Global Scientific Publishing.

Singh, P., Kushwaha, R., Kushwaha, J., & Singh, A. K. (2024b). A Qualitative Study to Rank Non-Conventional Energy Sources for Industrial Sustainability and Energy Management Decisions Using MoSCoW Prioritization Method. *Industrial and Manufacturing Designs: Quantitative and Qualitative Analysis*, 105-137. https://doi.org/DOI: 10.1002/9781394212668.ch4

Singh, K. A., Patra, F., Ghosh, T., Mahnot, N. K., Dutta, H., & Duary, R. K. (2025). Advancing food systems with industry 5.0: A systematic review of smart technologies, sustainability, and resource optimization. *Sustainable Futures : An Applied Journal ofTechnology, Environment and Society*, *9*, 100694. DOI: 10.1016/j.sftr.2025.100694

Singh, M., Srivastava, R., Fuenmayor, E., Kuts, V., Qiao, Y., Murray, N., & Devine, D. (2022). Applications of Digital Twin across Industries: A Review. *Applied Sciences (Basel, Switzerland), 12*(11), 5727. DOI: 10.3390/app12115727

Singh, N., & Bishnoi, S. (2024). Navigating GDPR Compliance: The Intersection of Data Governance, Accountability, and Organizational Culture. *International Journal of Innovative Research in Engineering & Multidisciplinary Physical Sciences, 12*(4), 230875. Advance online publication. DOI: 10.37082/IJIRMPS.v12.i4.230875

Singh, P. (2024). a. Weather index insurance viability in mitigation of climate change impact risk: A systematic review and future agenda. *Journal of Science and Technology Policy Management, 15*(1), 142–163. DOI: 10.1108/JSTPM-07-2021-0102

Singh, P., & Agrawal, G. (2019). Efficacy of weather index insurance for mitigation of weather risks in agriculture: An integrative review. *International Journal of Ethics and Systems, 35*(4), 584–616. DOI: 10.1108/IJOES-09-2018-0132

Singh, P., & Agrawal, G. (2020). Development, present status and performance analysis of agriculture insurance schemes in India: Review of evidence. *International Journal of Social Economics, 47*(4), 461–481. DOI: 10.1108/IJSE-02-2019-0119

Singh, P., & Agrawal, G. (2022). Mapping the customer centric weather index insurance service design using quality function deployment. *The TQM Journal, 34*(6), 1800–1822. DOI: 10.1108/TQM-08-2021-0236

Sinni, G. (2017). "Participatory Design for Public Services. Innovation in Public Administration." *Design Journal, 20*(sup1), 120-138. DOI: 10.1080/14606925.2017.1352841

Smith, M., Torres-Agüero, A., Grossman, R., Sen, P., Chen, Y., & Borcea, C. (2024). A Study of GDPR Compliance Under the Transparency and Consent Framework. *Proceedings of the ACM Conference.* https://doi.org/DOI: 10.1145/3589334.3645618

Snatkin, A., Karjust, K., Majaki, J., Aruvali, T., & Eiskop, T. (2013). Real time production tracking system in SME. *Estonian Journal of Engineering, 19*(1), 32–75. DOI: 10.3176/eng.2013.1.06

Soluk, J., Kammerlander, N., & Darwin, S. (2021). Digital entrepreneurship in developing countries: The role of institutional voids. *Technological Forecasting and Social Change, 170*, 120876. DOI: 10.1016/j.techfore.2021.120876

Song, F., Shi, N., Shan, F., Zhang, Z., Shen, J., Lu, H., Ling, Y., Jiang, Y., & Shi, Y. (2020). Emerging 2019 novel coronavirus (2019-nCoV) pneumonia. *Radiology, 295*(1), 210–217. DOI: 10.1148/radiol.2020200274 PMID: 32027573

Soori, M., Arezoo, B., & Dastres, R. (2023). Internet of things for smart factories in industry 4.0, a review. *Internet of Things and Cyber-Physical Systems*, *3*, 192–204. DOI: 10.1016/j.iotcps.2023.04.006

Srivastava, M. (2020). Digital economy expected to account for 20% of Indian GDP by 2025. https://kr-asia.com/digital-economy-expected-to-account-for-20-of-indian-gdp-by-2025

Stanciu, A., Titu, A. M., & Deac-Şuteu, D. V. (2021, July). Driving digital transformation of knowledge-based organizations through artificial intelligence enabled data centric, consumption based, As-a-service models. In *2021 13th international conference on electronics, computers and artificial intelligence (ECAI)* (pp. 1-8). IEEE.

Steiber, A., Alänge, S., Ghosh, S., & Goncalves, D. (2021). Digital transformation of industrial firms: An innovation diffusion perspective. *European Journal of Innovation Management*, *24*(3), 799–819. DOI: 10.1108/EJIM-01-2020-0018

Strielkowski, W., Rausser, G., & Kuzmin, E. (2022). Digital revolution in the energy sector: effects of using digital twin technology. In *Digital transformation in industry: Digital twins and new business models* (pp. 43–55). Springer. DOI: 10.1007/978-3-030-94617-3_4

Subasi, O., Bel, O., Manzano, J., & Barker, K. (2023). The landscape of modern machine learning: A review of machine, distributed and federated learning. *arXiv preprint arXiv:2312.03120*.

Sun, T. et al., (2022). The Digital Twin in Medicine: A Key to the Future of Healthcare?. *Frontiers in Medicine,* Volum 9.

Sun, T., He, X. & Li, Z., (2023). Digital twin in healthcare: Recent updates and challenges.. *Digital Health,* Volum 9.

Sundararajan, A. (2022). The sharing economy: The end of employment and the rise of crowd-based capitalism. *The Journal of Economic Perspectives*, *36*(2), 93–112. DOI: 10.1257/jep.36.2.93

Surendran, P. (2012). Technology acceptance model: A survey of literature. International journal of business and social research, 2(4), 175-178.

Taber, M. M. (2014). Use of Mobile Technology Among Museum Visitors: A Case Study. Use of Mobile Technology Among Museum Visitors, 1-50.

Tan, P., Li, X., & Zhang, J. (2021). AI-enhanced public transport efficiency: Case study of Singapore. *Journal of Intelligent Transport Systems*, *25*(4), 315–329.

Tao, F.. (2018). Digital twin-driven product design, manufacturing and service with big data. *Int J Adv Manuf Technol. Volum, 94*, 3563–3576.

Tao, F., Zhang, H., Liu, A., & Nee, A. Y. C. (2019). Digital twin in industry: State-of-the-art. *IEEE Transactions on Industrial Informatics, 15*(4), 2405–2415. DOI: 10.1109/TII.2018.2873186

Taormina, F., & Baraldi, S. B. (2022). Museums and digital technology: A literature review on organizational issues. *European Planning Studies, 30*(9), 1676–1694. DOI: 10.1080/09654313.2021.2023110

Tapscott, D., & Tapscott, A. (2016). *Die Blockchain-Revolution: Wie die Technologie hinter Bitcoin nicht nur das Finanzsystem, sondern die ganze Welt verändert.* Plassen Verlag.

Tapscott, D., & Williams, A. D. (2010). *Macrowikinomics: Rebooting Business and the World*. Portfolio.

Tariq, N., Asim, M., Al-Obeidat, F., Zubair Farooqi, M., Baker, T., Hammoudeh, M., & Ghafir, I. (2019). The security of big data in fog-enabled IoT applications including blockchain: A survey. *Sensors (Basel), 19*(8), 1788. DOI: 10.3390/s19081788 PMID: 31013993

Thames, L., & Schaefer, D. (2016). Software defined cloud manufacturing for Industry 4.0. *Procedia CIRP, 52*, 12–17. DOI: 10.1016/j.procir.2016.07.041

Thuan, P. Q., Khuong, N. V., Anh, N. D. C., Hanh, N. T. X., Thi, V. H. A., Tram, T. N. B., & Han, C. G. (2022). The determinants of the usage of accounting information systems toward operational efficiency in industrial revolution 4.0: Evidence from an emerging economy. *Economies, 10*(4), 83. DOI: 10.3390/economies10040083

Timm, I. J., & Lorig, F. (2015). Logistics 4.0: A challenge for simulation. In *2015 Winter Simulation Conference (WSC)* (pp. 3118–3119). IEEE. https://doi.org/DOI: 10.1109/WSC.2015.7408428

Tom Dieck, M. C., & Jung, T. H. (2017). Value of augmented reality at cultural heritage sites: A stakeholder approach. *Journal of Destination Marketing & Management, 6*(2), 110–117. DOI: 10.1016/j.jdmm.2017.03.002

Tóth, A., Nagy, L., Kennedy, R., Bohuš, B., Abonyi, J., & Ruppert, T. (2023). The human-centric Industry 5.0 collaboration architecture. *MethodsX, 11*, 102260. DOI: 10.1016/j.mex.2023.102260 PMID: 37388166

Trinkley, K. E., An, R., Maw, A. M., Glasgow, R. E., & Brownson, R. C. (2024). Leveraging artificial intelligence to advance implementation science: Potential opportunities and cautions. *Implementation Science : IS, 19*(1), 17. DOI: 10.1186/s13012-024-01346-y PMID: 38383393

Trischler, J., Røhnebæk, M., Edvardsson, B., & Tronvoll, B. (2023). Advancing Public Service Logic: Moving towards an ecosystemic framework for value creation in the public service context. *Public Management Review*, •••, 1–29. DOI: 10.1080/14719037.2023.2229836

Trischler, J., & Trischler, J. W. (2022). Design for experience – a public service design approach in the age of digitalization. *Public Management Review, 24*(8), 1251–1270. DOI: 10.1080/14719037.2021.1899272

Turab, M., & Jamil, S. (2023). A Comprehensive Survey of Digital Twins in Healthcare in the Era of Metaverse. *BioMedInformatics, 3*(3), 563–584. DOI: 10.3390/biomedinformatics3030039

Udayakumar, R., Devi, J., & Suganya, S. (2023). Integrated SVM-FFNN for fraud detection in banking financial transactions. *Journal of Internet Services and Information Security, 13*(4), 12–25. DOI: 10.58346/JISIS.2023.I4.002

Ukeyima, A. E. (2024). The importance of project management in underdeveloped countries. Journal of African Sustainable Development.

United Nations. (2019). *Sustainable Development Goals Report 2019*. United Nations Publications.

United Nations. (2020). *The impact of digital technologies*. https://www.un.org/en/un75/impact-digital-technologies

Uwaoma, P. U., Eboigbe, E. O., Eyo-Udo, N. L., Ijiga, A. C., Kaggwa, S., & Daraojimba, D. O. (2023). The fourth industrial revolution and its impact on agricultural economics: Preparing for the future in developing countries. *International Journal of Advanced Economics, 5*(9), 258–270. DOI: 10.51594/ijae.v5i9.646

Vallée, A., (2023). Digital twin for healthcare systems. *Frontiers in Digital Health,* Volum 5.

van der Aalst, W. M. P. (2021). *Process mining: Data science in action* (2nd ed., Vol. 1). Springer.

van der Aalst, W. M. P., Bichler, M., & Heinzl, A. (2018). Robotic process automation. *Business & Information Systems Engineering, 60*(4), 269–272. DOI: 10.1007/s12599-018-0542-4

Van der Meer, F. (2017). The Role of Technology in Public Administration: A Review of the Literature. *Public Administration Review*, *77*(4), 1–10.

van Erp, T., Carvalho, N. G. P., Gerolamo, M. C., Gonçalves, R., Rytter, N. G. M., & Gladysz, B. (2024). Industry 5.0: A new strategy framework for sustainability management and beyond. *Journal of Cleaner Production*, *461*, 142271. DOI: 10.1016/j.jclepro.2024.142271

Vanhoeyveld, J., Martens, D., & Peeters, B. (2020). Value-added tax fraud detection with scalable anomaly detection techniques. *Applied Soft Computing*, *86*, 105895. DOI: 10.1016/j.asoc.2019.105895

Vargo, S. L., & Lusch, R. F. (2008). Service-dominant logic: Continuing the evolution. *Journal of the Academy of Marketing Science*, *36*(1), 1–10. DOI: 10.1007/s11747-007-0069-6

Vargo, S. L., & Lusch, R. F. (2016). Institutions and axioms: An extension and update of service-dominant logic. *Journal of the Academy of Marketing Science*, *44*(1), 5–23. DOI: 10.1007/s11747-015-0456-3

Vasundhara, S., Venkatesh, K. S., Manimegalai, V., Sundharesalingam, P., Sathyakala, S., & Boopathi, S. (2024). AI-powered marketing revolutionizing customer engagement through innovative strategies. In *Cases on AI Ethics in Business* (pp. 21–46). IGI Global. DOI: 10.4018/979-8-3693-2643-5.ch002

Vial, G. (2019). Understanding digital transformation: A review and a research agenda. *The Journal of Strategic Information Systems*, *28*(2), 118–144. DOI: 10.1016/j.jsis.2019.01.003

Vidu, C., Zbuchea, A., & Pinzaru, F. (2021). Old Meets New: Integrating Artificial Intelligence in Museums' Management Practices. Strategica. Shaping the Future of Business and Economy, 830-844.

Vijayan, V., Connolly, J. P., Condell, J., McKelvey, N., & Gardiner, P. (2021). Review of Wearable Devices and Data Collection Considerations for Connected Health. *Sensors (Basel)*, *21*(16), 5589. DOI: 10.3390/s21165589 PMID: 34451032

Vorobyeva, K. I., Belous, S., Savchenko, N. V., Smirnova, L. M., Nikitina, S. A., & Zhdanov, S. P. (2025). Personalized learning through AI: Pedagogical approaches and critical insights. *Contemporary Educational Technology*, *17*(2), ep574. DOI: 10.30935/cedtech/16108

Wache, H., Hübner, E. J., Hönigsberg, S., & Dinter, B. (2022). Closing the Implementation Gap of Digital Twins. In *AMCIS*.

Wang, D., Zhang, Z., Zhang, M., Fu, M., Li, J., Cai, S., Zhang, C., & Chen, X. (2021). The role of digital twin in optical communication: Fault management, hardware configuration, and transmission simulation. *IEEE Communications Magazine*, *59*(1), 133–139. DOI: 10.1109/MCOM.001.2000727

Wang, N., Yang, W., Wang, X., Wu, L., Guan, Z., Du, X., & Guizani, M. (2024). A blockchain based privacy-preserving federated learning scheme for Internet of Vehicles. *Digital Communications and Networks*, *10*(1), 126–134. DOI: 10.1016/j.dcan.2022.05.020

Wang, S., Torngren, M., & Onori, M. (2015). Current status and advancement of cyber-physical systems in manufacturing. *Journal of Manufacturing Systems*, *37*, 517–527. DOI: 10.1016/j.jmsy.2015.04.008

Wang, S., Wan, J., Zhang, D., Li, D., & Zhang, C. (2016). Towards smart factory for Industry 4.0: A self-organized multi-agent system with big data based feedback and coordination. *Computer Networks*, *101*, 158–168. DOI: 10.1016/j.comnet.2015.12.017

Wang, S., Zhang, J., Wang, P., Law, J., Calinescu, R., & Mihaylova, L. (2024). A deep learning-enhanced Digital Twin framework for improving safety and reliability in human–robot collaborative manufacturing. *Robotics and Computer-integrated Manufacturing*, *85*, 102608. DOI: 10.1016/j.rcim.2023.102608

Wang, S., Zhang, Y., & Zhang, X. (2019). Machine learning applications in infrastructure project management: A review. *Automation in Construction*, *98*, 1–13.

Wang, X., Kim, J., Kim, J., & Koh, Y. (2024). Application of Natural-Resource-Based View to Nature-Based Tourism Destinations. *Sustainability (Basel)*, *16*(6), 2375. DOI: 10.3390/su16062375

Whitford, A. B., & Meier, K. J. (2013). Methodological Innovations in Public Administration: A Symposium. *Journal of Public Administration: Research and Theory*, *23*(2), 231–246. DOI: 10.1093/jopart/mut012

Wirtz, B. W., & Göttel, V. (2016). Business Model Innovation: A Review of the Literature. *International Journal of Management Reviews*, *18*(3), 1–20.

Wirtz, B. W., & Piehler, R. (2016). eGovernment Applications and public personnel acceptance: An empirical analysis of the public servant perspective. *International Journal of Public Administration*, *39*(3), 238–247. DOI: 10.1080/01900692.2014.1003384

Wirtz, B. W., Weyerer, J. C., & Geyer, C. (2019). Artificial intelligence and the public sector—Applications and challenges. *International Journal of Public Administration*, *42*(7), 596–615. DOI: 10.1080/01900692.2018.1498103

World Bank. (2020). *Infrastructure for Development: World Bank Report on Global Infrastructure*. World Bank Publications.

World Economic Forum. (2021). *The global risks report 2021* (16th ed.).

Wu, C. et al., (2022). Integrating mechanism-based modeling with biomedical imaging to build practical digital twins for clinical oncology.. 3(2), p. Biophysics Reviews.

Wu, J., Xue, Y., Zhang, Q., & Liu, Y. (2024). Digital economy, financial development, and corporate green technology innovation. *Finance Research Letters, 66*, 105552. DOI: 10.1016/j.frl.2024.105552

Xames, M., & Topcu, T. (2024). A Systematic Literature Review of Digital Twin Research for Healthcare Systems: Research Trends, Gaps, and Realization Challenges. *IEEE Access. IEEE Access : Practical Innovations, Open Solutions, 12*, 4099–4126. DOI: 10.1109/ACCESS.2023.3349379

Xu, L. D. (2011). Enterprise systems: State-of-the-art and future trends. *IEEE Transactions on Industrial Informatics*, *7*(4), 630–641. DOI: 10.1109/TII.2011.2167156

Xu, L. D., He, W., & Li, S. (2014). Internet of Things in industries: A survey. *IEEE Transactions on Industrial Informatics*, *10*(4), 2233–2243. DOI: 10.1109/TII.2014.2300753

Xu, L. D., Xu, E. L., & Li, L. (2018). Industry 4.0: State of the art and future trends. *International Journal of Production Research*, *56*(8), 2941–2962. DOI: 10.1080/00207543.2018.1444806

Xu, X., Lu, Y., Vogel-Heuser, B., & Wang, L. (2021). Industry 4.0 and Industry 5.0—Inception, conception and perception. *Journal of Manufacturing Systems*, *61*, 530–535. DOI: 10.1016/j.jmsy.2021.10.006

Yadav, D. R. T. C., Kala, K., Roy Kolachina, R. I., Kanneganti, M. C., & Pasupuleti, S. S. (2024). *Data Privacy Concerns and Their Impact on Consumer Trust in Digital Marketing*. Indian Scientific Journal of Research in Engineering and Management., DOI: 10.55041/IJSREM38555

Yang, J. P., Wang, W. L., & Zhou, S. K. (2019). A design of integrated quality management system based on artificial intelligence (AI) technology. *Destech Transactions on Computer Science and Engineering*, 15–20.

Yang, H., Kumara, S., Bukkapatnam, S. T. S., & Tsung, F. (2019). The Internet of Things for smart manufacturing: A review. *IISE Transactions*, *51*(11), 1190–1216. DOI: 10.1080/24725854.2018.1555383

Yang, J., Liu, Y., & Morgan, P. L. (2024). Human–machine interaction towards Industry 5.0: Human-centric smart manufacturing. *Digital Engineering*, *2*, 100013. DOI: 10.1016/j.dte.2024.100013

Yin, R. K. (2009). *Case study research: Design and methods* (Vol. 5). Sage Publications.

Yin, R. K. (2013). Validity and generalization in future case study evaluations. *Evaluation*, *19*(3), 321–332. DOI: 10.1177/1356389013497081

Yoo, Y., Boland, R. J.Jr, Lyytinen, K., & Majchrzak, A. (2012). Organizing for innovation in the digitized world. *Organization Science*, *23*(5), 1398–1408. DOI: 10.1287/orsc.1120.0771

Young, T., Hazarika, D., Poria, S., & Cambria, E. (2018). Recent trends in deep learning-based natural language processing. *IEEE Computational Intelligence Magazine*, *13*(3), 55–75. DOI: 10.1109/MCI.2018.2840738

Yousaf, Z., Radulescu, M., Sinisi, C. I., Serbanescu, L., & Păunescu, L. M. (2021). Towards sustainable digital innovation of SMEs from the developing countries in the context of the digital economy and frugal environment. *Sustainability (Basel)*, *13*(10), 5715. DOI: 10.3390/su13105715

Yu, C. S., & Tao, Y. H. (2009). Understanding business-level innovation technology adoption. *Technovation*, *29*(2), 92–109. DOI: 10.1016/j.technovation.2008.07.007

Yu, Y., Si, X., Hu, C., & Zhang, J. (2019). A review of recurrent neural networks: LSTM cells and network architectures. *Neural Computation*, *31*(7), 1235–1270. DOI: 10.1162/neco_a_01199 PMID: 31113301

Zafar, M. H., Langås, E. F., & Sanfilippo, F. (2024). Exploring the synergies between collaborative robotics, digital twins, augmentation, and industry 5.0 for smart manufacturing: A state-of-the-art review. *Robotics and Computer-integrated Manufacturing*, *89*, 102769. DOI: 10.1016/j.rcim.2024.102769

Zahid, A., Leclaire, P., Hammadi, L., Costa-Affonso, R., & El Ballouti, A. (2025). Exploring the potential of Industry 4.0 in manufacturing and supply chain systems: Insights and emerging trends from bibliometric analysis. *Supply Chain Analytics. Article*, *100108*, 100108. Advance online publication. DOI: 10.1016/j.sca.2025.100108

Zahidah, S., & Al-haimi, B. (2024). Telemedicine: Transforming healthcare accessibility and quality with sustainable technological advancement. *Navigating the Intersection of Business, Sustainability and Technology*, 137-149.

Zhang, Y., Wang, H., & Liu, J. (2020). IoT and AI for smart infrastructure: A sustainable approach. *Sustainable Cities and Society*, *50*, 101678.

Zhang, Z., Wen, F., Sun, Z., Guo, X., He, T., & Lee, C. (2022). Artificial intelligence-enabled sensing technologies in the 5G/internet of things era: From virtual reality/augmented reality to the digital twin. *Advanced Intelligent Systems*, *4*(7), 2100228. DOI: 10.1002/aisy.202100228

Zhao, W., Zhou, X., Li, J., Tang, B., Wang, Y., Duan, Y., & Liu, T. (2023). A survey of large language models. *arXiv preprint arXiv:2303.18223.*

Zheng, Q., Xu, Y., Liu, H., Shi, B., Wang, J., & Dong, B. (2024). A Survey of Tax Risk Detection Using Data Mining Techniques. *Engineering (Beijing)*, *34*, 43–59. DOI: 10.1016/j.eng.2023.07.014

Zhong, D., Xia, Z., Zhu, Y., & Duan, J. (2023). Overview of predictive maintenance based on digital twin technology. *Heliyon*, *9*(4), e14534. DOI: 10.1016/j.heliyon.2023.e14534 PMID: 37025897

Zizic, M. C., Mladineo, M., Gjeldum, N., & Celent, L. (2022). From Industry 4.0 towards Industry 5.0: A Review and Analysis of Paradigm Shift for the People, Organization and Technology. *Energies 2022, Vol. 15, Page 5221, 15*(14), 5221. DOI: 10.3390/en15145221

Zreik, M. (2024). Harnessing the Power of Digital Transformation and Sustainability: The Chinese Experience. In *Building Smart and Sustainable Businesses with Transformative Technologies* (pp. 247-266). IGI Global Scientific Publishing.

Zuboff, S. (2023). The age of surveillance capitalism: The fight for a human future at the new frontier of power. *Business Ethics Quarterly*, *33*(1), 156–182. DOI: 10.1017/beq.2022.41

Zyzak, B., & Martinussen, P. E. (2024). OK computer: Applying the public service logic on digital health services. *Public Management Review*, *26*(11), 3342–3364. DOI: 10.1080/14719037.2024.2337080

# About the Contributors

**Arshi Naim** is recognized among the top 2% of Global Scientists (Stanford University and Elsevier 2024). She is a distinguished scholar with a Ph.D. in Business Management and an MBA in Marketing. Her academic journey, rooted in excellence, began with First Class Honors in Economics from Aligarh Muslim University. As an e-learning expert and certified peer reviewer with Quality Matters, U.S., Dr. Naim has further enhanced her expertise with certificates in Digital Marketing and Social Media Analytics from Accenture, UK, and entrepreneurship from Harvard University. A prolific researcher, she has authored over 77 Scopus and Web of Science-indexed publications and penned more than 10 acclaimed books with prestigious publishers like Taylor & Francis, Nova Sciences and IGI Global. Her work seamlessly blends theory and practice, inspiring future generations of scholars. Through her commitment to mentoring doctoral students and her extensive contributions to global research, Dr. Naim continues to shape the landscape of academic excellence.

**Anuj Kumar** works as Head of Research and Associate Professor, at Rushford Business School, Switzerland. He is also associated with INTI International University, Malaysia as a Research Fellow (Honorary Role). His significant thrust areas are marketing, international business, and general management. He has completed an M.Sc. in International Business from University College Dublin, Michael Smurfit Graduate Business School, Ireland (Ranked among the top 200 colleges of the world – QS World ranking). He has completed PhD from Aligarh Muslim University (Central University). He has published research papers in reputed SCOPUS/Web of Science/SCI/ABDC/UGC Care Journals. Dr. Kumar has over eighty publications (Scopus- 48, ABDC- 25, Web of Science- 26, SCI-3). He also has three e-books and five book chapters to his credit (two are Scopus-indexed). He has published several cases in Emerald Emerging Case Studies (Scopus), International Social Science Journal (ABDC-B and Scopus), Small Enterprise Research (Scopus and Web of Science), Case Centre-UK, and other reputed Scopus, ABDC, and Web of Science journals. He has three internationally granted patents by South Africa and one patent granted by the Indian Government.

***

**Shaista Anwar** is at the Department of Finance & Accounting at Liwa College of Abu Dhabi, UAE. She completed her MBA in Finance and Master's in Commerce and earned a Ph.D. in Finance. She holds various professional certifications including; Decentralized Finance for Investment Professionals (CFA), and a Certificate in identifying predatory and Cloned journals in academics (CIRI). Her recent research articles on Blockchain-based finance gained wide visibility and she is an active speaker on Fintech and BlockchainTechnology

**Tiago Jordão Cardoso** is a Specialized Technical Officer at the General Secretariat of the Environment in Portugal, where he focuses on the integration of artificial intelligence and chatbots within public administration. Currently completing a master's in public administration, his research centers on technological acceptance, particularly in the public sector's use of AI to enhance sustainability and resilience. With strong involvement in climate change initiatives, Tiago explores the intersection of AI-driven innovation and sustainable practices, co-authoring articles on these topics and preparing additional proposals on AI's role in environmental resilience.

**Anita Gehlot** is associated with Uttaranchal University as Professor and Head – Research & Innovation with more than sixteen years of experience. Her area of expertise includes embedded systems, wireless sensor networks, Internet of Things, Artificial Intelligence and Automation. She has been featured among top ten inventors for ten years 2010-2020, by Clarivate Analytics in "India's Innovation Synopsis" in March 2021. She has filed more than five hundred patents including more than hundered Indian & International patent grants, 5 PCT and published more than three hundred and fifty research papers in SCI/Scopus journals. Five scholars have been awarded with PhD under her supervision. She has been featured by Indian and International media for smart systems and devices designed by her including-OBDAS, E-Parirakshak, Kawach, 20Sec4Life, Ally, Alithis, CT scan Diagnosis, among which five are designed to prevent COVID-19 as per WHO guidelines. She has authored/edited forty four books in the area of Embedded Systems, AI, Internet of Things. She has been awarded with "Gandhian Young Technological Innovation (GYTI) Award", at Rashtrapati Bahawan, India in 2018. She has been honoured with "Appreciation Letter" for contribution in the field of science & technology in Uttarakhand state by SPECS. She has been honored with "Certificate of Excellence" from 3rd faculty branding awards-15, Organized by EET CRS research wing for excellence in professional education and Industry, for the category "Young Researcher", 2015.

**Pedro Pinheiro Gomes** is a Technical Adviser at the Secretary General for the Environment in Portugal, where he currently focuses on project management and communication within the Environment Programme of the EEA Grants. He holds a master's degree in Environmental Engineering, specializing in environmental systems and policy. His professional journey also includes extensive experience in statistics and data analysis within the European Comparison Programme (ECP) for Purchasing Power Parities. In addition to his core roles, Pedro has authored multiple publications, including chapters in scientific books on environmental management, policy, and the integration of Artificial Intelligence in sustainability. He has contributed to high-impact projects, such as the development of mobile applications for coastal and tourism analysis aligned with the Inspire Directive, as well as tools for natural disaster alerts and forecasting.

**Gesualda Iodice**, PhD Candidate in Management at the Department of Economics, Management and Institutions at the University of Naples Federico II, deals with research, benchmarking and strategic analysis in the social and cultural sector. Her main research interests concern social entrepreneurship and Benefit Corporations driven by arts and culture; the use of disruptive technologies in cultural value propositions and cultural accessibility. She is the author of scientific articles of national and international relevance, speaker at international conferences and reviewer for international journals

**Mohammad Irfan** is presently working as a Professor at School of Business and Commerce, Chandigarh University, Lucknow, Unnao, India. Dr. Irfan has done his Ph.D. from the Central University of Haryana. He is MBA (Finance), M.Com (Account and Law), and MA (Economics). He has qualified UGC-JRF/SRF/NET in Management and Commerce. Dr. Irfan certified NSEs (NCFM) and BSEs certification. He has experience of more than sixteen years in the area of SAPM, Artificial Intelligence, Machine Learning, Blockchain, Cryptocurrency, Financial Engineering, Fintech, Green Finance, and Alternative Finance. He has to his credit more than 82+ Scopus Indexed articles, includes The Journal of Economic Cooperation and Development (Q2), International Journal of Business Excellence (IJBEX), International Journal of Economics and Management (IJEM), Montenegrin Journal of Economics (Q2), Cogent Business & Management (Taylor & Francis) (Q2), Indian Journal of Finance (IJF) and Journal of Islamic Monetary Economics and Finance (JIMF). His citations reached 494+ along with 13 H-index. Dr. Irfan has published 10 books in Springer, IGI Global Publication (Scopus indexed).

**Satvik Jain** is an accomplished academic and researcher, currently serving as an Assistant Professor at Kalinga University in Raipur. He holds a Ph.D. in Business and Commerce, with a strong focus on entrepreneurship, financial markets, and emerging technologies. Throughout his academic career, Dr. Jain has demonstrated a deep commitment to advancing knowledge in his fields of expertise. His research delves into the dynamics of entrepreneurship, the intricacies of financial markets, and the impact of emerging technologies on commerce. This multidisciplinary approach underscores his dedication to understanding and teaching the complexities of the modern business environment. In his role at Kalinga University, Dr. Jain is actively involved in mentoring students, fostering a learning environment that encourages critical thinking and innovation. His contributions to academia not only enhance the university's reputation but also prepare students to navigate and contribute to the evolving landscape of business and technology. Dr. Jain's work exemplifies the integration of academic rigor with practical insights, making him a valuable asset to both the academic community and the broader field of business and commerce.

**Navin Kumar** is an Assistant Professor at Bharat College of Law, Kurukshetra, India, with a Doctorate in Law from Chaudhary Devi Lal University (CDLU), Sirsa, Haryana (2022). He holds an LL.M. in Criminal Law from Kurukshetra University (2012) and a B.A. L.L.B. from the same institution (2007). Dr. Kumar is NET-qualified (2012) and has been teaching since 2017, specializing in criminal law, evidence law, and human rights. His academic contributions are complemented by significant research, with publications in esteemed journals, including the International Journal for Multidisciplinary Research and Indian Journal of Law and Legal Research. His research covers diverse topics, such as the compensation and rehabilitation of rape victims, the legal aspects of acid attacks, and judicial responses to Khap Panchayats. Dr. Kumar has presented papers at various national conferences on topics like India-South Africa relations, water conservation laws, and judicial attitudes towards Khap Panchayats. As a dedicated mentor, he guides students in their academic and research endeavors, encouraging critical thinking and fostering a deep understanding of legal issues. With his commitment to legal education, research, and student development, Dr. Kumar continues to inspire and contribute to the advancement of the legal field. Dr. Navin Kumar Assistant Professor Bharat College of Law Kurukshetra, Haryana, India.

**Anitha Kumari** is an Assistant Professor in the Department of Information Systems at GITAM School of Business, GITAM University, Bengaluru. She completed her Ph.D. in Information Systems at the Department of Business Administration, Hindustan Institute of Technology and Science, India. Her Bachelor's degree was in Computer Applications, while she obtained a Master's in Business Administration with a specialisation in finance and systems. Her current research focuses on information systems, blockchain, Fintech, business analytics, AI, and technological innovation. She has 6 years of teaching and research experience.

**Hesham Magd** has more than twenty-five years of combined experience in traditional and nontraditional higher education teaching, training, consultancy, community development, academic administration, curricula design, organizational change and development, distinguished research and scholarly writing, resulting in Honors, awards and recognition for academic excellence and outstanding achievement. Hesham is currently heavily involved in International accreditation and reshaping the aviation education in Oman and offer strategic direction on the direction of the sector In his previous appointments throughout the world in British, European, and American oriented Curriculum (Middle East, USA, and UK), he has been the driving force behind strategic institutional development during the time of profound change in the Higher Education in the Middle East. Hesham has broad knowledge of the UK, USA, Middle East University Systems, and Quality and accreditation systems (AACSB, AABI, OAAA, NCAAA…).

**Fadillah Mansor** is an Associate Professor at the Department of Shariah and Management, Academy of Islamic Studies, University Malaya. She obtained her PhD (in Islamic finance) from the La Trobe University, Australia. She holds a MBA, specialize in Finance from Graduate Business School, Universiti Malaya and a Bachelor in Shariah (Honours) minor in economics, also from Universiti Malaya. She has written numerous articles in Islamic finance and her works have been published in International journals and book chapters. In her teaching tasks, she has taught several subjects in in her field encompass Islamic financial system, Islamic capital market, economics for muamalat, Islamic banking operations and many more. Her research interest focus on Islamic banking and finance including Islamic mutual fund investments, Islamic financial literacy and Islamic financial planning. She is also interested in research related to halal management. Nationally, she is a panel assessor of curriculum programme in the area of Islamic finance, Islamic management and takaful for Malaysian Qualifications Agency (MQA), an agency that responsible for programme accreditations before the programme is listed under Malaysian Qualifications Register. Internationally, she serves as an appointed reviewer for peer-reviewed journals. She is active in the Islamic banking and finance industry as a member of Shariah Committee of Kenanga Investment Bank Berhad since 2021.

**Mostafa Mohamad** Associate Professor of Digital Innovation & Analytics Zayed University, College of Interdisciplinary Studies University of Manchester, Alliance Manchester Business School mostafa.mohamad@zu.ac.ae Dr. Mo is a scholar and industry expert in Digital Transformation, AI for Business, FinTech, Digital Innovation, and Business Analytics. Before joining Zayed University, he had 20 years of academic experience at Manchester Business School, Newcastle Business School, Leeds Business School, and Salford Business School. Mo served as a Chair of external relations, director of PG programmes, a director of Admissions, and a deputy DBA director. He also led significant research funds for Virtual Reality, Machine Learning, Artificial Intelligence, FinTech and Technology transfer in the Energy, automotive, tourism & telecommunication sectors. He is an associate editor in Q1 journals in IS & Digital Marketing. Mo has a record of publications in Q1 and Q2 Scopus-listed journals.

**Keerthana N V** is an Assistant Professor with 8 Years of experience in teaching and 1-year experience in research (IUCAA LIGO Data Centre, Pune). Her area of interest is Network Security, Blockchain Security, and Machine Learning Algorithms. She is pursuing PhD at Anna University in Security in Blockchain. She has a working experience in High-Performance Computing (HTCondor) in LIGO INDIA Project. She has written many research articles in Network Security and assisted Scientists in analyzing Datasets in HTCondor.

**Konstantinos Peramatzis** is a prominent professional in the field of aviation engineering, currently serving as an EASA-approved B2 Instructor in Aviation Engineering at the Australian University (AU) since September 2019. A proud Greek scholar, Konstantinos started his academic journey with a Bachelor degree in Aviation Engineering from the Technical Academy of the Greek Military Air Force, focusing on Telecommunication-navigation and autopilot systems. He later earned a Bachelor's degree in Physics from the Open University of Greece. Konstantinos' strong educational background, includes also a Part-66 B2 EASA License & EASA-Approved B2 Instructor, Examiner, Exam Invigilator, and Practical Assessor from the Greek Civil Aviation Authority and the UK's Civil Aviation Authority, a Train the Trainer degree from the Greek Military Air Force, and degrees in Computer Programming (C++ & JAVA), degree in support and maintenance on computer networks and also degree in Quantum Physics. Konstantinos has held key administrative roles, notably as the Head of the Avionics Workshop for aircrafts C-130 and EMBRAER, including also a significant role in the Greek government VIP squadron for 10 years. His teaching experience spans over 20 years, with roles as an instructor, examiner, and exam invigilator at respected institutions such as KLMUK College in Norwich, ΩMEGA AVIATION ACADEMY, and the Greek

Aerospace Industry. In addition to his teaching and administrative roles, Konstantinos is actively involved in research, focusing on human factors that influence aviation safety. His research aims to improve safety practices within the aviation industry.

**Carmine Sergianni**, PhD student at the Department of Economics, Management, and Institutions of the University of Naples Federico II. His main focus is the PSL – Public Service Logic – (Osborne, 2018, 2020), which identifies and analyses value creation in public services. His research interest analyses the customer journey from a PSL perspective, in order to support the improvement of the value proposition in the public sector and the impact of new technologies on the customer journey in public services.

**Ahmad Shaheen** is an Assistant Professor of Marketing at the Arab Academy for Science and Technology and Maritime Transport, Alexandria, Egypt. He holds a Ph.D. in Marketing from an AACSB-accredited institution, with a research focus on digital marketing, consumer behavior, AI-driven marketing strategies, and marketing analytics. Dr. Shaheen has extensive experience in teaching, research, and industry consulting, specializing in the integration of artificial intelligence, data science, and marketing automation into business practices. His research explores emerging trends in influencer marketing, customer engagement, and market dynamics. He has published in peer-reviewed journals and actively collaborates on interdisciplinary projects related to data-driven decision-making and strategic marketing management.

**Ravi Kant Sharma** is currently working as A Professor& (Division Chair-Marketing) in School of Business. He is associated since long in marketing department. He has total experience of more than 26 years of teaching and industry. He has been associated with Galgotias University since Nov.2022. He has participated in numerous Intl. and National Conferences and Faculty Development Programs (FDP).

**Pankaj Singh** is an Associate Professor of Marketing Science in School of Business at Galgotias University, Greater Noida, India. He completed his doctoral research in marketing domain at ABV-Indian Institute of Information Technology and Management, Gwalior, Madhya Pradesh. He holds a masters' in International Business Management and M. Phil. in Marketing and his particular research interest in the area of Service Marketing, Digital Marketing, Social Media and Public Policy. He published several Scopus indexed, ABDC listed research articles, IEEE conference proceedings and book chapters in all leading publishers i.e., Springer, IGI Global, CRC Press, Willey and Emerald publications. He has more than fourteen years of experience in academics and industry. In previous assignments, Dr. Singh is associated with SRM University, Sonipat and GLA University, Mathura as a faculty of Marketing Management.

**Mahendrakumar Subramaniam** received the Bachelors degree in Electronics andCommunication Engineering in 2006, the Masters degree in Communication Systemsin 2009 and received Ph.D - Information and Communication Engineering in 2019from Anna University, Chennai, India. He is currently working as Professor inthe department of Electronics and Communication Engineering, Velalar College ofEngineering and Technology since 2009. He is an Anna University recognized Ph.Dsupervisor and eight scholars are pursuing Ph.D under his supervision. Hisresearch interests include Antenna, Microwave, Wireless Network, MAC andNetwork Protocols. He has published 32 papers in international journals. He isa life member of ISTE and IEICE.

**T. Nithya** presently serves as an Assistant Professor in the Department of Information Technology at Velalar College of Engineering and Technology in Erode, Tamil Nadu. She earned his B.Tech. in II, followed by an M.E. in Computer Science and Engineering from Anna University Chennai. With an extensive 18 years of experience in Academic, Administration, Teaching, and Research, she is a highly accomplished professional in the field of Information Technology, specializing in Data Analysis and Data Mining. She has published more than 20 research papers in National and International journals and presented more than 50 papers in National and International conferences across India. She published 2 Indian Patents. She actively supervises students undertaking their final year projects at Anna University Chennai. Additionally, she holds an active life membership in the Professional Society of ISTE. Her research interests encompass Data Analysis, Machine Learning and Deep Learning.

# Index

## E



## F



## G



## H



## I

## M



## O



## P



## R



## S

## T



## V

www.ingramcontent.com/pod-product-compliance
Lightning Source LLC
LaVergne TN
LVHW060007031026
849161LV00025B/2246